D1145222
WITHDRAWN
DE MONTFORT UNIVERSITY
LIBRARY

Transform Techniques in Chemistry

MODERN ANALYTICAL CHEMISTRY

Series Editor: David Hercules
University of Pittsburgh

ANALYTICAL ATOMIC SPECTROSCOPY
By William G. Schrenk

PHOTOELECTRON AND AUGER SPECTROSCOPY
By Thomas A. Carlson

MODERN FLUORESCENCE SPECTROSCOPY, VOLUME 1
Edited by E. L. Wehry

MODERN FLUORESCENCE SPECTROSCOPY, VOLUME 2
Edited by E. L. Wehry

APPLIED ATOMIC SPECTROSCOPY, VOLUME 1
Edited by E. L. Grove

APPLIED ATOMIC SPECTROSCOPY, VOLUME 2
Edited by E. L. Grove

TRANSFORM TECHNIQUES IN CHEMISTRY
Edited by Peter R. Griffiths

ION-SELECTIVE ELECTRODES IN ANALYTICAL CHEMISTRY
Edited by Henry Freiser

Transform Techniques in Chemistry

Edited by
Peter R. Griffiths
Ohio University, Athens

Plenum Press · New York and London

Library of Congress Cataloging in Publication Data

Main entry under title:

Transform techniques in chemistry.

(Modern analytical chemistry)
Includes bibliographical references and index.
1. Fourier transform spectroscopy.
I. Griffiths, Peter R., 1942-
QD96.F68T7 543'.085 77-29271
ISBN 0-306-31070-8

A Division of Plenum Publishing Corporation
227 West 17th Street, New York, N.Y. 10011

Printed in the United States of America

Contributors

Michael B. Comisarow, Chemistry Department, University of British Columbia, Vancouver, B.C., Canada V6T 1W5

James W. Cooper, Department of Chemistry, Tufts University, Medford, Massachusetts 02155

Thomas C. Farrar, Chemistry Division, National Science Foundation, Washington, D.C., 20550

Charles T. Foskett, Digilab, Inc., Cambridge, Massachusetts 02139

Peter R. Griffiths, Department of Chemistry, Ohio University, Athens, Ohio 45701

Martin Harwit, Center for Radiophysics and Space Research, Cornell University, Ithaca, New York 14853

Peter C. Jurs, Department of Chemistry, The Pennsylvania State University, University Park, Pennsylvania 16802

Russell D. Larsen, Department of Chemistry, University of Nevada—Reno, Reno, Nevada 89507. *Present address*: Department of Chemistry, University of Michigan, Ann Arbor, Michigan 48109

John O. Lephardt, Phillip Morris U.S.A., Research Center, P. O. Box 26583, Richmond, Virginia 23261

Alan G. Marshall, Chemistry Department, University of British Columbia, Vancouver, B.C., Canada V6T 1W5

Charles L. Wilkins, Department of Chemistry, University of Nebraska, Lincoln, Nebraska 68501

Preface

The application of the Fourier transform is being seen to an increasing extent in all branches of chemistry, but it is in the area of chemical analysis that the greatest activity is taking place. Fourier transform infrared and nuclear magnetic resonance spectrometry are already routine methods for obtaining high-sensitivity IR and NMR spectra. Analogous methods are now being developed for mass spectrometry (Fourier transform ion cyclotron resonance spectrometry) and microwave spectroscopy, and Fourier transform techniques have been successfully applied in several areas of electrochemistry. In addition the fast Fourier transform algorithm has been used for smoothing, interpolation, and more efficient storage of data, and has been studied as a potential method for more efficient identification of samples using pattern recognition techniques.

Linear transforms have also been shown to be useful in analytical chemistry. Probably the most important of these is the Hadamard transform, which has been applied in alternative methods for obtaining IR and NMR data at high sensitivity. Even though *measurements* involving this algorithm will probably not be applied as universally as their Fourier transform analogs, in the area of pattern recognition application of the Hadamard transform will in all probability prove more important than application of the Fourier transform.

In this book, distinguished investigators in the various fields mentioned above have written on their area of expertise at a level that should be understandable to graduate analytical chemists and to the advanced undergraduate, as well as the professional maintaining and updating research skills. It is hoped that the similarities between the various spectroscopic and data manipulation techniques will become evident throughout the book. We have omitted treatment of crystallographic applications because they seem outside the mainstream of analytical interests.

After a brief look at the history of transform techniques in chemistry and an editorial forecast of their applications in the future (Chapter 1), the mathematical basis of the Fourier transform is introduced by Charles Foskett of Digilab, Inc., in Chapter 2. In Chapter 3, Alan Marshall and Melvin Comisarow of the University of British Columbia discuss the foundation of multiplex methods in spectroscopy, showing the origin of the advantages of instruments that do not measure a spectrum directly, but rather generate a signal that is related to the spectrum through the Fourier or Hadamard transform. In Chapter 4, James Cooper of Tufts University discusses the nature of the data-handling and computer capabilities required for on-line Fourier transform spectrometry.

The next six chapters describe the theory, instrumentation, and applications of several different types of multiplex spectroscopy. In Chapters 5 and 6 the editor describes Fourier transform infrared spectrometry, and in Chapter 7 Martin Harwit of Cornell University describes Hadamard transform infrared spectrometry, including how this technique may be used to multiplex information both spectrally and spatially. In Chapter 8, Thomas Farrar of the National Science Foundation introduces Fourier transform–NMR spectrometry, and some of the more recent advances in this subject are described in the subsequent chapter (Chapter 9) by James Cooper. In Chapter 10, the nature of Fourier transform ion cyclotron resonance spectrometry is introduced by Melvin Comisarow.

Several of the more important applications of the Fourier transform in data processing are discussed in Chapter 11 by John Lephardt of Philip Morris U.S.A., and in Chapter 12, the application of transform techniques in pattern recognition is described by Charles Wilkins of the University of Nebraska and Peter Jurs of Pennsylvania State University. Chapter 13, by Russell Larsen of the University of Nevada at Reno, describes potential applications of binary transforms for very rapid data processing; to a greater extent than the previous chapters, this represents a look into the future and is written to a slightly more advanced audience than the earlier chapters. We believe it represents an important new transform technique of the future and merits a more advanced treatment since there is little readily available reference material on this subject. Finally, the editor has summarized the applications of the Fourier transform in electrochemistry in a chapter that illustrates not only the sensitivity advantage obtained through acquiring data at several frequencies simultaneously but also shows how the information content of different types of electrochemical data may be enhanced by the application of the Fourier transform independently of the manner in which the data were acquired.

That a volume such as this may be compiled is a tribute to the many pioneers in all the areas covered in this book. The fact that so many of the techniques that are described here are now available in the market place is similarly a tribute to the individuals and companies who had faith that transform techniques were of sufficient value to be developed commercially. On a more personal level, the editor and many of the authors would like to thank the various agencies that, in such timely fashion, supported their research and the many co-workers without whose cooperation and hard work many of the results described in this book would not have been found. Finally, the secretarial assistance so valuable in preparing the manuscripts of these chapters is gratefully acknowledged.

Peter R. Griffiths

Contents

Chapter 1

Transform Techniques in Chemistry: Past, Present, and Future

Peter R. Griffiths

Chapter 2

The Fourier Transform and Related Concepts: A First Look

Charles T. Foskett

Chapter 3

Multichannel Methods in Spectroscopy

Alan G. Marshall and Melvin B. Comisarow

Chapter 4

Data Handling in Fourier Transform Spectroscopy

James W. Cooper

Chapter 5

Fourier Transform Infrared Spectrometry: Theory and Instrumentation

Peter R. Griffiths

Chapter 6

Infrared Fourier Transform Spectrometry: Applications to Analytical Chemistry

Peter R. Griffiths

Chapter 7

Hadamard Transform Analytical Systems

Martin Harwit

Chapter 8

Pulsed and Fourier Transform NMR Spectroscopy

Thomas C. Farrar

Chapter 9

Advanced Techniques in Fourier Transform NMR

James W. Cooper

Chapter 10

Fourier Transform Ion Cyclotron Resonance Spectroscopy

Melvin B. Comisarow

Chapter 11

Fourier Domain Processing of General Data Arrays

John O. Lephardt

Chapter 12

Fourier and Hadamard Transforms in Pattern Recognition

Charles L. Wilkins and Peter C. Jurs

Chapter 13

Spectral Representations for Quantized Chemical Signals

Russell D. Larsen

Chapter 14

Applications of the FFT in Electrochemistry

Peter R. Griffiths

Chapter 1

Transform Techniques in Chemistry: Past, Present, and Future

Peter R. Griffiths

1.1. THE PAST

1.1.1. Optical Spectroscopy

Although the use of transform techniques in analytical chemistry and applied spectroscopy has only become widespread in the past five years, the history of this subject can be traced back to the middle of the nineteenth century when the effect of the interference of light was first used to derive spectroscopic information. In 1862, Fizeau[1] used Newton's rings to show that the yellow sodium radiation was a doublet whose separation was 1/980 of their average wavelength. At the end of the century Michelson designed the *interferometer*, which now bears his name.[2,3] The initial uses of this instrument for spectroscopic purposes concerned the determination of spectral profiles through the use of the visibility technique,[4] which is essentially a study of the envelope of what we now call the *interferogram*. Rayleigh[5] pointed out that a unique spectral distribution cannot be found from the visibility curve itself, and the Fourier transform of the interferogram is needed to calculate the spectrum unequivocally.

The actual calculation of a digital Fourier transform was beyond the technical resources available at the turn of the century and, in a remarkable

Peter R. Griffiths • Department of Chemistry, Ohio University, Athens, Ohio 45701

attempt to circumvent this limitation, Michelson developed a harmonic synthesizer that was designed to output the Fourier transform of an input signal. This was an 80-channel device consisting of 80 gears driving 80 wheels to rotate at speeds proportional to the integers 1 through 80. Each wheel rocked a lever, which in turn generated a simple harmonic motion in an adjustable arm. Each arm was linked by springs to an axle, and the composite motion of all 80 arms moved a pen. There is no doubt that this was the first Fourier transform computer; however, there is no record of Michelson ever using this device to obtain a spectrum from an optical signal. He did successfully reinvert synthetic signals, which in itself is a remarkable feat since it was achieved more than 40 years before the same operation was performed on a digital computer.

Michelson was able to show that the red Balmer line of hydrogen is a doublet and that the red line of cadmium is exceptionally narrow. He proposed this line as a wavelength calibrant, and it was used as the standard of length until 1960, when it was supplanted by the orange line of krypton produced by a lamp operating at the triple point of nitrogen. He also showed that the green line of natural mercury is a complicated multiplet, which he was unable to resolve completely with his 80-channel harmonic synthesizer.

The first true interferogram was published by Rubens and Wood in 1911[6]; they were investigating the far-infrared radiation emitted by a Welsbach mantle, and chose to use an interferometer because a quartz prism with sufficient dispersion absorbed too much of the incident radiation. They did not use a Michelson interferometer, but rather one that worked using the same principles as the Newton's rings apparatus employed by Fizeau.[1] Rubens and his co-workers, in this and subsequent work, guessed a spectral distribution, calculated the Fourier transform, and then adjusted the estimate to try to make the calculated and observed interferograms match. No reason was ever given why this method was preferred to direct Fourier transformation. In Rubens' work the multiplex gain (*vide infra*) was realized, but there is no evidence that it was ever appreciated.

Several refinements and applications of interferometry were made in the next forty years, but it remained a tool for high-resolution spectroscopy until Jacquinot in France and Fellgett in England recognized two important advantages of interferometers for the measurement of spectra. Jacquinot[7,8] recognized that the optical energy *throughput* (the product of the area and solid angle of a beam at its focus) of a Michelson interferometer used for spectroscopy is greater than that of a monochromator used for spectral measurements at the same resolution. Fellgett[9] not only published the first numerically transformed interferogram, but also recognized that an interferometer gave a fundamental advantage over a scanning monochromator, that of *multiplexing* the spectral information. The multiplex, or Fellgett, advantage is the basis for several types of spectrochemical and electrochemical methods described in this book.

The gain in signal-to-noise ratio resulting from the application of the multiplex principle may be appreciated intuitively on the basis that the signal integrates in direct proportion to T, the time of observation, whereas the noise integrates in proportion to $T^{1/2}$. If a total observation time T is available for the exploration of M spectral elements, they may be investigated sequentially or simultaneously, provided that, in the latter case, they may be *decoded* at the end of the measurement. In sequential investigations, each element is observed for an average time T/M, with a noise level proportional to $(T/M)^{1/2}$, so that the signal-to-noise ratio is proportional to $(T/M)^{1/2}$. In the simultaneous investigation, each element is observed for a time T, and the signal-to-noise ratio is proportional to $T^{1/2}$, indicating a gain of $M^{1/2}$ over the sequential case. For spectra measured with equal signal-to-noise ratios on each type of spectrometer, the observation time required for the sequential measurement is M times longer than the simultaneous measurement.

The time, effort, and cost involved in decoding the spectral information from the output of a multiplex spectrometer was a principal reason for the reluctance of chemists to use multiplex methods for infrared spectroscopy, especially in view of the fact that most infrared spectrochemical data could be obtained using a scanning monochromator (albeit at less than the optimum signal-to-noise ratio). Only those scientists who could not obtain acceptable spectra using a monochromator because of the weakness of their sources had a strong interest in developing multiplex methods. It is therefore not surprising that many of the pioneers of Fourier transform infrared (FT–IR) spectrometry were astronomers and far-infrared spectroscopists. Connes and Mertz, who led the development of high-resolution and rapid-scanning low-resolution interferometers, respectively, are both astronomical spectroscopists. Gebbie and Strong, who pioneered the development of Michelson and lamellar grating interferometers, respectively, were both interested in far-infrared measurements. The first Michelson interferometer sold commercially was designed for measurements in the far-infrared region.

It is surprising that, despite the simplicity of Michelson interferometers designed for far-infrared spectroscopy, the first commercial instrument (manufactured by Research and Industrial Instruments Corporation in England) was not delivered until 1964, well over a decade after Jacquinot and Fellgett showed the fundamental advantages of this type of instrument. Since that time, the number of FT–IR spectrometers has steadily increased. The development of the fast Fourier transform (FFT) algorithm has substantially reduced the time to compute a spectrum, to the point that it is rarely the rate-limiting step in spectroscopic measurements. The development of small, relatively inexpensive data systems has greatly increased the flexibility and ease with which these instruments can be used. Now that the advantages of FT–IR spectrometry are becoming appreciated by chemical spectroscopists, several new applications for infrared spectroscopy are finally being used routinely in the analytical laboratory.

1.1.2. NMR Spectroscopy

To a certain extent the history of Fourier transform nuclear magnetic resonance (FT–NMR) spectroscopy parallels that of FT–IR, but in a much shorter timeframe. The year 1976 marked the thirtieth anniversary of the discovery of nuclear magnetic resonance in bulk materials.[10,11] Bloch *et al.* showed that NMR can be observed in several ways, two of which have proved to be most important. The *slow passage* experiment consists of slowly sweeping the radio frequency (RF) applied to a sample in a fixed magnetic field (or, alternatively, slowly sweeping the field with a fixed RF). When the RF perturbs an ensemble of like nuclei resonance occurs, producing an absorption line in the spectrum. This method of measuring NMR spectra is commonly known as a continuous wave (CW) technique, since the RF is applied continuously while the spectrum is observed.

A second method for observing magnetic resonance, also suggested by Bloch *et al.* in 1946 and put into practice initially by Torrey in 1949[12] and Hahn in 1950[13], makes use of short *pulses* of RF power at a discrete frequency. The observation of the nuclear spin system is made after the RF is turned off. Most of the early work in pulsed NMR concerned studies of nuclear relaxation rather than the measurement of spectra. In 1957 Lowe and Norberg[14] showed that the free induction decay signal obtained in pulsed NMR experiments and the absorption spectrum obtained in a CW NMR experiment constitute a Fourier transform pair, but it was only a decade ago that Ernst and Anderson[15] recognized the multiplex advantage for pulsed NMR over CW methods.

In the last decade, FT–NMR spectrometers have evolved from the large, intricate, expensive, and very touchy first-generation instruments to turn-key third-generation spectrometers, which are smaller, less expensive, much more reliable, and simpler to operate than the first-generation instruments. FT–NMR spectrometers have opened up a completely new area of investigation, that of ^{13}C NMR spectrometry without isotopically enriching the sample. ^{13}C spectra of samples with carbon at its natural isotopic abundance can now be measured in minutes rather than days, and are being found very useful in assigning the molecular structure of a compound. The great sensitivity of FT–NMR has also been used for the measurement of proton magnetic resonance spectra of samples at very low concentration and, as we will see in Section 1.2, NMR spectra with other nuclei, such as ^{19}F and ^{31}P.

1.1.3. Data Processing

Data processing using the Fourier transform algorithm has been used extensively in the past by electrical engineers. However only recently, possibly alerted by the nature of data processing that can be carried out on interferograms and free induction decay signals *before* the FFT, have chemists begun to realize what can be done if data measured on *conventional* spectrometers are first subjected to a transform algorithm and then treated in an analogous fashion.

Most of this work has been published since 1973 and is rightly treated in Section 1.2. The current status of the application of transform techniques to analytical chemistry and applied spectroscopy will be illustrated in Section 1.2 and covered in more detail in the remainder of this book.

1.2. THE PRESENT

Any survey of papers published in chemical journals over the past few years would show that the Fourier transform algorithm is being applied with increasing frequency to a variety of problems in analytical chemistry and applied spectroscopy. Other transform algorithms, such as the Hadamard and Walsh transforms, are also being applied to chemical problems, albeit less frequently than the Fourier transform. As a demonstration of the relevance of transform techniques to the field of analytical chemistry today, a survey of the feature articles and contributed papers published in the last 60 issues of the journal *Analytical Chemistry* was made. The next 42 references[16-57] are all to papers whose principal subject matter is an application of the Fourier (primarily) or Hadamard transform.

The *Instrumentation* feature articles published in *Analytical Chemistry* are indicative of the more important developments in instrumental analysis; in the past five years, papers have appeared in this section on Fourier and Hadamard transform methods in spectroscopy in general,[16] Hadamard transform infrared spectroscopy,[17] Fourier transform infrared spectroscopy,[18] atomic spectrochemical measurements with a Fourier transform spectrometer,[19] ^{13}C nuclear magnetic resonance spectroscopy by Fourier transform techniques,[20] and the application of Fourier transform methods to electrochemistry.[21,22] Apparently the editors of this journal consider transform techniques important enough to occupy over 10% of the *Instrumentation* feature articles.

The research papers contributed to *Analytical Chemistry* since 1972 demonstrate the wide variety of areas to which transform techniques are currently being applied. They fall into two categories, either describing *measurements* made using an instrument whose output is the transform of the conventional spectrum or *applications* in which data measured in conventional fashion have been manipulated through the use of a transform algorithm. Most of the former types of measurement are possible because of the multiplex advantage of the spectrometer, and the number of papers describing such measurements currently exceeds the number of papers in which data manipulation using transform techniques is described. Because of the benefits accrued through the application of transform techniques to experiments of both categories, it is highly probable that the next few years will see an increase in the number of analytical problems that will be investigated using transform techniques.

Perhaps the area to which transform techniques have been most widely applied in chemistry is nuclear magnetic resonance spectrometry; papers on FT–NMR have appeared in *Analytical Chemistry* not only describing studies on such "conventional" nuclei as ^{1}H[23-25] and ^{13}C,[26-29] but also less commonly used nuclei, such as ^{19}F[30] and ^{31}P,[31,32] and even such unusual nuclei as ^{7}Li, ^{3}H, and ^{3}He.[33] FT–IR ranks second in popularity among "transform spectroscopies" to FT–NMR. Among the topics studied by FT–IR that have been reported in *Analytical Chemistry* are the infrared identification of components separated by gas chromatography[34] and thin-layer chromatography,[35] measurement of the spectra of small quantities of materials,[36] and the quantitative determination of closely related compounds.[37-39] Analogous methods have been applied to the measurement of ultraviolet–visible spectra,[40,41] but with less marked success because the multiplex advantage is offset by the fact that the noise level of a photomultiplier detector increases with the incident signal.

It may be noted that only a small proportion of the published work in the area of FT–NMR and FT–IR has appeared in *Analytical Chemistry*. Applications of FT–NMR have been reported in many places, from general journals, such as the *Journal of the American Chemical Society*, to highly specialized periodicals such as *Organic Magnetic Resonance*. FT–IR applications can also be found in many different journals, but by far the greatest proportion of papers on recent instrumental developments in FT–IR has appeared in *Applied Spectroscopy*.

It is not only in the areas of spectroscopy *per se* that Fourier transform techniques have been applied to obtain new and important data. The work of Smith at Northwestern University[42-45] illustrates how *electrochemical* data can also be acquired with the multiplex advantage.

The types of data manipulation using transform techniques that have been described in recent editions of *Analytical Chemistry* include methods for more efficient data acquisition,[46] data conversion,[47,48] storage,[49] smoothing,[50,51] interpolation,[52] and deconvolution.[53] These methods have been applied in such different areas as electron spin resonance spectroscopy,[49] AC polarography,[50] circular dichroism spectroscopy,[51] photodiode array spectrometry,[52] and steric exclusion chromatography.[53]

One of the most important new ways of analyzing chemical data is through the use of pattern recognition techniques and it will come as no surprise to hear that the Fourier and Hadamard transform algorithms have also been applied in this field.[54-57]

1.3. THE FUTURE

A prediction of the future of transform techniques in analytical chemistry is difficult, if not impossible. Nevertheless, in this section we shall attempt such a prediction, even though in five years time this section will probably have been proven to be far from accurate!

FT–NMR spectrometry is now becoming a routine analytical tool for measuring natural abundance ^{13}C NMR spectra. The cost of special-purpose ^{13}C FT–NMR spectrometers has dropped dramatically in the last few years, but will probably not come down much further. However, it is highly unlikely that the popularity of the technique will decrease, the reverse being far more probable. Microsampling techniques for ^{1}H NMR are still being developed, and the application of proton magnetic resonance for trace analysis will probably increase significantly.

FT–IR spectrometry has been accepted far more slowly than FT–NMR; however, the reliability and flexibility of these instruments is now so good that one can predict a continued increase in their number, especially for mid-infrared spectroscopy. One application that will be of particular significance is the on-line measurement of the infrared spectra of peaks eluting from gas chromatographs and high-performance liquid chromatographs using rapid-scanning FT–IR spectrometers. The detection limit of on-line GC–IR measurements is now about 100 ng and this may be forecast to drop by about an order of magnitude, but little more, in the next few years. Tunable diode lasers will assume an increasingly important role in high-resolution infrared spectroscopy, but the small ranges over which these lasers may conveniently be tuned suggest that FT–IR spectrometers will still be used for general-purpose measurements over wide spectral ranges if resolution widths narrower than 0.05 cm^{-1} are not required.

Hadamard transform infrared spectrometers will probably not find general acceptance for routine analytical spectroscopy, since they do not appear to be as sensitive as FT–IR spectrometers. However their use for two-dimensional spectroscopy (spectral imaging), as discussed by Harwit in Chapter 7 of this volume, will certainly increase, since there are almost no other ways to obtain these data. The routine use of either Fourier or Hadamard transform spectrometers for the measurement of ultraviolet–visible spectra is quite unlikely in view of the fact that photomultipliers are photon noise limited.

Little has been mentioned in this chapter about two other types of spectroscopy whose sensitivity has been increased by multiplexing the output. Fourier transform ion cyclotron resonance (FT–ICR) spectrometry is being developed both in academic research laboratories and commercially, and is described in Chapter 10 of this book. It is certain that application of this technique will enable mass spectra to be measured very rapidly and at very high sensitivity, and it is highly probable that much more will be heard about FT–ICR in the future.

During the preparation of the manuscript of this volume, the first pulsed microwave Fourier transform spectrometer has been described.[(58)] This technique is somewhat analogous to pulsed NMR, but differs in that a high-power pulse train is sent through the absorption cell, and the transient emission of excited transitions in a band up to 50 MHz around the carrier frequency is observed after each pulse using a superheterodyne detection system. This system shows a marked improvement in signal-to-noise ratio and resolution over conventional spectrometers. However, conventional microwave spectrometers are themselves very sensitive and specific, but for some reason have not yet been accepted by the chemical community as a useful analytical technique. It is doubtful in the mind of this author that the increased sensitivity obtained by pulsing the microwave source will change this situation appreciably.

In terms of data manipulation operations, the next few years will almost certainly see more applications of transforms of the type reported in references 50–54. It is very probable that transform methods for signal-to-noise enhancement, interpolation, convolution (and deconvolution), and correlation will become increasingly important in many types of instrumental analysis.

REFERENCES

1. H. Fizeau, *Ann. Chim. Phys.* (3), **66**, 429 (1862).
2. A. A. Michelson, *Phil. Mag.* (5), **31**, 256 (1891).
3. A. A. Michelson, *Phil. Mag.* (5), **34**, 280 (1892).
4. M. Born and E. Wolf, *Principles of Optics,* 2nd ed., pp. 320ff, The Macmillan Co., New York, 1964.
5. Lord Rayleigh, *Phil. Mag.* (5), **34**, 407 (1892).
6. H. Rubens and R. W. Wood, *Phil. Mag.* **21**, 249 (1911).
7. P. Jacquinot and C. J. Dufour, *J. Rech. C.N.R.S.* **6**, 91 (1948).
8. P. Jacquinot, *Rep. Prog. Phys.* **23**, 267 (1960).
9. P. B. Fellgett, Ph.D. thesis, University of Cambridge, 1951.
10. F. Bloch, *Phys. Rev.* **70**, 460 (1946).
11. F. Bloch, W. W. Hansen, and M. Packard, *Phys. Rev.* **69**, 127 (1946).
12. H. C. Torrey, *Phys. Rev.* **76**, 1059 (1949).
13. E. L. Hahn, *Phys. Rev.* **80**, 580 (1950).
14. I. J. Lowe and R. E. Norberg, *Phys. Rev.* **107**, 46 (1957).
15. R. R. Ernst and W. A. Anderson, *Rev. Sci. Instrum.* **37**, 93 (1966).
16. A. G. Marshall and M. B. Comisarow, *Anal. Chem.* **47**, 491A (1975).
17. J. A. Decker, *Anal. Chem.* **44**, 127A (1972).
18. P. R. Griffiths, *Anal. Chem.* **46**, 645 A (1974).
19. G. Horlick and W. K. Yuen, *Anal. Chem.* **47**, 775A (1975).
20. G. A. Gray, *Anal. Chem.* **47**, 547A (1975).
21. D. E. Smith, *Anal. Chem.* **48**, 221A (1976).
22. D. E. Smith, *Anal. Chem.* **48**, 517A (1976).
23. J. E. Sarneski and C. N. Reilley, *Anal. Chem.* **48**, 1303 (1976).
24. M. L. Lee, M. Novotny, and K. D. Bartle, *Anal. Chem.* **48**, 1566 (1976).
25. H. C. Dorn and D. L. Wooton, *Anal. Chem.* **48**, 2146 (1976).
26. J. E. Sarneski, H. L. Surprenant, F. K. Molen, and C. N. Reilley, *Anal. Chem.* **47**, 2116 (1975).
27. O. A. Subbartin and N. M. Sergeyev, *Anal. Chem.* **48**, 545 (1976).
28. D. W. Vidrine and P. E. Petersen, *Anal. Chem.* **48**, 1301 (1976).
29. J. N. Shoolery and W. L. Budde, *Anal. Chem.* **48**, 1458 (1976).

30. E. G. Brame and F. W. Yeager, *Anal. Chem.* **48**, 709 (1976).
31. T. W. Gurley and W. M. Ritchey, *Anal. Chem.* **47**, 1444 (1975).
32. T. W. Gurley and W. M. Ritchey, *Anal. Chem.* **48**, 1137 (1976).
33. A. Attalla and R. C. Bowman, *Anal. Chem.* **47**, 728 (1975).
34. J. O. Lephardt and B. J. Bulkin, *Anal. Chem.* **45**, 706 (1973).
35. C. J. Percival and P. R. Griffiths, *Anal. Chem.* **47**, 154 (1975).
36. D. H. Anderson and T. E. Wilson, *Anal. Chem.* **47**, 2482 (1975).
37. R. T. Yang and M. J. D. Low, *Anal. Chem.* **45**, 2014 (1973).
38. R. M. Gendreau, P. R. Griffiths, L. E. Ellis, and J. R. Anfinsen, *Anal. Chem.* **48**, 1907 (1976).
39. R. M. Gendreau and P. R. Griffiths, *Anal. Chem.* **48**, 1910 (1976).
40. F. W. Plankey, T. H. Glenn, L. P. Hart, and J. D. Winefordner, *Anal. Chem.* **46**, 1000 (1974).
41. T. L. Chester, J. J. Fitzgerald, and J. D. Winefordner, *Anal. Chem.* **48**, 779 (1976).
42. D. E. Glover and D. E. Smith, *Anal. Chem.* **45**, 1869 (1973).
43. S. C. Creason and D. E. Smith, *Anal. Chem.* **45**, 2401 (1973).
44. K. R. Bullock and D. E. Smith, *Anal. Chem.* **46**, 1069 (1974).
45. K. R. Bullock and D. E. Smith, *Anal. Chem.* **46**, 1567 (1974).
46. P. C. Kelley and G. Horlick, *Anal. Chem.* **45**, 518 (1973).
47. R. C. Williams, R. M. Swanson, and C. L. Wilkins, *Anal. Chem.* **46**, 1803 (1974).
48. R. C. Williams and F. D. Crary, *Anal. Chem.*, **48**, 1150 (1976).
49. T. Nishikawa and K. Someno, *Anal. Chem.* **47**, 1290 (1975).
50. J. W. Hayes, D. E. Glover, D. E. Smith, and M. W. Overton, *Anal. Chem.* **45**, 277 (1973).
51. C. A. Bush, *Anal. Chem.* **46**, 890 (1974).
52. G. Horlick and W. K. Yuen, *Anal. Chem.* **48**, 1643 (1976).
53. T. A. Maldacker, J. E. Davis, and L. B. Rogers, *Anal. Chem.* **46**, 647 (1974).
54. B. R. Kowalski and C. F. Bender, *Anal. Chem.* **45**, 2234 (1973).
55. J. B. Justice and T. L. Isenhour, *Anal. Chem.* **46**, 223 (1974).
56. T. R. Brunner, R. C. Williams, C. L. Wilkins, and P. J. McCombie, *Anal. Chem.* **46**, 1798 (1974).
57. T. R. Brunner, C. L. Wilkins, R. C. Williams, and P. J. McCombie, *Anal. Chem.* **47**, 662 (1975).
58. J. Ekkers and W. H. Flygare, *Rev. Sci. Instrum.* **47**, 448 (1976).

Chapter 2

The Fourier Transform and Related Concepts: A First Look

Charles T. Foskett

2.1. INTRODUCTION: GUITAR TUNING

The Fourier Transform is one of the most common transformations occurring in nature. Certain features associated with this transform are found used by man in a variety of occupations and applications. For example, Fourier transforms are used in encephalography, X-ray crystallography, radar, network design, and chemical Fourier transform spectroscopy in both nuclear magnetic resonance and infrared analysis. One example of a physical Fourier transform is far-field or Fraunhofer diffraction; this optical phenomenon occurs with narrow slits in dispersive spectroscopy.

Certain features of the method are used in tuning pianos and guitars. Consider this method of tuning a guitar. The player strikes an open string, let us say the lower E-string (Figure 2.1) and adjusts it until it is in the proper tune. He then depresses the E-string at the fifth fret and strikes it (this is now a properly tuned A-note). He then strikes the open A-string and adjusts the tension on the A-string until it is in tune with the depressed (fifth-fret) E-string.

One criterion for judging whether or not the A-string is properly tuned is whether or not one hears two notes, a beat note and sum note, or just one note, the A-note. The further apart the two notes are, the higher the frequency of the beat note. The closer they are, the less frequently the beats occur.

Charles T. Foskett • Digilab, Inc., 237 Putnam Avenue, Cambridge, Massachusetts 02139

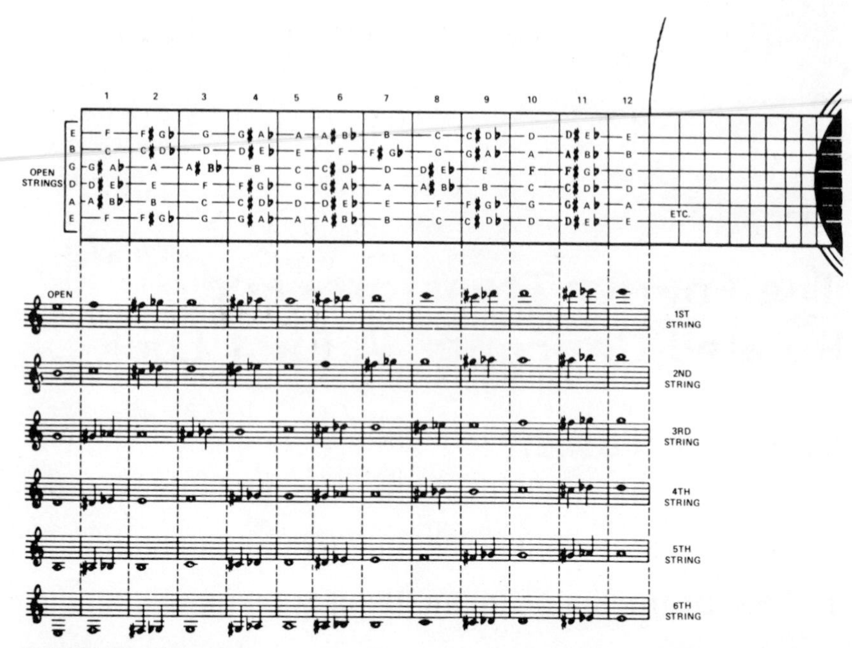

FIGURE 2.1. The scales on a guitar. The basic six notes are E, A, D, G, B, E. The intermediate notes on the full scale are attained by depressing the strings tightly against the frets indicated, thus altering the resonant frequency of the string.

This method of tuning a string instrument illustrates two important concepts of Fourier transforms: the linear combination of signals and resolution. Consider the following idealized expression for a single note of frequency f_1:

$$h_1(t) = a_1 \cos 2\pi f_1 t \tag{2.1}$$

where a_1 is the amplitude of the time-varying sinusoid. The second note may be expressed in time as

$$h_2(t) = a_2 \cos 2\pi f_2 t \tag{2.2}$$

The linear combination of tones h_1 and h_2 we will express as

$$h = a_1 \cos 2\pi f_1 t + a_2 \cos 2\pi f_2 t \tag{2.3}$$

and we will assume that the player struck each string with equal intensity so that $a_1 = a_2$. At this point we recall a fundamental trigonometric identity:

$$\cos X + \cos Y = 2 \cos \left[\tfrac{1}{2}(X + Y)\right] \cos \left[\tfrac{1}{2}(X - Y)\right] \tag{2.4}$$

By substitution of the two notes in (2.1) and (2.2) into this identity,

$$h = 2a\cos\left[2\pi\left(\frac{f_1+f_2}{2}\right)t\right]\cos\left[2\pi\left(\frac{f_1-f_2}{2}\right)t\right] \tag{2.5}$$

we can identify the beat frequency as one-half the difference between f_1 and f_2, and the sum note as the average value of f_1 and f_2. This is shown in Figure 2.2. Now what does the guitar player hear as he adjusts his string tension? As his string gets closer to the correct tension, the sum frequency $\frac{1}{2}(f_1 + f_2)$ gets closer and closer to f_1, the A-note. The difference or beat frequency $\frac{1}{2}(f_1 - f_2)$ gets lower and lower, and the time between the beats, t_B, gets greater. When f_2 is identically tuned to f_1 the beats disappear, i.e., t_B is infinite.

In tuning his guitar, the player has linearly combined two signals and analyzed their relative position by the use of beats. He intuitively knows

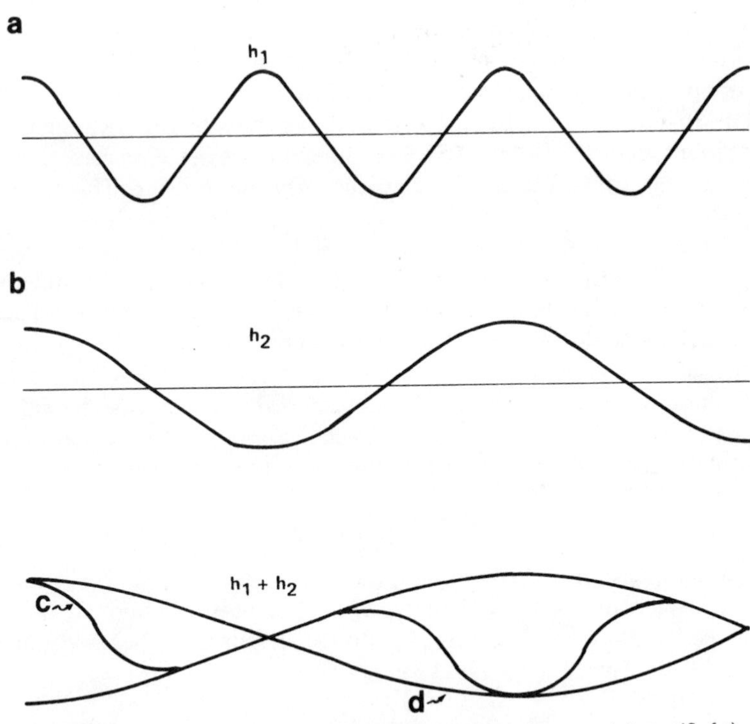

FIGURE 2.2. The signals h_1 and h_2 are schematically shown at (a) $\cos(2\pi f_1 t)$ and (b) $\cos(2\pi f_2 t)$. The linear combinations of h_1 and h_2 are shown producing the sum note (c) $\cos\{2\pi[(f_1+f_2)/2]t\}$, and the beat note (d) $\cos\{2\pi[(f_1-f_2)/2]t\}$.

a varietal expression of Heisenberg's quantum mechanical uncertainty principle, which is that his ability to resolve the difference between f_1 and f_2 is proportional to his patience in measuring the time between beats:

$$(f_1 - f_2)\, t_B \geq \tfrac{1}{2} \tag{2.6}$$

2.2. DIFFERENCES IN SPACE AND TIME: PHASE

The guitar player described in Section 2.1 generated a signal that is time dependent. A child creating a wave by dropping a stone in water creates a similar sinusoidal signal, but one that at any instant in time is space dependent:

$$h(r) = a \cos 2\pi\sigma r \tag{2.7}$$

where r is the radial distance from the stone and σ the "spatial frequency" of the wave. Most of our discussion in this chapter will use the variables f and t for frequency and time, but there will be separate cause to use variables for other coordinate systems as well, such as distance as above, and the pair of optical retardation x and wavenumber ν.

In any of these coordinate systems, it is possible for two signals to exist simultaneously, but to have had different origins. Consider that our guitar player strikes *first* the E-string and *then* the A-string. This is shown in Figure 2.3.

In the jargon of signal analysts, h_1 *leads* h_2 by $t_2 - t_1$, or h_2 *lags* h_1 by $t_2 - t_1$. This time lead or lag can be translated into a phase difference. Phase is generally expressed as an angle and refers to the number of degrees or radians a sine or cosine function has progressed from its origin. Consequently, at $t = 0$, $\cos 2\pi f_1 t$ has zero phase. However, at $t = 4/f_1$, $\cos 2\pi f_1 t$ has a phase of $\pi/2$ or 90°. Similarly, a cosine *leads* a sine wave by $\pi/2$ or a sine wave *lags* a cosine wave by $\pi/2$. In the case of our two signals above, h_1 is out of phase with respect to h_2 at time t_2 by the phase *angle*

$$\phi = 2\pi f_1 (t_2 - t_1) \tag{2.8}$$

Certainly we may ask why phase is important. One reason is that it enables us to take account of the different origin of signals. Suppose we were to introduce the phase of equation (2.8) into the signals described in equations (2.1) and (2.2). Then $h_1(t)$ would be expressed as

$$h_1(t) = a \cos (2\pi f_1 t + \phi) \tag{2.9}$$

If t_2 and t_1 were such that $t_2 - t_1$ were equal to $2/f_1$, then ϕ would be exactly equal to π. This phase difference, although highly improbable in practice,

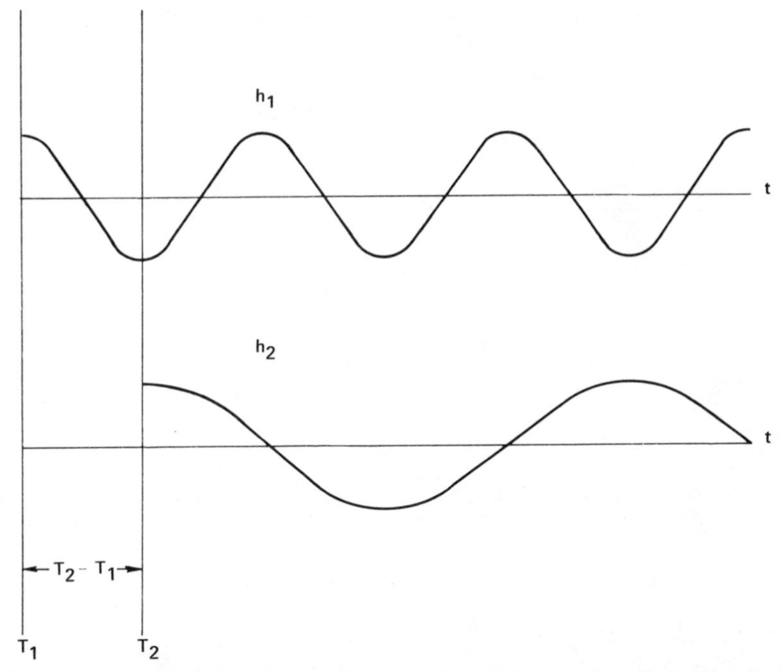

FIGURE 2.3. The phase lag between h_2 and h_1 is shown, since h_1 begins at time T_1 and h_2 begins at a later time T_2.

would cause the sum of the two signals to vanish once the guitar was in tune. That is, a phase difference of π is an inversion of the sign of the first cosine function, causing it to be *out of phase* with respect to the second.

2.3. SUMS, INTEGRALS, AND ORTHOGONALITY

Linear combination is an element critical to Fourier analysis. That is, the quantities or signals must be combined or summed in direct proportion to their amplitudes, not quadratically or cubically. In Section 2.1 we combined signals h_1 and h_2 to form the signal h, which can be generalized in a linear sum:

$$h = \sum_{l=0}^{N-1} h_l \tag{2.10}$$

or

$$h(t) = \sum_{l=0}^{N-1} a_l \cos 2\pi f_l t \tag{2.11}$$

In both of these expressions the combination is completely linear. In the limit of large N (and given certain restrictive conditions on the a_l that may be ignored here), we can write

$$h(t) = \frac{1}{2\pi}\int a(f)\cos 2\pi ft\, df \tag{2.12}$$

and again, each amplitude $a(f)$ and each cosine function is linearly weighted with respect to all the others.

Another form of a linear sum is the vector **R**, which describes a point in three-dimensional space,

$$\mathbf{R} = \mathrm{X}\hat{\mathbf{i}} + \mathrm{Y}\hat{\mathbf{j}} + \mathrm{Z}\hat{\mathbf{k}} \tag{2.13}$$

We may recall that the length of this vector is given by

$$R_l = \mathbf{R}_l \cdot \mathbf{R}_l \tag{2.14}$$

While the length R_l is *not unique* (it describes a sphere in Euclidean space) $\mathbf{R}_l$ is a unique point because the unit vectors $\hat{\mathbf{i}}$, $\hat{\mathbf{j}}$, and $\hat{\mathbf{k}}$ are orthogonal, i.e.,

$$\hat{\mathbf{i}}\cdot\hat{\mathbf{j}} = \hat{\mathbf{k}}\cdot\hat{\mathbf{j}} = \hat{\mathbf{i}}\cdot\hat{\mathbf{k}} = 0 \tag{2.15}$$

and

$$\hat{\mathbf{i}}\cdot\hat{\mathbf{i}} = \hat{\mathbf{k}}\cdot\hat{\mathbf{k}} = \hat{\mathbf{i}}\cdot\hat{\mathbf{i}} = 1 \tag{2.16}$$

Similar to the unit vectors shown here, sines and cosines are orthogonal functions, either in summation or under integration. This may be briefly indicated by reference to the following trigonometric identity similar to the one used in our discussion of beats:

$$\cos x \cos y = \tfrac{1}{2}[\cos(x + y) + \cos(x - y)] \tag{2.17}$$

If we view the integration of two cosine functions as analogous to the inner or scalar product of equations (2.15) and (2.16), we see that

$$\int \cos 2\pi f_1 t \cos 2\pi f_2 t\, dt = \int \{\cos[2\pi(f_1 + f_2)t] + \cos[2\pi(f_1 - f_2)t]\}\, dt \tag{2.18}$$

The first term on the right-hand side of (2.18) is always zero, because the average value of a sinusoid is always zero; there is as much area above zero as below zero. The second term is likewise zero except when $f_1 = f_2$. Therefore, $\cos 2\pi f_1 t$ is said to be *orthogonal* to $\cos 2\pi f_2 t$.

Equations (2.11) and (2.12) are two different ways of expressing a linear combination of orthogonal functions. The orthogonal functions are the cosine functions, and the weighting values are the a_l or the $a(f)$. These are two basic expressions of a Fourier cosine series or a Fourier cosine transform.

2.4. VARIOUS EXPRESSIONS OF FOURIER TRANSFORM RELATIONSHIPS

Equation (2.12) is an expression of the *Fourier cosine* transform. Similarly, one may write the *Fourier sine* transform,

$$S(t) = \frac{1}{2\pi} \int_{-\infty}^{\infty} a(f) \sin 2\pi ft \, df \tag{2.19a}$$

The cosine transform allows us to synthesize (or analyze) functions of *even symmetry* about the origin. Even functions (of which the cosine is an example) are identical when the negative independent variable is substituted for the position:

$$h_e(t) = h_e(-t) \tag{2.19b}$$

Odd functions (e.g., the sine function) change sign when the independent variable changes sign:

$$h_o(-t) = -h_o(t) \tag{2.20}$$

The combination of Fourier sine and cosine transforms permits us to synthesize (and analyze) time functions of arbitrary symmetry, in other words, time functions whose spectra have arbitrary phase. More formally, we can express the Fourier cosine transform as

$$H_c(f) = \frac{1}{2\pi} \int_{-\infty}^{\infty} h_e(t) \cos 2\pi ft \, dt \tag{2.21}$$

and its inverse as

$$h_e(t) = \int_{-\infty}^{\infty} H_c(f) \cos 2\pi ft \, df \tag{2.22}$$

Similarly, the Fourier sine transform is

$$H_s(f) = \frac{1}{2\pi} \int_{-\infty}^{\infty} h_o(t) \sin 2\pi ft \, dt \tag{2.23}$$

and its inverse

$$h_o(t) = \int_{-\infty}^{\infty} H_s(f) \sin 2\pi ft \, df \tag{2.24}$$

A function that has purely even symmetry or purely odd symmetry tends to be exceptional in the practice of physics, chemistry, or engineering. However, all functions of arbitrary symmetry can be expressed as a combination of an even function and an odd function. Consider the function $h(t)$. By defining a function $h_e(t)$ as

$$h_e(t) = h(t) + h(-t) \tag{2.25}$$

and

$$h_o(t) = h(t) - h(-t) \tag{2.26}$$

we can readily see that each function obeys the rules of even and odd symmetry, respectively, and also that

$$h(t) = \tfrac{1}{2}[h_e(t) + h_o(t)] \tag{2.27}$$

This statement has in no way restricted the definition of the symmetry of $h(t)$. If we recall that the products of an even function by an even function and an odd function by an odd function are both even, while the product of an odd function by an even function is odd, we can then understand that

$$\int_{-\infty}^{\infty} h_e(t) \cos 2\pi ft \, dt = \int_{-\infty}^{\infty} h(t) \cos 2\pi ft \, dt \tag{2.28}$$

and

$$\int_{-\infty}^{\infty} h_o(t) \sin 2\pi ft \, dt = \int_{-\infty}^{\infty} h(t) \sin 2\pi ft \, dt \tag{2.29}$$

A function of arbitrary symmetry has both a sine transform and a cosine transform, while a function of even symmetry has only a cosine transform and a function of odd symmetry has only a sine transform. In that most practical functions are of arbitrary symmetry, the above notation would have become very cumbersome had not mathematicians invented a remarkable shorthand that combines the concepts of phase, orthogonality, and symmetry in one fell swoop. This shorthand is the *exponential notation* for transcendental circular functions.

A sine function may be expressed by Taylor expansion as

$$\sin x = x - \frac{x^3}{3!} + \frac{x^5}{5!} + \cdots \tag{2.30}$$

and a cosine function as

$$\cos x = 1 - \frac{x^2}{2!} + \frac{x^4}{4!} + \cdots \tag{2.31}$$

Also, the exponential function is

$$\exp(x) = 1 + x + \frac{x^2}{2!} + \frac{x^3}{3!} + \cdots \tag{2.32}$$

The shorthand notation defines $Z = i\theta$, and further defines

$$\exp(Z) = \exp(i\theta) = \cos\theta + i\sin\theta \tag{2.33}$$

where i is the "imaginary number" $\sqrt{-1}$. The fundamental properties of i, $i^2 = -1$, and $i^{-1} = -i$ allow the shorthand notation to accommodate the symmetry of the sine and cosine functions and simultaneously preserve the orthogonality between the sine and cosine. This leads to the shorthand notation used for the combined sine and cosine transform of an arbitrary function:

$$\int_{-\infty}^{\infty} h(t)\exp(-i2\pi ft)\,dt = \int_{-\infty}^{\infty} h(t)(\cos 2\pi ft - i\sin 2\pi ft)\,dt \tag{2.34}$$

The expression on the left-hand side of equation (2.34) is known as the complex Fourier transform of $h(t)$ and is normally noted* as $\hat{h}(t)$ or $H(f)$,

$$H(f) = \int_{-\infty}^{\infty} h(t)\exp(-i2\pi ft)\,dt \tag{2.35}$$

This relationship between $H(f)$ and $h(t)$ may be further shortened to

$$H(f) \longleftrightarrow \mathrm{h}(t) \tag{2.36}$$

where the double-headed arrow symbolizes the complex Fourier transform. $h(t)$ is recovered through the inverse complex Fourier transform

$$h(t) = \frac{1}{2\pi}\int_{-\infty}^{\infty} H(f)\exp(i2\pi ft)\,df \tag{2.37}$$

2.5. CONCEPTS AND COROLLARIES FOR FOURIER TRANSFORMS

"The spirit killeth, but the letter giveth life"—T. S. Eliot.

To realize the beauty of Fourier transforms a certain level of intimacy with them must be attained. That is the purpose of this section.

The best way to know how the Fourier transform operates is to see its operation on some simple functions. One such simple function is the "delta-function" $\delta(x)$, which has the property that it is identically zero everywhere, except where $x = 0$, and at that point it is infinite. However,

* The selection of the sign in the exponential kernel of equation (2.35) and the assignment of the normalization constant $(2\pi)^{-1}$ to the right-hand side of equation (2.37) is a matter of selecting one of several conventions and is not presented here with any particular physical argument. This is the convention used by Papoulis,[1] and is method 2 in Bracewell.[2]

its width is infinitely narrow, so that the area under the function is unity; hence,

$$\int_{-\infty}^{\infty} \delta(x)\, dx = 1 \tag{2.38}$$

and this has the further interesting property that

$$\int_{-\infty}^{\infty} f(x)\, \delta(x - x_0)\, dx = f(x_0) \tag{2.39}$$

which is sometimes called the "sifting" property of a δ-function.

This function has some physical analogies; for example, a laser is zero everywhere in the spectrum, except at ν_L, its lasing wavenumber, where its output may be so strong it could be said to be infinite. Similarly, the resonant frequency of a piano string or an organ pipe is analogous in the acoustic spectrum to the δ-function, for if these resonated at more than one frequency, then their value as generators of pure tones would be lost to music lovers. A slit in a dispersive spectrometer is a δ-function in that the entire image plane is blocked except at the slit, where relatively the transmission is infinite. A camera shutter is a δ-function in both space and time. In nuclear magnetic resonance spectroscopy an isolated narrow resonance may approximate a δ-function in frequency space.

The Fourier transform of a δ-function situated at the origin in time is

$$\int_{-\infty}^{\infty} \exp(-i2\pi ft)\, \delta(t)\, dt = 1 \tag{2.40}$$

and for a δ-function at other than the origin, we have

$$\int_{-\infty}^{\infty} \exp(-i2\pi ft)\, \delta(t - t_0)\, dt = \exp(-i2\pi ft_0) \tag{2.41}$$

for a δ-function in frequency space, we have

$$\frac{1}{2\pi}\int_{-\infty}^{\infty} \exp(i2\pi ft)\, \delta(f)\, df = \frac{1}{2\pi} \tag{2.42}$$

and

$$\frac{1}{2\pi}\int_{-\infty}^{\infty} \exp(i2\pi ft)\, \delta(f - f_0)\, df = \frac{1}{2\pi}\exp(i2\pi f_0 t) \tag{2.43}$$

This shows that a constant in one space transforms into a δ-function at the origin in the other space, and that a "pure tone" in one space transforms into a δ-function in the other space. An operational (but not closed-form) definition of the δ-function is

$$\delta(t - t_0) = \frac{1}{2\pi}\int_{-\infty}^{\infty} \exp[i2\pi f(t - t_0)]\, df \tag{2.44}$$

and

$$\delta(f - f_0) = \int_{-\infty}^{\infty} \exp[i2\pi t(f - f_0)]\, dt \tag{2.45}$$

These relationships are depicted schematically in Figure 2.4. If one tries to evaluate the right-hand side of (2.44) or (2.45) one finds that the only thing simple about a δ-function is the schematic in Figure 2.4.

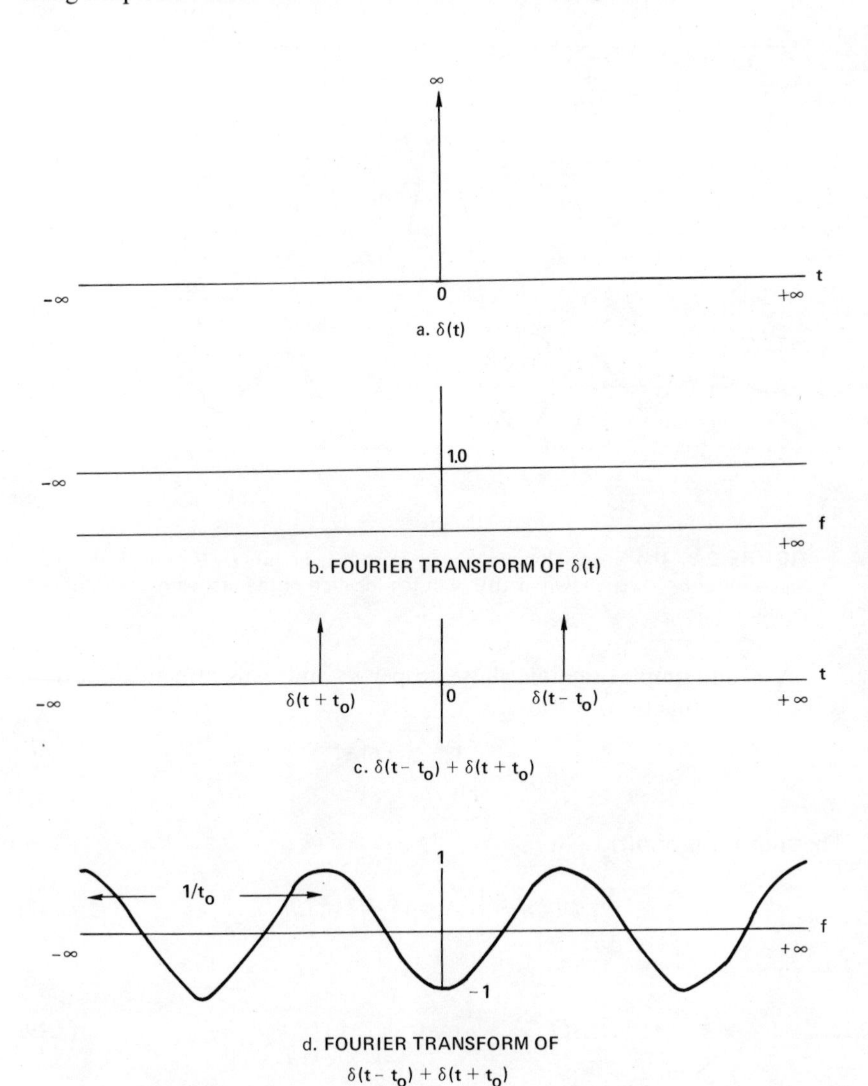

FIGURE 2.4. δ-functions at the origin (a), and at points $\pm t_0$ (c), are shown along with their respective Fourier transforms (b) and (d).

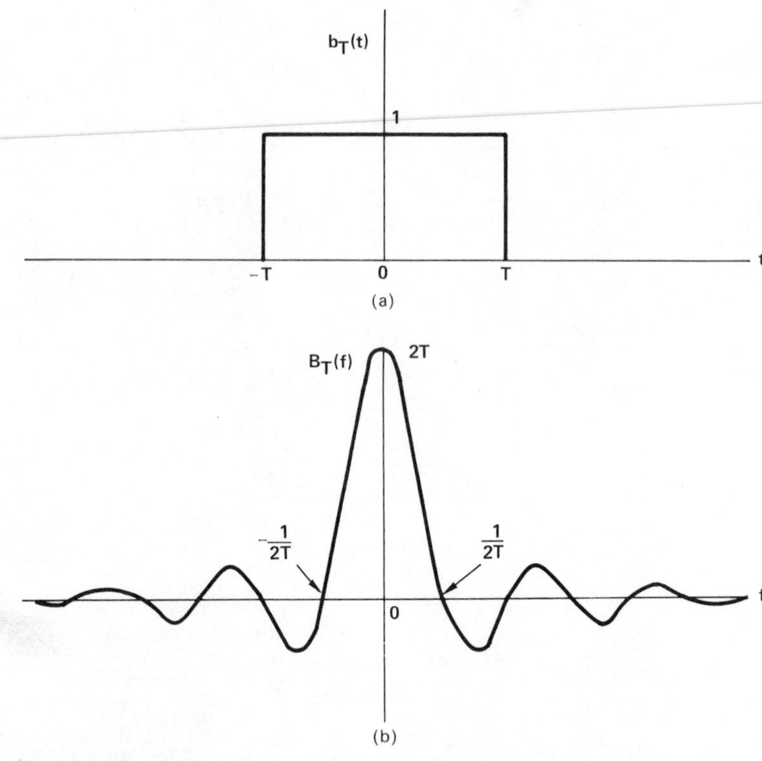

FIGURE 2.5. The boxcar function $b_T(t)$ is shown in (a). $B_T(f)$, the Fourier transform of $b_T(t)$, is shown in (b), with the location of the first zero-crossings noted.

A second simple function whose properties and transform are important is the boxcar function,

$$b(t) = \begin{cases} 1, & |t| \le \mathrm{T} \quad (2.46) \\ 0, & |t| > \mathrm{T} \quad (2.47) \end{cases}$$

The transform of $b(t)$ is $B(f)$,

$$B(f) = \int_{-T}^{T} \exp(-i2\pi ft)\, dt \tag{2.48}$$

or

$$B(f) = \frac{-i}{2\pi f} \exp(-i2\pi ft) \bigg|_{-T}^{T} \tag{2.49}$$

and since the cosine function is even,

$$B(f) = 2T \sin 2\pi f T / 2\pi f T \tag{2.50}$$

which is often called a "(sin x)/x function" or a "sinc function," shown in Figure 2.5.

Two other simple functions that deserve mention because they are commonly encountered in chemical problems are the Gaussian function, $\exp(-\alpha t^2)$, and the exponential decay, $\exp(-\alpha|t|)$. In the first instance,

$$\int_{-\infty}^{\infty} \exp(-\alpha t^2)\exp(-i2\pi ft) = \left(\frac{\pi}{\alpha}\right)^{\frac{1}{2}} \exp\left[\frac{-(2\pi f)^2}{4\alpha}\right] \tag{2.51}$$

we see that the Fourier transform of a Gaussian is another Gaussian, which is a rather unique property. Note that its width in one space is proportional to the inverse of its width in the other. In the second instance, we transform to a form recognized as Lorentzian,

$$\int_{-\infty}^{\infty} \exp(-\alpha|t|)\exp(-i2\pi ft)\,dt = \frac{2\alpha}{\alpha^2 + (2\pi f)^2} \tag{2.52}$$

There are five fundamental theorems related to the Fourier transform that need to be formally stated, and described, before ready application of the theory to physical systems can be realized. These theorems are (1) linearity (linear combination), (2) time shifting, (3) frequency shifting, (4) differentiation, and (5) convolution. Additional theorems can be found in the references, but will not be addressed here.

Theorem 1. Linearity.

The Fourier transform of a sum is equal to the sum of the Fourier transforms. This was highlighted in our early discussion on guitar tuning:

$$g(t) = ag_1(t) + bg_2(t) \tag{2.53}$$

and

$$g_1(t) \leftrightarrow G_1(f) \tag{2.54a}$$

$$g_2(t) \leftrightarrow G_2(f) \tag{2.54b}$$

$$g(t) \leftrightarrow G(f) \tag{2.54c}$$

then

$$G(f) = aG_1(f) + bG_2(f) \tag{2.55}$$

Theorem 2. Time Shifting.

The Fourier transform of a time-shifted function is modulated by a complex phase function. If $g(t)$ is a function such that

$$g(t) \leftrightarrow G(f) \tag{2.56}$$

then

$$g(f - t_0) \leftrightarrow G(f) \exp(-i2\pi f t_0) \tag{2.57}$$

Theorem 3. Frequency Shifting.

The inverse Fourier transform of a frequency-shifted function is modulated in time by a complex phase function. That is, if

$$G(f) \leftrightarrow g(t) \tag{2.58}$$

then

$$G(f - f_1) \leftrightarrow g(t) \exp(i2\pi f_0 t) \tag{2.59}$$

Theorem 4. Differentiation.

The nth derivative of a function is proportional to the Fourier transform of the product of its Fourier transform and the nth power of the independent variable:

$$(d/df)^n\, g(t) \leftrightarrow (-i2\pi f)^n\, G(f) \tag{2.60}$$

Theorem 5. Convolution.

The convolution theorem describes what happens under Fourier transformation to two functions in product. Their Fourier transform is equal to the area under the curves of the Fourier transform of the multiplicand with the folded form of the Fourier transform of the multiplier as a function of the shifting distance between their origins. That is,

$$\int_{-\infty}^{\infty} g(t)\, h(t) \exp(-i2\pi f t)\, dt = \int_{-\infty}^{\infty} G(f')\, H(f - f')\, df' \tag{2.61}$$

and

$$\frac{1}{2\pi} \int_{-\infty}^{\infty} G(f)\, H(f) \exp(i2\pi f t)\, df = \int_{-\infty}^{\infty} g(t')\, h(t - t')\, dt' \tag{2.62}$$

Each of the above theorems is important in the understanding of the Fourier transform, but the most important is the convolution theorem. Since these operations, convolution and multiplication, are very common in physical systems, a complete understanding of the theorem is invaluable.

Let us consider, for example, the problem of "boxcar" versus "triangular" weighting in infrared Fourier transform spectroscopy. In this case, the pertinent Fourier transform pairs are the interferogram $g(x)$ and the spectrum $G(\nu)$ such that

$$G(\nu) = \int_{-\infty}^{\infty} g(x) \exp(-i2\pi \nu x)\, dx \tag{2.63}$$

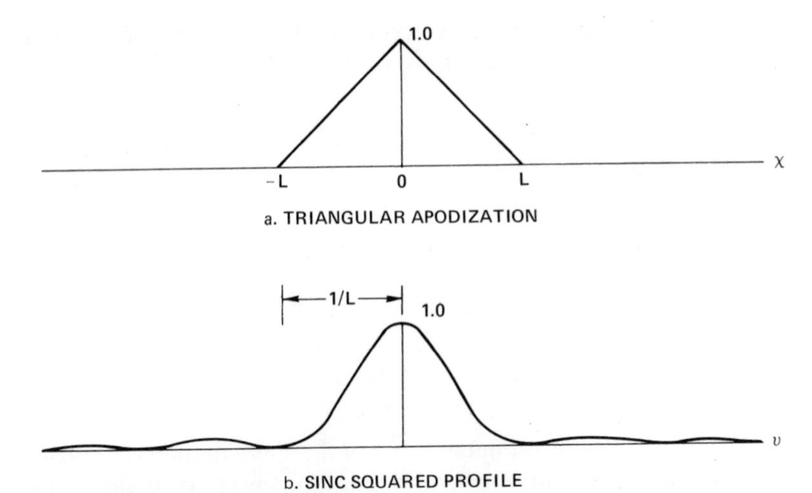

FIGURE 2.6. When an interferogram is weighted with the triangular function shown in (a), the resulting spectra are convolved with the profile shown in (b). The distance of the first zero-crossing from the center of the profile is noted.

In the case where the interferometer moving mirror travels over x from $-L$ to $+L$, we have simple boxcar truncation, because the interferogram is weighted with the boxcar function $b_L(x)$. Hence the observed spectrum is given by

$$G'(x) = \int_{-\infty}^{\infty} g(x)\, b_L(x) \exp(-i2\pi x v)\, dx \tag{2.64}$$

and by the convolution theorem

$$G'(x) = \int_{-\infty}^{\infty} G(v')\, B(v - v')\, dv' \tag{2.65}$$

$B(v)$ was previously defined as [see equation (2.50)]

$$B_L(v) = 2L \sin(2\pi v L)/2\pi v L \tag{2.66}$$

Boxcar truncation produces undesirable side effects or sidelobes. Often, in order to suppress this effect, triangular *apodization* is employed, wherein the function

$$a(x) = 1 - |x|/L$$

is used to weight the interferogram. Then the observed spectrum is

$$G'(v) = \int_{-\infty}^{\infty} g(x)\, a(x) \exp(-i2\pi v x)\, dx \tag{2.67}$$

The question becomes "How does $a(x)$ affect the spectrum?" By using the convolution theorem twice, we can see that

$$a(x) = \int_{-\infty}^{\infty} b_{L/2}(x^1)\, b_{L/2}(x - x^1)\, dx^1 \tag{2.68}$$

and

$$A(f) = B_{L/2}(f)\, B_{L/2}(f) \tag{2.69}$$

so that the spectrum observed is

$$G'(v) = \int_{-\infty}^{\infty} G(v^1)\, B_{L/2}^2(v - v^1)\, dv^1 \tag{2.70}$$

In other words, as shown in Figure 2.6, when the interferogram is triangularly apodized, the spectrum in convolved with a $(\sin x/x)^2$ or *sinc squared* function whose full width at halfheight is twice that described in (2.66). The relative amplitude of the sidelobes is decreased by a power of 2, but the resolution is also degraded by a factor of approximately 2.

2.6. MORE ON PHASE AND PHASE CORRECTION

Both Fourier transform infrared spectroscopy and Fourier transform nuclear magnetic resonance spectroscopy have peculiarities that result in a need for phase correction.

In FT–NMR, the free induction decay is a causal signal, i.e., until the spin system is perturbed by the pulse, there is no signal. A single line FID may be characterized by

$$g(t) = \begin{cases} 0, & t < 0 \quad (2.71a) \\ \exp(-t/\tau)\cos 2\pi f_0 t, & t \geq 0 \quad (2.71b) \end{cases}$$

This signal is clearly asymmetric and may be considered the sum of an even function and an odd function:

$$g_e(t) = \tfrac{1}{2}\exp(-|t|\alpha)\cos 2\pi f_0 t \tag{2.72a}$$

$$g_o(t) = \frac{1}{2}\left[\underset{t>0}{\exp(-t\alpha)} - \underset{t<0}{\exp(t\alpha)}\right]\cos 2\pi f_0 t \tag{2.72b}$$

The cosine Fourier transform of the even part of this function leads to an absorption spectral line

$$\int_{-\infty}^{\infty} g_e(t)\cos 2\pi f t\, dt = \frac{2/\alpha}{(f - f_0)^2 + \alpha^2} \tag{2.73}$$

and the sine Fourier transform of the odd portion leads to the dispersion line

$$\int_{-\infty}^{\infty} g_o(t) \sin 2\pi f t \, dt = \frac{(2/\alpha)(f - f_0)}{(f - f_0)^2 + \alpha^2} \tag{2.74}$$

The complex spectrum is expressed as

$$G(f) = \frac{[1 + i(f - f_0)]^2/\alpha}{(f + f_0)^2 + \alpha^2} \tag{2.75}$$

In practice, delays due to electronic filters, phase detectors, and other effects cause generally constant or linear phase shifts that result in the type of spectral line profile shown in Figure 2.7. This is easily corrected by introducing a correcting factor

$$\exp[-i(2\pi k f + \varepsilon)] \tag{2.76}$$

and rotating the spectra back into the form of equation (2.75).

In Fourier transform infrared spectroscopy the situation is somewhat different. Typically the signal is not causal in the same sense as is the NMR signal because the moving mirror may be scanned over both sides of the point of zero path difference. Because the signal is detected with a square-law detector, the ideal infrared interferogram is described as

$$h(x) = 2\varepsilon \int_{-\infty}^{\infty} H(v) \cos 2\pi v x \, dv \tag{2.77}$$

but in practice the radiation undergoes phase shifts due to beamsplitter characteristics, signal processing delays, refraction effects in materials, etc., and therefore must be expressed as

$$h(x) = 2\varepsilon \int_{0}^{\infty} H(v) \cos[2\pi v x + \theta(v)] \, dv \tag{2.78}$$

where $H(v)$ is an arbitrary phase function, and E an efficiency factor. Simple trigonometry gives us

$$h(x) = 2\varepsilon \left[\int_{0}^{\infty} H(v) \cos 2\pi v x \cos \theta \, dv - \int_{0}^{\infty} H(v) \sin 2\pi v x \sin \theta \, dv \right] \tag{2.79}$$

or

$$h(x) = 2\varepsilon \int H(v) \cos 2\pi v x \cos \theta \, dv - 2\varepsilon \int H(v) \sin 2\pi v x \sin \theta \, dv \tag{2.80}$$

Because $\theta(v)$ is slowly varying,

$$h(x) \simeq 2\varepsilon \overline{\cos \theta} \int H(v) \cos 2\pi v x \, dv - 2\varepsilon \overline{\sin \theta} \int H(v) \sin 2\pi v x \, dv \tag{2.81}$$

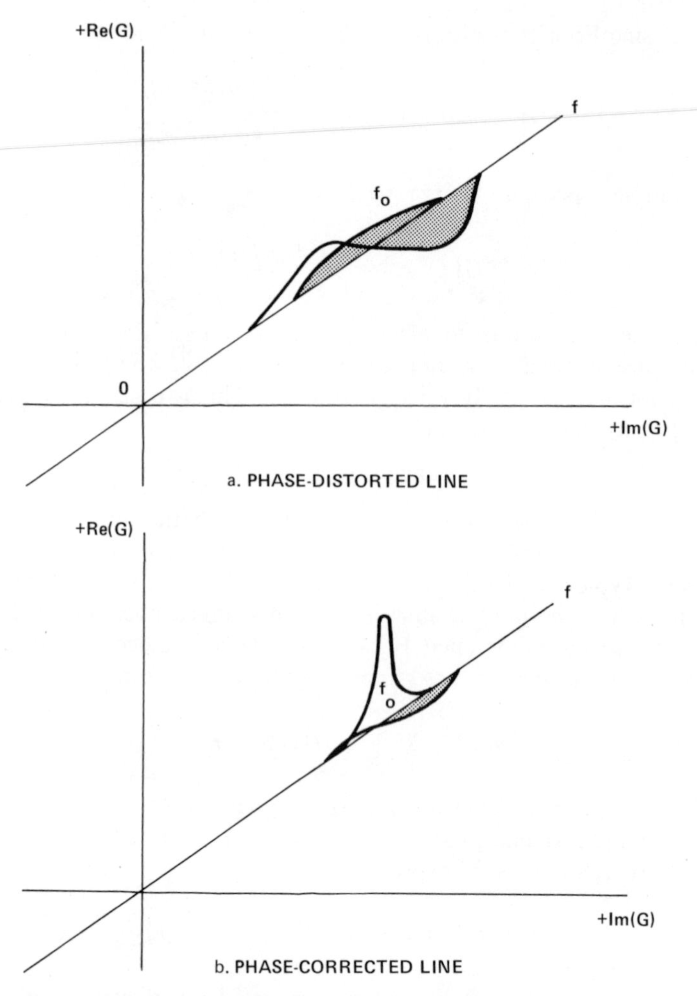

FIGURE 2.7. A phase-distorted Lorentzian line in (a) and a phase-corrected line in (b) are shown in three-dimensional complex space.

Here we can see rather clearly how the phase angle θ controls the relative effect of the cosine transform or the sine transform on $h(x)$. As θ approaches 2π, $h(v)$ becomes an even function. As θ approaches $\pi/2$, $h(x)$ becomes an odd function. The spectrum is recovered from the interferogram by the Fourier transform, that is,

$$H'(v) = \int_{-\infty}^{\infty} h(x) \exp(-i2\pi vx)\, dx \tag{2.82}$$

The first term in (2.81) only has a cosine transform contribution to the recovery, and the second term has only a sine transform contribution. Hence,

$$H'(v) = 2\varepsilon \iint H(v')[\cos\theta \cos 2\pi v x + \sin\theta \sin 2\pi v' x] \times \exp(-i2\pi v x)\, dv'\, dx \tag{2.83}$$

and interchanging the order of integration and integrating gives us

$$H'(v) = 2\varepsilon \int \{H(v')\cos\theta[\delta(v - v') + \delta(v' - v)] - iH(v')\sin\theta[\delta(v' - v) + \delta(v - v')]\}\, dv' \tag{2.84}$$

and so the recovered spectrum is $H'(v)$:

$$H'(v) = 2\varepsilon H(v)\exp[-i\theta(v)] \tag{2.85}$$

This is the original spectrum rotated in complex space by the phase vector $\exp[-i\theta(v)]$. $H(v)$ can only be fully recovered if $\theta(v)$ is known. One method of learning $H(v)$ in practice is to assume that $\theta(v)$ is slowly varying as a function of v. (This is generally true.) Then a low-resolution Fourier transform of $h(x)$ will yield a low-resolution function of v, $H_L'(v)$, where

$$H'_L(v) = H_L(v)\exp[-i\theta(v)] \tag{2.86}$$

and

$$\exp[-i\theta(v)] = H'_L(v)/|H_L(v)| \tag{2.87}$$

The phase-corrected high-resolution spectrum is then recovered by

$$H(v) = H'(v)\, H'_L(v)^*/|H'_L(v)| \tag{2.88}$$

This method was first suggested by Mertz.[3]

The fundamental assumption in the correct determination of $H(v)$ is that $\theta(v)$ is very slowly varying with wavenumber. It is not required that $\theta(v)$ be linear. Basically, if $\theta(v)$ is expressed as a series

$$\theta(v) = K_0 + K_1 v + K_2 v^2 + K_3 v^3 + \cdots \tag{2.89}$$

it is necessary that K_1 and K_2 be small, and that terms higher than the third effectively vanish. There are methods for phase-correcting spectra with higher-order phase terms (from "chirped" interferograms) but they tend to be tedious and sometimes difficult to implement.

2.7. APODIZATION AND RESOLUTION ENHANCEMENT

Apodization and resolution enhancement, in the first approximation, are similar techniques. In the signal-processing nomenclature the term "windowing" is often used to describe the technique of apodization. In the following discussion we shall use the FT–IR interferogram. However, a similar argument may be made for the FT–NMR free induction decay simply by changing variables and viewing the pulse origin (i.e., $t = 0$) as equivalent to the interferometric point of zero retardation (or stationary phase, i.e., where all the energy has approximately the same phase and produces the "centerburst").

The term *apodization* derives from the Greek expression meaning "to take off the feet." The implication is best seen by making reference to the boxcar function described in equation (2.46), expressed here in retardation coordinates:

$$b(x) = \begin{cases} 1, & |x| \leq L \\ 0, & |x| > L \end{cases} \tag{2.90}$$

This function has the Fourier transform

$$B(\nu) = \sin 2\pi\nu L/\pi\nu \tag{2.91}$$

The boxcar function and its Fourier transform are shown in Figure 2.5. This boxcar function is the "window" or "weighting" function for the unapodized Fourier transform. $B(\nu)$ is the spectral line profile of an isolated δ-function line. It has sidelobes (feet) that diminish inversely with distance from the line center; the first negative side lobe is twenty-one percent of the line center height.

The existence of these sidelobes is related to the idea of "Gibbs phenomena" in Fourier series, or to the approximation of the calculated spectrum to the real spectrum. Similar features are seen when one tries to approximate a curve (for example, a spectrum or observed set of data points) with a finite polynomial. The computed curve often oscillates about the real data, even when a least squares method is used to achieve a good fit. (It can be mathematically demonstrated that the best fit to a curve, in the least squares sense, is given by the Fourier coefficients of the curve.)

Because the window is finite ($-L$ to L) and because the weighting function is abrupt, i.e., a boxcar, we will see the phenomena of sidelobes on our spectra. Apodization is the technique used to reduce those sidelobes. The technique consists of weighting the interferogram by a function that monotonically and continuously reduces the value of the interferogram so that it smoothly reaches zero at its maximum domain L. One typical technique[4] used is shown in Figure 2.6. As with all apodization techniques there are three effects:

(1) The apparent signal-to-noise ratio may be increased.
(2) The effect of the sidelobes is reduced.
(3) The resolution is reduced, i.e., the full width at half-maximum is increased.

The normal interferometric definition of resolution is that $\Delta\nu = 1/L$. In this definition, $\Delta\nu$ is the distance from the line center at which the spectral line function has its first zero crossing. There is a strong similarity here to Lord Rayleigh's description of same. For the case of boxcar apodization this zero crossing occurs at $\Delta\nu' = 1/2L$, and hence the total wavenumber distance from zero crossing to zero crossing is $1/L$. If triangular apodization is used, as was described earlier, the first zero crossing occurs at a distance of $1/L$ from the line center, and the resolution of the line is degraded by a factor of 2.

Norton and Beer[(5)] have examined the classes of functions used for apodization and have shown that the preferred set of functions produces an optimum compromise between sidelobe suppression and loss of resolution. This set of functions is given by

$$F(u) = \sum_{i=0}^{n} C_i(1 - u^2)^i, \quad \sum_{i=0}^{n} C_i = 1, \qquad n = 0, 1, 2, 3 \tag{2.92}$$

where C_i is an empirical coefficient and U a normalized path difference.

Resolution enhancement may be achieved by the use of weighting functions on the interferogram or free induction decay. In all cases it is obtained only at the expense of signal-to-noise. A straightforward method is to weight the interferogram directly rather than inversely with distance. The Fourier transform of this function provides the first derivative of the spectrum, i.e.,

$$xg(x) \leftrightarrow \frac{i}{2\pi}\frac{d}{d\nu}G(\nu) \tag{2.93}$$

and the second derivative is provided by a quadratic weighting,

$$x^2g(x) \leftrightarrow \frac{-1}{(2\pi)^2}\frac{d^2}{d\nu^2}G(\nu) \tag{2.94}$$

Very often one obtains a "resolution-enhanced" spectrum by using the difference between the second and first derivatives, namely,

$$G_{\mathrm{RE}}(\nu) = G''(\nu) - G'(\nu) \tag{2.95}$$

From the prior equations one can see that these derivatives are available from Fourier transformation. One can also see that the noise is increased via the two effective differentiations and the subtraction.

An alternative and somewhat less tractable form of resolution enhancement is available. If we recall the early example of the guitar tuner, as the guitar became more tuned the time between beat notes became larger. In FT–IR, the resolution is a function

$$\Delta\nu \geq 1/L \tag{2.96}$$

and we again see an expression similar to the quantum mechanical uncertainty principle.

Consider the interferogram of finite extent ($-L$ to $+L$) shown in Figure 2.8. If that interferogram were of infinite signal-to-noise ratio in the region ($-L$, L) then via analytic continuation, by a Taylor expansion, the interferogram could be specified outside the region ($-L$, L). That is, for the interferogram $g(x)$ in the region of $x = x_0$,

$$g(x) = \sum_{n=0}^{n} \frac{g^{(n)}(x_0)}{n!} (x - x_0)^n + R_n \tag{2.97}$$

Clearly, since $g(x)$ is known around $x = L$, the above expression could be

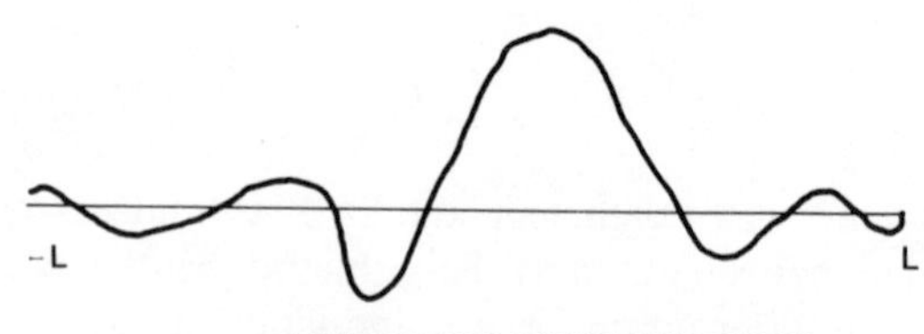

a. ASYMMETRIC INTERFEROGRAM

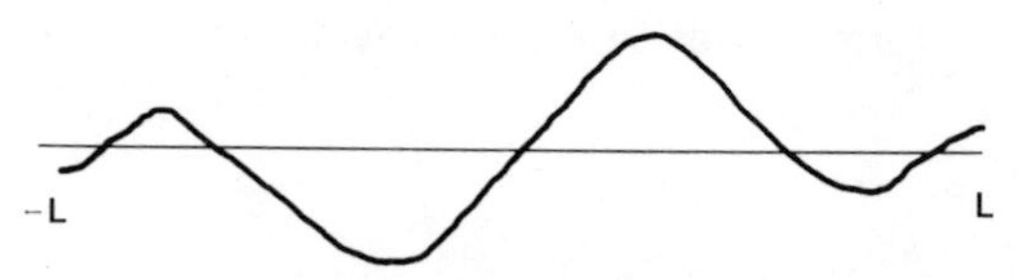

b. ANTISYMMETRIC INTERFEROGRAM

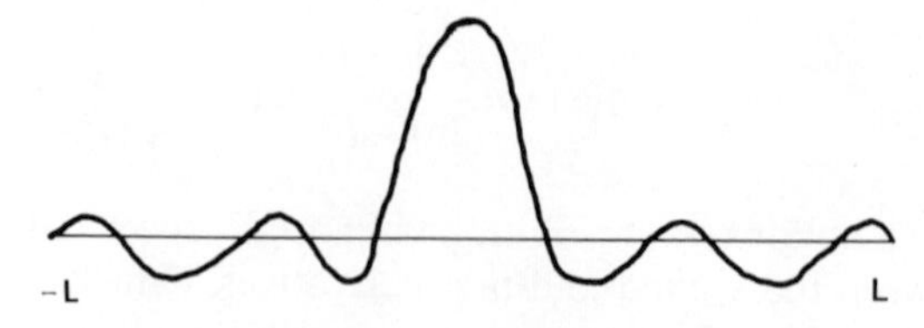

c. SYMMETRIC INTERFEROGRAM

FIGURE 2.8. The effect of phase distortion on an interferogram is shown. In (a) the phase angle (θ) is arbitrary, in (b) it is $\pi/2$, and in (c) it is zero. A value of π would provide an inversion of C.

used to describe $g(x)$ just outside the region $(-L, L)$, for example at $x = L+\varepsilon$. Then this new value of $g(L+\varepsilon)$ could be similarly used to describe the function at $L+2\varepsilon$, etc. There are two problems with this technique. The first is that the computation is tedious at the very least. The second is that the relative noise in $g(x_0)$ is

$$\text{noise} = \Delta g/g(x_0) \tag{2.98}$$

and the error in g is

$$\Delta g = (dg/dx)\,\Delta x \tag{2.99}$$

where

$$\frac{dg}{dx} = \sum_{n=0}^{n} \frac{g^{(n+1)}(x_0)}{(n+1)\,!}(x-x_0)^{n+1} + R_{n+1} \tag{2.100}$$

That is, the correction to $g(x_0)$ that is used in the computation of $g(x+\varepsilon)$ is directly proportional to the noise in $g(x_0)$ divided by the incremental distance Δx.

A more elegant and tractable approach to this problem has been proposed by Landau *et al.*[6] This approach involves the use of prolate spheroidal wave functions, a set of functions that enjoy the unique characteristic of orthogonality over two domains, the domain $(-1, 1)$ and $(-\infty, \infty)$. Expansion of the known interferogram $g(x)$ in terms of the coefficients of the functions enables the expansion of the function outside the region $(-L, L)$.

2.8. THE DISCRETE FOURIER TRANSFORM

We have previously defined the Fourier transform of a function $h(t)$ as

$$H(f) = \int_{-\infty}^{\infty} h(t)\exp(-i2\pi ft)\,dt \tag{2.101}$$

and the inverse transform as

$$h(t) = \frac{1}{2\pi}\int_{-\infty}^{\infty} H(f)\exp(i2\pi ft)\,df \tag{2.102}$$

Now, in order to heuristically derive the relationships[7] for a discrete Fourier transform, we will restrict ourselves to thinking of functions of a certain class, known as "band-limited functions." These are functions whose spectra are zero valued above a certain maximum frequency f_m. If $h(t)$ is a time-varying function whose spectrum is band-limited, then we can sample each frequency in $h(t)$ uniquely by sampling $h(t)$ at equidistant time points Δt such that

$$\Delta t = 1/2f_{\mathrm{m}} \tag{2.103}$$

Therefore, we can express a series h_n such that

$$(h_n) = \sum_{n=0}^{N-1} h(n\,\Delta t)\,\delta(t - n\,\Delta t) \tag{2.104}$$

By the prudent choice of N, we determine for how long we observe $h(t)$, and thus determine our ability to resolve frequencies in $H(f)$, giving us the elemental resolution Δf in $H(f)$,

$$\Delta f = 1/N\,\Delta t \tag{2.105}$$

The Fourier transform of (h_n) may be simply stated by substituting (2.104) into (2.101) and interchanging integration and summation:

$$H(f) = \sum_{n=0}^{N-1} h_n \exp(-i2\pi f n\,\Delta t) \tag{2.106a}$$

where h_n is an abbreviated form of (h_n). Since the elemental resolution in f is Δf, $H(f)$ may be approximated by a series of $f_{\mathrm{m}}/\Delta f$ points Δf apart:

$$H(k\,\Delta f) = \sum_{n=0}^{n-1} h_n \exp(-i2\pi kn\,\Delta t\,\Delta f) \tag{2.106b}$$

Thus

$$H_k = \sum_{n=0}^{N-1} h_n \exp(-i2\pi kn/N) \tag{2.107}$$

An analogous argument will produce the inverse discrete Fourier transform, similar in form to equation (2.102):

$$h_n = \frac{1}{N} \sum_{n=0}^{N-1} H_k \exp(i2\pi kn/N) \tag{2.108}$$

Without proof we will describe several properties of the discrete Fourier transform. First, we will examine some distinctions between the discrete Fourier transform and the continuous Fourier transform (CFT) and then examine some similarities.

Distinctions

Distinction 1. The continuous forward and inverse Fourier transform exists over the real axis from $-\infty$ to $+\infty$. The domain of the forward and inverse DFT is the unit circle from 0 to $N - 1$. [So defined because it is described by $\exp(i2\pi kn/N)$.]

Distinction 2. The DFT and the IDFT are periodic in N.

$$h_n = h_0, \quad h_{n+1} = h_1, \quad \text{etc.,} \qquad \text{as } H_1 = H_{N+1}$$

Similarities

These properties of the DFT are similar to the CFT.[8]

Similarity 1. Convolution:

$$X_1^{(n)} \cdot X_2^{(n)} \leftrightarrow X_1(k) * X_2(k) \tag{2.109}$$

Similarity 2. Shifting:

$$X(n)\exp(i2\pi k_0 n/N) \leftrightarrow X(k - k_0) \tag{2.110}$$

Similarity 3. Differentiation:

$$(-iN/2\pi)\, X'(n) \leftrightarrow kX(k) \tag{2.111}$$

2.9. WALSH AND HADAMARD TRANSFORMS

Fourier analysis is a technique used to expand arbitrary functions into coefficients of the trigonometric functions sine and cosine. The sine and cosine functions, which vary from -1 to 1 *continuously*, can be used to form complete sets. There are other types of periodic functions that also have been used to form complete sets.

In 1922, in Hamburg, Rademacher discovered a set of functions on the interval [0, 1] that take on the value ± 1, and in 1923, in Cambridge, Massachusetts, Walsh[9] discovered a similar set. While the Rademacher functions were not a complete and orthogonal set, the Walsh functions were. Moreover, they exhibited oscillation, recursiveness, and many of the other attributes of the trigonometric functions.

Walsh functions[10] are functions that have a visual similarity to "square waves," vary between -1 and $+1$, and behave according to strict relations similar to those found in the trigonometric functions. Instead of an argument of frequency (the number of cycles per unit time), these functions use the argument *sequency* (defined as "one-half the average number of zero crossings per unit time"). Figure 2.9 shows some example orthogonal cosine functions, while on the right are shown basic orthogonal Walsh elements. The expressions "Cal" and "Sal" denote the Walsh analogs of cosine and sine. It is beyond the scope of this chapter to pursue Walsh functions further, but the theory of orthogonal functions may be completely applied to them. A recommended reference on the subject is by Harmuth.[10]

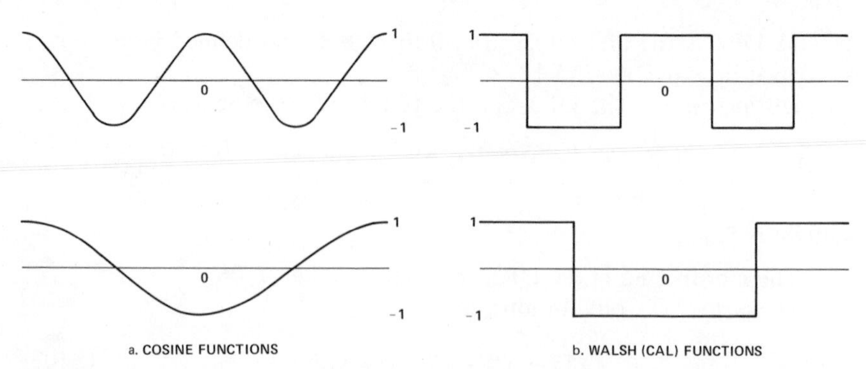

FIGURE 2.9. Two cosine functions (a) and their analogous Walsh (Cal) functions (b).

Walsh functions may be used in a method analogous to the method of Fourier or trigonometric functions in the DFT. This technique is known as the Walsh–Hadamard transform (WHT). Specifically, if $\mathbf{x}$ is a $1 \times n$ matrix, then the $1 \times n$ matrix is given by the WHT of $\mathbf{x}$:

$$\mathbf{X} = \mathbf{H} \cdot \mathbf{x} \tag{2.112}$$

The operand $\mathbf{H}$ has rows and columns made up of values ± 1 and has an inverse $\mathbf{H}^{-1}$ that recovers $\mathbf{x}$,

$$\mathbf{x} = \mathbf{H}^{-1} \cdot \mathbf{X} \tag{2.113}$$

An application of this theory has been realized by Decker[11] and Harwit.[12] In their case $\mathbf{x}$ is the image plane of a dispersive spectrometer and H is a Hadamard mask of 0's and 1's. $\mathbf{X}$ is the series of n observations of the image plane with different mask positions (corresponding to different rows in the operand mask $\mathbf{H}$). After the n observations have been collected, mathematical operation of $\mathbf{H}^{-1}$ on $\mathbf{X}$ recovers the spectrum $\mathbf{x}$. In the degenerate case of $\mathbf{H}$ being the diagonal matrix $\mathbf{1}$, we have the case of the scanning slit spectrometer. In all other cases of $\mathbf{H}$ some degree of the multiplex advantage is achieved since many resolution elements are being simultaneously observed.

2.10. SUMMARY

An understanding of the Fourier transform may be achieved through basic understanding of the relationship between time and frequency. Arithmetic operations change under Fourier transformation (e.g., multiplication transforms into convolution) and a firm grasp of how these operations change under transformation simplifies the application of the theory to real systems. The principle of uncertainty is fundamental to Fourier transforms theory; the width of the observation in one space varies inversely as the width in the other. Higher spectral resolution at constant signal-to-noise ratio is always obtained at the expense of measurement time.

REFERENCES

1. A. Papoulis, *The Fourier Integral and Its Applications,* McGraw–Hill, New York, 1962.
2. R. Bracewell, *The Fourier Transform and Its Applications,* McGraw–Hill, New York, 1965.
3. L. Mertz, *Transformations in Optics,* John Wiley & Sons, New York, 1965.
4. P. R. Griffiths, C. T. Foskett, and R. Curbelo, *Appl. Spectroscopy Reviews* **6,** 31 (1972).
5. R. Norton and R. Beer, *J. Opt. Soc. Am.* **66**, 3 (1976).
6. H. J. Landau, H. O. Pollack, and D. Slepian, *Bell Syst. Tech. J.* **40**, 1, 43–64 (1961); **40**, 1, 65–84 (1961); **41**, 4, 1295–1336 (1962).
7. B. Gold and C. Rader, *Digital Processing of Signals,* McGraw–Hill, New York, 1969.
8. A. Oppenheim and R. Schafer, *Digital Signal Processing,* Prentice–Hall, Englewood Cliffs, New Jersey, 1975.
9. J. Walsh, *Proc. Symp. Workshop Applications of Walsh Functions 1970,* vii, 1970 (AD-707-431).
10. H. Harmuth, *Transmission of Information by Orthogonal Functions,* Springer–Verlag, New York, 1970.
11. J. Decker, *Proc. Applications of Walsh Functions,* p. 101, 1973 (AD-763-000).
12. M. Harwit, *Proc. Applications of Walsh Functions,* p. 108, 1973 (AD-763-000).

Chapter 3

Multichannel Methods in Spectroscopy*

Alan G. Marshall† and Melvin B. Comisarow

3.1. INTRODUCTION

This chapter is intended to introduce, using a minimum of instrumental detail, the basic principles behind the advantages of Fourier and Hadamard transform methods in spectroscopy. Applications of these principles to each of several branches of spectroscopy, along with illustrative examples, may be found in succeeding chapters.

This chapter begins with a discussion of single-channel (scanning-type) spectrometer sources and detectors. It is noted that spectrometer sources are either inherently monochromatic (e.g., tunable laser or radio frequency oscillator) or are inherently broad-band frequency sources that may be made monochromatic by a monochromator. However, when a broad-band radiation source is available, it is in principle possible and highly advantageous to detect the whole spectrum at once (*multichannel* detection).

The most obvious multichannel method is based on use of an array of individual detectors, each of which is arranged to observe a small part of the spectral range (*multidetector* spectrometer). While the multidetector principle is desirable for spectroscopy in any wavelength region, its present feasibility is usually limited to optical spectra. Alternatively, Hadamard and Fourier (*multiplex*) methods are based on initial encoding of the individ-

* Work supported by grants (to A.G.M. and M.B.C.) from the National Research Council of Canada.

† Alfred P. Sloan Research Fellow, 1976–1978.

Alan G. Marshall and Melvin B. Comisarow • Department of Chemistry, University of British Columbia, Vancouver, B.C., Canada V6T 1W5

ual spectral elements (intensities or amplitudes) prior to detection, and subsequent decoding of the response from a *single* broad-band detector to produce the desired broad-band spectrum.

Examination of the optimal use of a simple pan balance is then used to provide an analogy that simultaneously accounts for the advantages (time-saving or improved signal-to-noise ratio) of multichannel methods, and also leads to several encoding–decoding schemes for multiplex operation. Multidetector and multiplex solutions for the three-channel case are discussed in sufficient detail to provide for generalization to the N-channel case. The advantages of Hadamard transform spectroscopy follow immediately from the balance analogy.

The advantages of coherent radiation sources and detectors (e.g., high accuracy in frequency determination, conveniently variable spectral resolution) are presented. The balance analogy is then invoked to account for the advantages of the Fourier encoding–decoding scheme. Fourier analysis of the broad-band detector response then provides a description of several limiting spectral line shapes. Fourier theory is then used to explain the formation of a broad-band coherent radiation source from a monochromatic radiation pulse.

The chapter concludes with a unified classification of single-channel and multichannel spectrometers, developed from the preceding discussion.

3.2. SPECTROMETER SOURCES AND DETECTORS

3.2.1. Terminology

A *spectrometer* may be defined as a device that provides and records the *spectrum* (intensity or amplitude of absorption or emission of energy as a function of incident or emitted energy, frequency, or wavelength) from a sample of material. In this chapter, *absorption* spectrometers are classified according to whether or not the whole spectrum is *excited* at once and, independently, whether or not the whole spectrum is *detected* at once. (Extension to *emission* spectrometers will be obvious.) *Channel* will denote a specified narrow-energy range for excitation or detection; a full spectrum thus consists of an intensity measurement for each of a large number of adjacent channels.

3.2.2. Single-Channel (Scanning-Type) Spectrometer

Figure 3.1 shows the combination of a single detector with either an inherently monochromatic source (Figure 3.1a) or an inherently broad-band frequency source that has been made monochromatic by dispersing the radiation according to frequency (e.g., prism or grating for visible radiation) and then removing the radiation from all but one narrow-energy

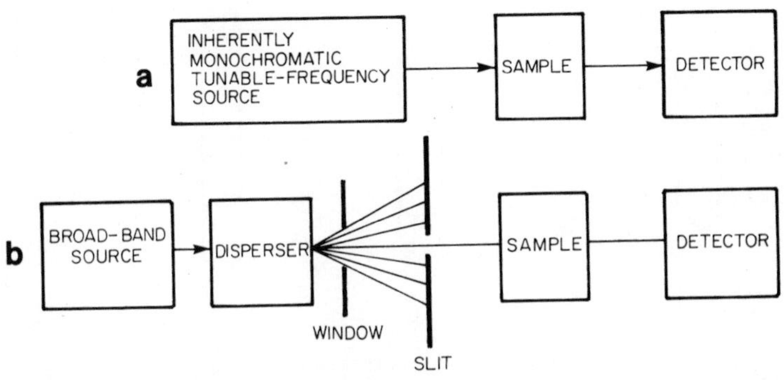

FIGURE 3.1. Schematic diagrams of single-channel (scanning-type) spectrometers. (a) Source provides inherently monochromatic radiation. (b) Source produces inherently broad-band radiation, which is made monochromatic by passage through a disperser–window-slit combination as shown. (Spectrometer operation is described in Section 3.2.2.) [While the single detector may be either *broad-band* (detects radiation anywhere in the overall spectral range) or *narrow-band* (detects radiation only in the single channel emitted by the monochromatic source), it is usual to associate a broad-band detector with a broad-band source (Figure 3.1b) and a narrow-band detector with a narrow-band source (Figure 3.1a); see Section 3.5 for more details.]

channel (Figure 3.1b) by use of (for example) a slit. In either case, the detector measures the intensity of radiation absorbed for a given channel, the source frequency is then shifted to an adjacent channel, and the measurement is repeated channel by channel until the entire spectrum has been obtained. The devices in Figure 3.1 are called *scanning-type* spectrometers, because it is necessary to scan one channel at a time over many individual energy (frequency) channels in order to obtain the final spectrum.

3.2.3. Multidetector Spectrometer

When a broad-band energy source is available, a direct and obvious improvement is to retain the disperser of Figure 3.1b, but remove the slit, thereby opening up the spectral "window" from just one channel to the entire multichannel spectrum. The single detector of Figure 3.1b is then replaced by an array of detectors, so that each member of the array detects one channel of the dispersed spectrum, as shown in Figure 3.2. The *multidetector* spectrometer of Figure 3.2 thus detects the whole spectrum (all N channels) at once, or more precisely, in the same observation time required for a single intensity measurement (one channel) by the scanning-type spectrometer of Figure 3.1. In other words, the multidetector spectrometer provides the same (whole) spectrum in a factor of $1/N$ as much time as the

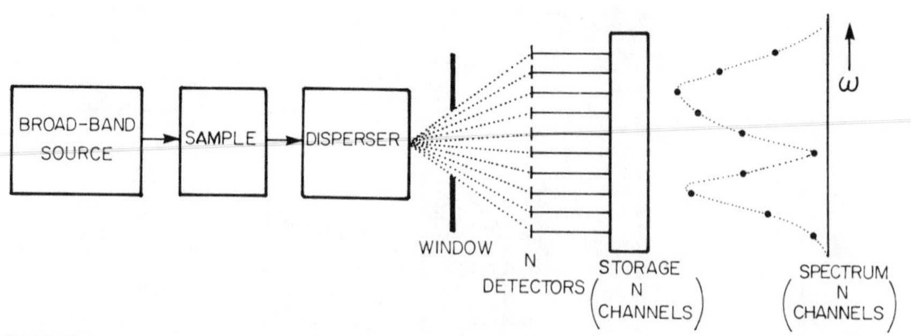

FIGURE 3.2. Schematic multidetector spectrometer. This spectrometer is similar to that of Figure 3.1b with the slit removed, so that the full N-channel spectrum impinges on N respective individual detectors, each positioned to detect the radiation from just one spectral channel. The spectral intensity in each channel accumulates in the storage unit, and is ultimately displayed in the point-by-point intensity spectrum shown at the far right of the figure. Spectrometer operation is discussed in Section 3.2.3.

scanning-type spectrometer. We shall return to discuss this $1/N$ advantage in Section 3.3.

Because of the conceptual *simplicity* of the multidetector spectrometer of Figure 3.2, it is logical to investigate its *feasibility*. It is desirable to be able to resolve spectral detail as narrow as the width of a typical spectral absorption line; therefore, the minimum number of channels that will be required in any multichannel spectrometer is simply the width of the *entire* spectral range of interest, divided by the width of a *single* spectral line. The resultant necessary number of channels for various forms of spectroscopy is shown in Table 3.1.

From Table 3.1 it would appear that electronic (visible–uv) spectroscopy is the least likely candidate for success with a multidetector spectrometer, but in fact, multidetector recording of visible–uv radiation is readily accomplished photographically. The resolution of a fine-grain photographic plate is sufficient to provide the huge number of required channels, since the desired spectrum may be dispersed over the necessary distance (a few meters) without undue effort. When the required number of channels is greatly reduced (say, to a few thousand), as for applications requiring lower resolution and/or more limited spectral range, any of several types of electronic multidetector devices[1] (e.g., vidicon or self-scanning photodiode array) can replace the photoplate. For example, in the vidicon (television camera) type, dispersed photons of different wavelength are simultaneously recorded at different regions of the focal plane of the camera. While it is true that the electronic multidetector has relatively few channels compared to the photoplate multidetector, the electronic device furnishes an immediate spectrum without the delays of photographic processing and densitometric reading of the developed photoplate.

TABLE 3.1. Minimum Number of Channels Required for Various Types of Multidetector Spectrometers

Type of spectroscopy	Largest usual frequency (Hz) [energy (eV)][a]	Typical spectral frequency (Hz) [energy (eV)] range	Width of one line [Hz (eV)]	Approximate minimum number of channels[b]
Mossbauer	6×10^{18} (2.4×10^{4})[c]	10^{8} (4×10^{-7})	10^{7} (4×10^{-8})	10
ESCA	3.5×10^{17} (1.5×10^{3})	10^{17} (3×10^{2})	10^{14} (3×10^{-1})	1,000
Photoelectron	5×10^{15} (20)	3×10^{15} (10)	10^{12} (3×10^{-3})	3,000
Electronic	1.5×10^{15} (6)	1.2×10^{15} (5)	10^{9} (4×10^{-6})	1,250,000
Vibrational	2×10^{14} (8×10^{-1})	1.5×10^{14} (6×10^{-1})	3×10^{9} (1.2×10^{-5})	50,000
Rotational	4×10^{10} (1.6×10^{-4})	3×10^{10} (1.2×10^{-4})	10^{5} (4×10^{-10})	300,000
^{13}C NMR	8×10^{7} (3×10^{-7})[d]	2×10^{4} (10^{-10})	4×10^{-1} (2×10^{-15})	50,000
ICR[e]	2×10^{6} (10^{-8})	2×10^{6} (10^{-8})	10^{2} (5×10^{-13})	20,000

[a] $1\ eV = 2.4 \times 10^{14}$ Hz.
[b] Number of channels is obtained by dividing the typical spectral range by the width of one line.
[c] ^{119m}Sn.
[d] Magnetic field strength approximately 75 kG.
[e] Ion cyclotron resonance; magnetic field strength approximately 20 kG; minimum ionic mass-to-charge ratio, 15.

In ESCA (electron spectroscopy for chemical analysis)[2] and photoelectron spectroscopy,[3] electrons are dislodged from atoms or molecules by X-ray or uv radiation, respectively, and the released electrons have a translational energy that depends on the energy of the bound state occupied by that electron in the original atom or molecule. By scanning the energy of the observed dislodged electrons, the energies of the original molecular electronic states may be determined. By passing the electrons through a perpendicular electric field, the dislodged electrons may be dispersed in space according to their velocity to achieve the arrangement shown schematically in Figure 3.2. In this case, an electronic multidetector can again be based on the vidicon, in which an arriving electron strikes a fluorescent screen on the face of a television camera. Since electrons of different velocity can be dispersed to strike different regions of the screen, their arrivals will be recorded independently by different elements of the television camera grid. Because of the small number of detector channels required (see Table 3.1), the multidetector spectrometer of Figure 3.2 is thus feasible[4] for ESCA and photoelectron spectroscopy.

For the other forms of spectroscopy listed in Table 3.1, direct multidetector methods are less attractive. For rotational (microwave) spectroscopy, for example, there is no broad-band radiation source available: a blackbody radiation source such as employed for other radiation energies (xenon or hydrogen discharge for uv, hot tungsten wire for visible, globar for near infrared and infrared, mercury vapor for far infrared) would have to be operated at an unreasonably high temperature in order to obtain sufficient radiation flux for use as a radiation source. Alternatively, it would be conceivable to construct an array of individual (narrow-band) microwave transmitters (for about $1000 each) as the "broad-band" radiation source, but Table 3.1 shows that the cost would be excessive. For infrared spectroscopy, on the other hand, the necessary broad-band source is available, but it would be necessary to disperse the spectrum over many meters in order to be able to resolve the desired spectral detail with existing individual (thermopile) detectors of about 1-mm width each, and at a cost of about $200 per detector, the total multidetector cost again becomes unmanageable. (Photographic detection does not extend beyond about 12,000 Å, and is thus unavailable.) Finally, for nuclear magnetic resonance (NMR) and ion cyclotron resonance (ICR) spectroscopy, broad-band sources are again available, but the cost of an array of tens of thousands of individual narrow-band mixer–filter detectors (see Section 3.5) is again unreasonably high. For infrared, microwave, and radio frequency spectrometers then, the multidetector approach of Figure 3.2 is just not feasible, either geometrically of financially.

3.3. WEIGHTS ON A BALANCE: THE MULTICHANNEL ADVANTAGE. MULTIPLEX METHODS

Hadamard and Fourier procedures as applied to spectroscopy are two particular examples of the *multiplex* principle. The multiplex advantage follows from the same concept that led us to the *multidetector* spectrometer, and the connection between these approaches may be expressed very simply by analogy to the best way of using an ordinary double-pan balance.

3.3.1. One-at-a-Time Weighing: The Scanning Spectrometer

Consider the hypothetical problem of determining the weights of three unknown objects, by use of the schematic two-pan balance shown in Figure 3.3. Conventionally, we would solve the problem by weighing the unknowns one at a time in (say) the left pan, by placing the appropriate known weight on the right, as shown schematically in Table 3.2. The obvious advantage of this procedure is that each of the measurements (y_1, y_2, or y_3) yields the weight of each unknown (x_1, x_2, or x_3) directly; therefore no data reduction is required:

$$y_1 = x_1, \quad y_2 = x_2, \quad y_3 = x_3 \tag{3.1}$$

TABLE 3.2

Observed weighing result	Unknown weight		
	x_1	x_2	x_3
y_1	1	0	0
y_2	0	1	0
y_3	0	0	1
Number of times each unknown is weighed	1	1	1

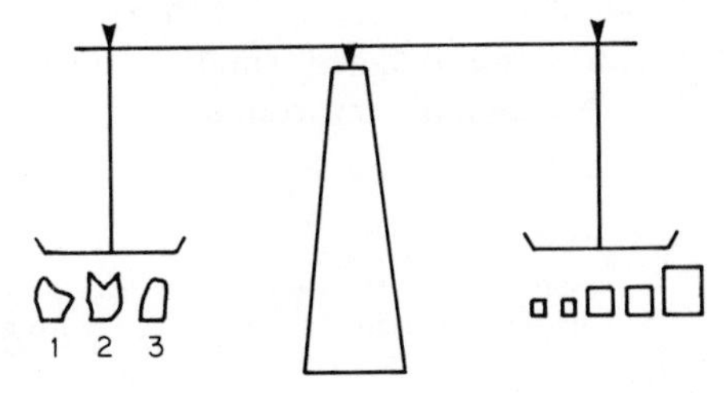

FIGURE 3.3. Schematic diagram of a double-pan balance, with three unknown weights on the left, and a set of standard (known) weights on the right. Uses of the balance model are discussed in Section 3.3.

The disadvantage, which we shall now discuss, is that each unknown has been weighed only once. The one-at-a-time weighing procedure is analogous to scanning a spectrum one channel at a time, as in the spectrometers of Figure 3.1.

3.3.2. Many Balances: The Multidetector Spectrometer

Returning to the balance problem, suppose that we could use three balances *simultaneously*, one for each unknown weight (i.e., the multidetector approach of Figure 3.2). Clearly, the weights of the three unknowns would be determined in 1/3 the time ($1/N$ the time for N unknowns) it would take to make the same determination with just one balance. Alternatively, we could *repeat* the three-balance experiment three times in the *same* time it would take to determine each unknown weight using a single balance.

In any experimental measurement characterized by a certain level of random imprecision or "noise," it is desirable to repeat the measurement many times in order to obtain a more accurate result. The *signal* (in this case the weight of a given unknown) will accumulate as the number of weighings, N. But, if the *noise* is random, its magnitude may be treated as a random walk about zero (the average noise level), and the average absolute distance away from zero after N steps of a random walk (more precisely, the root-mean-square distance), is proportional to $N^{1/2}$.* Thus, for the repeated measurement using the three-balance method illustrated above, the true measure of signal quality, the *signal-to-noise ratio*, will be better by a factor of $3/3^{1/2} = 3^{1/2}$ than for the one-balance method illustrated above, when the *same total observation time* is allotted for each method.

The above results are readily generalized to give a *multidetector* advantage that may be stated and realized in either of two common ways:

(a) For a specified signal-to-noise ratio in the final determined unknown weights (final N-channel spectrum), a multibalance (multidetector spectrometer) experiment will yield the desired weights of the N unknown objects (N-channel spectrum) in $1/N$ the time required for one-at-a-time weighing (scanning) with a single balance (single-channel spectrometer).

(b) A multibalance (multidetector spectrometer) experiment will yield the desired weights of N unknown objects (N-channel spectrum) with a factor of $N^{1/2}$ better signal-to-noise ratio than for one-at-a-time weighing with a single balance (single-channel spectrometer) experiment requiring the same total observation time.

* An exception is in situations where the noise is proportional to the signal strength (e.g., "modulation" or "scintillation" noise). See Appendix for a discussion of various types of noise in spectrometers.

It should be evident that the signal-to-noise ratio and time-saving advantages can be traded off against one another in a variety of intermediate ways, of which (a) and (b) are the extreme choices.

3.3.3. Half the Weights on the Balance at Once: Hadamard Multiplexing

Returning again to use of a single balance,[5] suppose that approximately *half* the unknown weights (in this case, two of the three total unknown weights) are placed on the balance in any one weighing, and that we repeat this procedure with N (in this case three) different linearly independent arrangements of unknown weights as shown in Table 3.3.

This time, the three unknown weights (x_1, x_2, x_3) are related to the three measurements (y_1, y_2, y_3) by three linear algebraic equations:

$$y_1 = x_1 + x_2 \tag{3.2a}$$

$$y_2 = x_1 + x_3 \tag{3.2b}$$

$$y_3 = x_2 + x_3 \tag{3.2c}$$

which may be solved to yield the desired unknown weights (see Section 3.4). However, since each unknown has now been weighed twice, the "signal" per unknown is two times larger than with one-at-a-time weighing; since there are three measurements, the "noise" is $3^{1/2}$ times larger than with one-at-a-time weighing, so that the signal-to-noise ratio of each calculated weight is now better by a factor of $2/3^{1/2}$ than with the conventional one-at-a-time weighing.*

* These statements are valid only when the noise in each weighing (spectral channel) is independent of the weight on the balance (spectral signal), as discussed more fully in the Appendix.

TABLE 3.3

Observed weighing result	Unknown weight		
	x_1	x_2	x_3
y_1	1	1	0
y_2	1	0	1
y_3	0	1	1
Number of times each unknown is weighed	2	2	2

Since the same total number of weighings (three) are required, the total observation time required for the new experiment (i.e., half the weights on the balance at once) is also the same as for the one-at-a-time method. It is clear that for an arbitrary number N of unknown weights, the improvement in signal-to-noise ratio for this encoding–decoding ("multiplex") scheme is a factor of $(N/2)/N^{1/2} = N^{1/2}/2$ for an experiment that takes no longer to carry out than N conventional single-object weighings. Although the experimental details of the spectroscopy experiment are completely different from those of the balance experiment, the preceding encoding–decoding scheme still applies and forms the basis for the signal-to-noise advantage of Hadamard "transform" spectroscopy (see Section 3.4).[5]

3.3.4. All the Weights on the Balance at Once: The Fourier Advantage

By logical extension of the preceding argument, one might think of putting all three weights on the balance in each weighing,[5] being careful to choose three linearly independent arrangements of weights such as shown in Table 3.4, and recording the (known) weight required to balance any particular arrangement of unknowns:

$$\begin{aligned} y_1 &= x_1 + x_2 - x_3 \\ y_2 &= -x_1 + x_2 + x_3 \\ y_3 &= x_1 - x_2 + x_3 \end{aligned} \tag{3.3}$$

Again, it is possible to extract the three desired individual unknown weights by straightforward solution of three coupled linear algebraic equations. However, since each unknown has now been weighed three times, and since there are again three weighings, the signal-to-noise ratio for each calculated weight will be improved by a factor of $3/3^{1/2} = 3^{1/2}$ compared to conventional one-at-a-time weighing experiment requiring the same total time for com-

TABLE 3.4

Measurement	Unknowns[a] in:	
	Left pan	Right pan
y_1	x_1, x_2	x_3
y_2	x_2, x_3	x_1
y_3	x_1, x_3	x_2

[a] Each unknown weighed three times.

pletion. For an arbitrary number of unknowns N, it follows that the general improvement in signal-to-noise ratio will be a factor of $N^{1/2}$, just as for the multidetector example given earlier. This $N^{1/2}$ improvement also applies to Fourier multiplexing described in Section 3.6.1.

By placing either half or all the unknown weights on the balance at once, the signal-to-noise ratio of each calculated unknown weight is improved by a factor of $N^{1/2}/2$ or $N^{1/2}$, respectively, for a total experimental observation time that is the *same* as in conventional one-at-a-time weighing. This improvement is called the Fellgett[6] advantage; all that remains is to show that we conventionally conduct spectroscopy as inefficiently as we conventionally use a double-pan balance, and to describe the particular Hadamard and Fourier multiplex schemes for exploiting the spectroscopic Fellgett advantage.

3.4. HADAMARD MULTIPLEXING OF SPATIALLY DISPERSED SPECTRA

Figure 3.4 shows the instrumental modification that allows for introduction of the Hadamard multiplex scheme: using the broad-band source, disperser, and single broad-band detector of the spectrometer of Figure 3.1b, a mask is interposed between the desired spectroscopic "window" and the detector. The mask is constructed so that its smallest opening is the width of the (narrow) slit width of the conventional scanning spectrometer (Figure 3.1b), but with approximately *half* the total possible positions open. The pattern of open and shut slits is pseudorandom (see Example).

Let the spectrum of intensities transmitted by the sample (Figure 3.2) be represented by spectral "elements" $x_1, x_2, \ldots, x_N$. When the mask of Figure 3.4 is in position, the total measured intensity at the detector, y, is composed of a sum of all the desired spectral elements, each weighted by a factor a_n of either zero or one, depending on whether that particular slit was shut or open, respectively:

$$y = a_1 x_1 + a_2 x_2 + \cdots + a_N x_N, \qquad a_n = 0 \text{ or } 1 \tag{3.4}$$

In other words, the detector has provided one observable (y), expressed in terms of N unknowns (the desired individual spectral intensities, x_1 to x_N), according to the "code" (a_1 to a_N) of equation (3.4). This situation is clearly parallel to that of putting half the unknown weights on the left pan of the balance, as already discussed. In order to recover the desired spectrum, x_1 to x_N, the next step is to replace the first mask with a second mask, which again has a pseudorandom arrangement of open and shut slits with approximately half the slits open, such that this second slit arrangement is linearly independent compared to the first arrangement. Proceeding in this way, by using N different masks having N linearly independent arrangements of

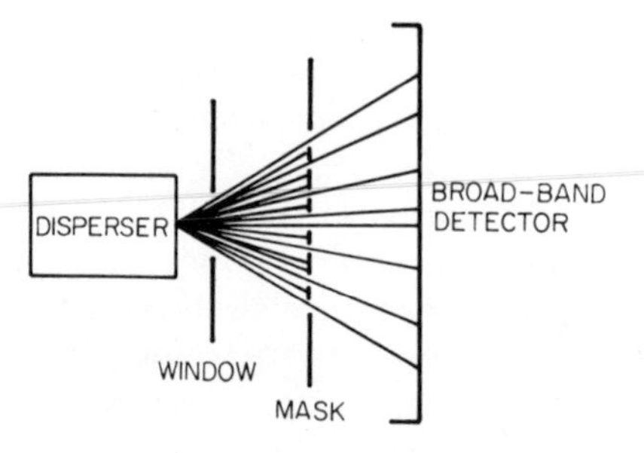

FIGURE 3.4. Encoding of spectral intensities of an *N*-channel spectrum, by insertion of a mask between the spectral window and the (single, broad-band) detector. The mask is constructed so that its smallest opening is the width of the (narrow) slit of the conventional scanning spectrometer of Figure 3.1b, but with approximately *half* the total possible slits open. The encoding process is discussed in Section 3.4.

open and shut slits, one readily obtains N *observables* (y_1 to y_N, the total detected intensity through each of the N respective masks), expressed in terms of N *unknowns* (x_1 to x_N, the desired spectral intensities), according to a "code" (a_{nm}) in which all the coefficients are either zero or unity, and roughly half the coefficients in any one row are zero:

$$\begin{aligned} y_1 &= a_{11}x_1 + a_{12}x_2 + \cdots + a_{N1}x_N \\ y_2 &= a_{21}x_1 + a_{22}x_2 + \cdots + a_{N2}x_N \\ &\vdots \\ y_N &= a_{N1}x_1 + a_{N2}x_2 + \cdots + a_{NN}x_N \end{aligned} \tag{3.5}$$

The "Hadamard" aspect of this scheme follows from a particular choice for the "code" coefficients,[7] and is most easily illustrated with an example.

Example. Hadamard multiplexing for $N = 3$.

Consider a three-slit mask with two of the three slits open in any one measurement, and let each of the three slit arrangements differ from the preceding one by *cyclic permutation*, as shown in equation (3.2). The advantage of this choice is that instead of three separate masks (namely, 110, 101, and 011), it suffices to construct just a single mask consisting of five potentially open slit positions (11011): then by placing the first three mask slits over the window, one achieves the first slit arrangement [110, equation (3.2a)], and by moving the mask by one slit width, one obtains the second slit arrangement [101, equation (3.2b)], etc. Figure 3.5 shows the seven linearly independent slit arrangements (each a cyclic permutation of the preceding one) derived from a single 13-slit mask (0101110010111) in analogous fashion. In general, the condition that each mask differ from the preceding one by cyclic permutation is sufficient to provide a single mask of $2N - 1$ slits rather than N separate masks. Thus, changing from one mask to another simply consists of translating the $(2N - 1)$-slit linear mask across the spectral window by one position per change.[8] The key to the tremendous mechanical simplification of the preceding scheme is that it be possible to choose slit arrangements that are cyclically permutable, linearly independent, and such that approximately half the possible slits are open in any one arrangement. It turns out that these conditions are

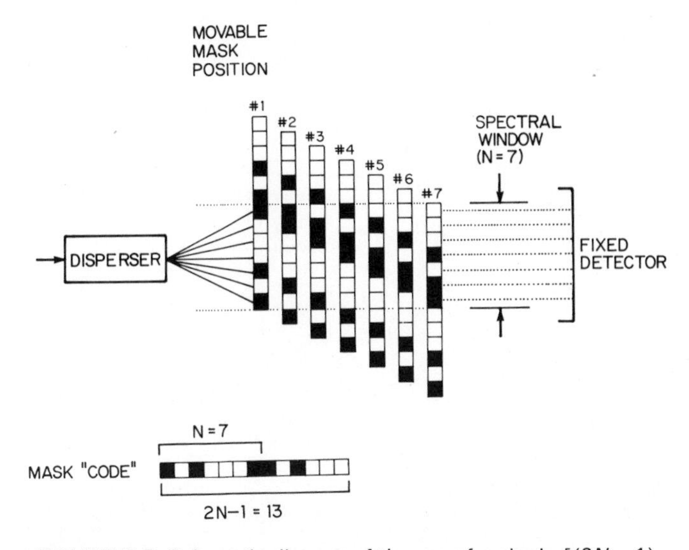

FIGURE 3.5. Schematic diagram of the use of a single [(2*N* − 1)-slit] mask to generate the *N* different linear combinations of spectral intensities from an *N*-channel spectrum, for the $N = 7$ case. The 13-slit mask is first placed between the spectral window and the detector as shown at the far left, to give a pattern of open and shut slits described by the sequence 0 1 0 1 1 1 0, and the transmitted total spectral intensity is measured by the detector. Next, the mask is translated by one slit position across the window, to generate the new pattern of open and shut slits described by the sequence 1 0 1 1 1 0 0, and the total transmitted intensity is again recorded. Proceeding in this way, one obtains $N = 7$ different observed total transmitted intensities, corresponding to seven linearly independent combinations of the (desired) unknown spectral intensities in each of the seven spectral channels. By the methods described in Section 3.4, these $N = 7$ equations in $N = 7$ unknowns may be solved to yield the desired *N*-channel intensity spectrum from the sample. The encoding–decoding scheme is called Hadamard "multiplexing," as discussed in the text.

compatible only for certain choices of N (for example, $N = 4$ will not satisfy all three conditions at once). Since Hadamard matrices can be used to generate the slit arrangements that satisfy all three conditions for the most common choices for N [namely, $N = 3, 7, 15, (2r - 1)$, where r is an integer], this particular encoding–decoding scheme is known as Hadamard multiplexing.[7]

We have until now avoided the problem of actually solving the N equations in N unknowns [equations (3.5)] required to recover the desired spectrum (x_1 to x_N) from the N measurements of the detector of Figure 3.4. In general, the numerical solution of N linear algebraic equations in N unknowns can be difficult or worse.[9] However, the solution is trivial for the Hadamard case, as is easily illustrated for the $N = 3$ case.

Expressing the N observables (y_1 to y_N) in terms of the N unknowns (x_1 to x_N)

for the $N = 3$ example of equation (3.2),

$$\begin{aligned} y_1 &= 1 \cdot x_1 + 1 \cdot x_2 + 0 \cdot x_3 \\ y_2 &= 1 \cdot x_1 + 0 \cdot x_2 + 1 \cdot x_3 \\ y_3 &= 0 \cdot x_1 + 1 \cdot x_2 + 1 \cdot x_3 \end{aligned} \tag{3.6}$$

or, in more compact matrix notation,

$$\mathbf{y} = \mathbf{A}\mathbf{x} \tag{3.7}$$

$$\begin{pmatrix} y_1 \\ y_2 \\ y_3 \end{pmatrix} = \begin{pmatrix} 1 & 1 & 0 \\ 1 & 0 & 1 \\ 0 & 1 & 1 \end{pmatrix} \begin{pmatrix} x_1 \\ x_2 \\ x_3 \end{pmatrix} \tag{3.8}$$

It is easily shown that the solution of equations (3.6) is

$$\begin{aligned} x_1 &= 1 \cdot y_1 + 1 \cdot y_2 - 1 \cdot y_3 \\ x_2 &= 1 \cdot y_1 - 1 \cdot y_2 + y_3 \\ x_3 &= - \ 1 \cdot y_1 + 1 \cdot y_2 + 1 \cdot y_3 \end{aligned} \tag{3.9}$$

or again, in more compact matrix notation,

$$\mathbf{A}^{-1}\mathbf{A}\mathbf{x} = \mathbf{x} = \mathbf{A}^{-1}\mathbf{y} \tag{3.10}$$

$$\begin{pmatrix} x_1 \\ x_2 \\ x_3 \end{pmatrix} = \begin{pmatrix} 1 & 1 & -1 \\ 1 & -1 & 1 \\ -1 & 1 & 1 \end{pmatrix} \begin{pmatrix} y_1 \\ y_2 \\ y_3 \end{pmatrix} \tag{3.11}$$

In other words, the solution of the original equations (3.8) is obtained immediately by replacing zero by -1 in the original coefficient matrix and then calculating the desired unknown spectral intensities, x_1 to x_N, from equation (3.11). It can readily be verified that cyclic permutation of the sequence, 0 1 0 1 1 1 0, produces a mask "code" having the same simple relation between the original code [e.g., (3.8)] and the "inverse" code [e.g., (3.11)], and the same is true for the larger values of N indicated above.[10]

By extrapolation from the $N = 3$ example just given, it is clear that the "Hadamard" code based on cyclically permuted successive slit arrangements offers substantial advantages on both mechanical [replacement of N different masks by a single $(2N - 1)$-slit mask] and computational (near-trivial data reduction) grounds. The signal-to-noise advantage of the Hadamard spectrometer of Figures 3.4 and 3.5 has been shown to be $2/3^{1/2}$ when $N = 3$; it becomes $4/7^{1/2}$ for $N = 7$ (four slits open in each measurement gives four times larger signal, while seven measurements increases noise by $7^{1/2}$);

and is in general given by

$$\text{Hadamard multiplex signal-to-noise advantage}^{(11)} = \frac{(N+1)/2}{N^{1/2}} \tag{3.12a}$$

or equivalently,[12]

$$\text{Hadamard multiplex signal-to-noise advantage}^{(11)} = \frac{N^{1/2}}{2 - 2/(N+1)} \tag{3.12b}$$

$$= \frac{N^{1/2}}{2} \quad \text{for large } N \tag{3.13}$$

Finally, as with the multidetector spectrometer of Section 3.2.3, the Hadamard advantage given in equations (3.12) and (3.13) may be expressed in either of two alternate ways:

(a) A Hadamard spectrometer will yield the desired N-point intensity spectrum of a given sample with a factor of $N^{1/2}/2$ better signal-to-noise ratio than for one-at-a-time scanning with a single-channel spectrometer experiment requiring the same total observation time.*

(b) A Hadamard spectrometer (Figures 3.4 and 3.5) experiment will yield the desired N-point intensity spectrum having the same signal-to-noise ratio in $4/N$ the time required in one-at-a-time scanning with a single-channel spectrometer.

3.5. ADVANTAGES OF COHERENT RADIATION IN SPECTROMETER DETECTION

In order to proceed to Fourier methods in spectroscopy, it is important to understand that the spectrometers discussed up to now (Figures 3.1, 3.2, and 3.4) can operate with an incoherent radiation source; that is, there is no necessary common phase relationship (see below) between the various radiation components issuing from the source. For such incoherent source spectrometers, Hadamard mask techniques provide a means for effectively opening up the slit width without sacrificing resolution (see Figure 3.11).

* Since approximately half the possible N slits are open at once in the Hadamard spectrometer, the Hadamard spectrometer effectively provides a spectrum having half the signal-to-noise ratio in each channel, in $1/N$ the time required to obtain the spectrum by a scanning (single-channel) spectrometer. Thus, in order to increase the Hadamard spectrometer signal-to-noise ratio by a factor of 2 (to make it the same as that from the scanning spectrometer), it is necessary to operate the Hadamard spectrometer for $(2^2)\cdot(1/N) = (4/N)$, the time required for the same signal-to-noise ratio per channel from the scanning spectrometer. (See Appendix for further examples and discussion.)

Apart from Fourier applications, a coherent radiation source and coherent detector in a single-channel (scanning) spectrometer (Figure 3.1a) provide two important advantages to the spectroscopist. First, since the frequency of the coherent radiation source is easily determined to very high accuracy by use of electronic counting techniques, the line positions in a spectrum may be determined very conveniently and very accurately, simply by measuring the frequency of the source as it is (slowly) scanned over the spectral window. Second, coherent radiation permits the implementation of electronic filtering techniques that can make the spectrometer resolution arbitrarily high. Thus, the spectral line shape determined by a coherent radiation spectrometer can be made characteristic of the sample, by making the instrumental line-broadening arbitrarily small.

The basic operation of a coherent source spectrometer is illustrated in Figure 3.6, which shows a hypothetical infrared laser-source spectrometer for use in vibrational spectroscopy. The radiation issuing from the source consists of a plane-polarized electric field whose magnitude varies sinusoidally with time. Upon encountering an electric dipole (i.e., a polar molecule), the electric field will force the dipole to oscillate at the frequency of the radiation, but the amplitude of that dipolar oscillation will be appreciable only when the electric field oscillation frequency is the same as ("in resonance with") the "natural" vibration frequency of the dipole. If the source radiation is coherent, then all the electric dipoles in a given region of space will oscillate together, forming a *macroscopic* oscillating electric dipole in the sample. That macroscopic oscillating electric dipole then induces an oscillating charge on the parallel plates of the capacitor enclosing the sample, and thus a corresponding oscillating voltage in the external circuit (see Figure 3.6). That induced oscillating voltage may then be amplified and, in the most important step, multiplied (in a "mixer") by the oscillating signal from the source and the product decomposed electronically into the sum and difference of the two sine wave frequencies, just as the product of two

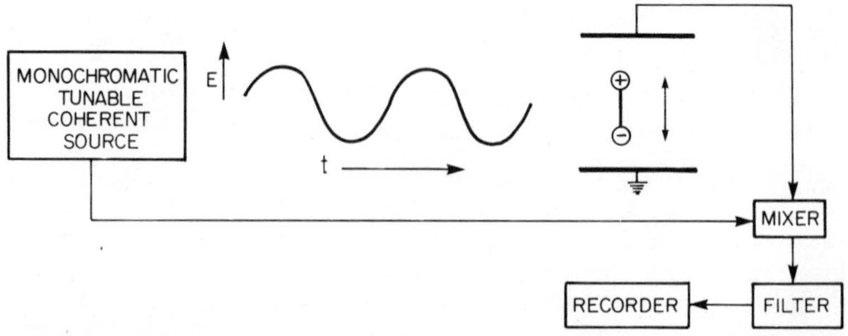

FIGURE 3.6. Hypothetical infrared laser source spectrometer. Operation of this device is discussed in Section 3.5.

sine waves may be decomposed algebraically (by a trigonometric identity) into sine waves of the sum and difference frequencies. The low-pass filter of Figure 3.6 rejects the (higher) "sum" frequency and passes the (lower) "difference" frequency, which is then recorded. The above mixing–filtering process effectively extracts a small spectral segment that is centered at the source frequency and whose width is determined by the bandwidth of the low-pass filter.

In the language of optical spectroscopy, this sort of spectrometer provides a slit *position* that is determined by the frequency of the source and a slit *width* that is determined by the bandwidth of the electrical low-pass filter: the slit width may therefore be made arbitrarily wide or narrow without any mechanical adjustment of the spectrometer geometry. Spectrometers in which a macroscopic change in a physical property of the sample is induced by incident radiation from a coherent source, and that macroscopic change is detected electronically, in the manner described above, have long been employed in nuclear magnetic resonance (NMR) spectroscopy,[13] and ion cyclotron resonance spectroscopy,[14] and have recently been introduced in microwave[15] and infrared[16,17] spectroscopy.

The coherent source and detector make possible another type of encoding–decoding scheme (Fourier multiplexing) for opening up the spectral window while preserving spectral resolution (see Figure 3.11). Finally, one reason that both Hadamard (incoherent source) and Fourier (coherent source) methods may be applied to infrared spectroscopy is that the Michelson interferometer can be thought of as a device that effectively converts incoherent to coherent radiation in the present context.

3.6. FOURIER METHODS

3.6.1. Fourier Multiplexing: The Multichannel Advantage

Fourier transform methods at first seem strange to our intuition, because we are prejudiced by our eyes and ears to analyze our surroundings in the *frequency* domain—we judge light by its color and sound by its pitch. It is, however, equally useful to analyze observations in the *time* domain, as we will now try to show.

Figure 3.7a shows two equivalent ways of representing a single-frequency sine wave signal (such as the electric field amplitude of monochromatic coherent radiation), using either time domain or frequency domain display, respectively. Figure 3.7b shows a somewhat more complicated signal consisting of the sum of two sine wave oscillations of different amplitude and different frequency; again, it is possible to represent the signal using either time or frequency domain display. In fact, it is possible to represent virtually any periodic time domain signal in terms of its frequency "components," as shown in Figure 3.8, where the time domain signal is

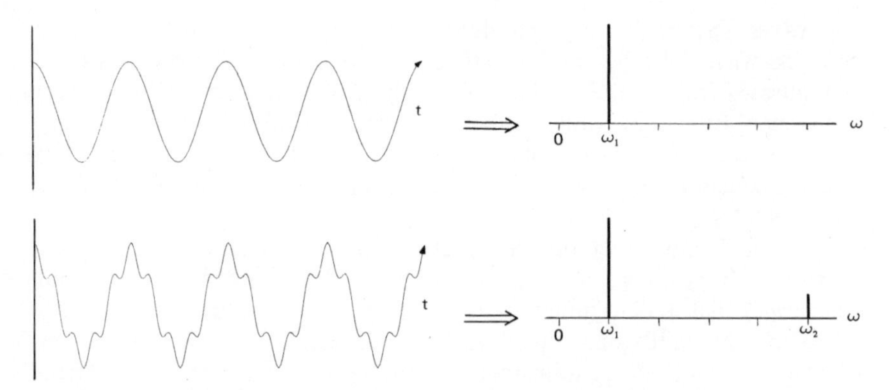

FIGURE 3.7. Time domain (left) and frequency domain (right) representations of a *single* cosine wave (top) and of a *sum* of two cosine wave oscillations of different amplitude and different frequency (bottom). While either representation contains the same information, the frequency domain picture is usually more familiar in spectroscopic applications (see Section 3.6.1).

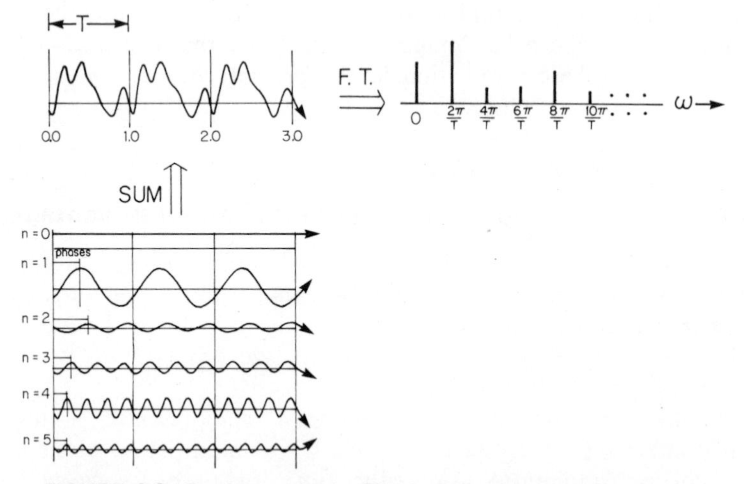

FIGURE 3.8. Fourier synthesis. The periodic time domain signal at the upper left may be constructed from a sum of sinusoidal oscillations of various amplitudes and phases, where the frequencies (in hertz) of the sinusoidal oscillations are chosen to be integral multiples of the reciprocal of the period of the time domain signal. These sinusoidal components may be represented either by their time domain behavior (bottom of diagram) or by their frequency domain components (upper right of diagram), as discussed in Section 3.6.1.

"synthesized" from a sum of sine waves whose frequencies are multiples of the reciprocal of the period of the time domain signal.

For the physical applications described in subsequent chapters, it is usual to *sample* the time domain signal (from, say, a spectrometer detector) at N equally spaced time intervals, $t = T/N, 2T/N, \ldots, T$, to give a set of N time domain data points (the amplitude of the time domain signal at each of the sampled times), $y(t_1), y(t_2), \ldots, y(t_N)$. As for the *continuous* time domain signals of Figures 3.7 and 3.8, it is possible to represent *each* time domain data point in terms of a *series* of frequency domain amplitudes, $x(\omega_1), x(\omega_2), \ldots, x(\omega_N)$, where the frequencies $\omega_1, \omega_2, \ldots, \omega_N$ are integral multiples of $1/T$:

$$\begin{aligned} y(t_1) &= a_{11}x(\omega_1) + a_{12}x(\omega_2) + \cdots + a_{1N}x(\omega_N) \\ y(t_2) &= a_{21}x(\omega_1) + a_{22}x(\omega_2) + \cdots + a_{2N}x(\omega_N) \\ &\vdots \\ y(t_N) &= a_{N1}x(\omega_1) + a_{N2}x(\omega_2) + \cdots + a_{NN}x(\omega_N) \end{aligned} \tag{3.14}$$

where

$$t_n = n(T/N), \qquad n = 1, 2, \ldots, N \tag{3.15}$$

$$\omega_m = (2\pi m/T), \qquad m = 1, 2, \ldots, N \tag{3.16}$$

$$a_{nm} = \exp[i\omega_m t_n] = \exp[2\pi inm/N] \tag{3.17}$$

where the a_{nm} represent the phases (see Figure 3.8) of each frequency component, using complex notation for brevity. [A phase factor of $\exp(i\theta)$ indicates a phase angle of θ rad, as illustrated in Figure 3.7. For example, $\theta = 0$ denotes a pure cosine wave whose phase factor is $\exp(0) = 1$, while $\theta = \pi/2$ denotes a pure sine wave whose phase factor is $\exp(i\pi/2) = i$; etc.]

Since there are N independent known sampled time domain amplitudes, $y(t_1)$ to $y(t_N)$, each expressed in terms of all N desired discrete frequency domain spectral amplitudes, $x(\omega_1)$ to $x(\omega_N)$, it is possible to "decode" the sampled time domain data to obtain the desired discrete frequency domain amplitude spectrum, simply by solving the N linear algebraic equations in N unknowns of equations (3.14). For the particular choice of *equally spaced* time domain samples and *equally spaced* frequency domain amplitudes [equations (3.15) and (3.16)],[18] the decoding procedure is called a (discrete) *Fourier transformation* (or *Fourier series*), and equations (3.14) may be solved rapidly and reliably by a digital computer.[19]

The multichannel advantage of the Fourier approach (sampling of a time domain signal, followed by Fourier transformation to give a discrete frequency domain spectrum) can now be understood. In contrast to the Hadamard technique, in which *half* the possible spectrum is detected in any given observation [i.e., *half* the a_{nm} in any one row of equations (3.5) are zero], the magnitude of *each* a_{nm} in any one row of equations (3.14) is *unity*:

$$|a_{nm}| = |\exp[2\pi i n m/N]| = 1 \tag{3.18}$$

so that in the Fourier experiment, it is as if *all* the possible slits were open during any one (time domain) signal measurement. Because all N possible channels or slits are open at the same time, it is now clear from the arguments previously used for the double-pan balance examples that detection and sampling of the *time*-domain response, followed by Fourier transformation (decoding) to obtain the discrete *frequency* domain spectrum, provides a frequency spectrum exhibiting either (i) signal-to-noise ratio improvement of a factor of $N^{1/2}$ in the same total observation period, or (ii) a spectrum having the same signal-to-noise ratio in a factor of $1/N$ as much time as required by a conventional spectrometer, which scans the spectrum one slit width at a time (see Appendix).

While equation (3.18) indicates how Fourier data reduction effectively accomplishes *detection* of the whole multichannel spectrum at once, we have not yet discussed a *source* of broad-band coherent radiation. The operation of the broad-band source becomes clear, once we have analyzed the detected signal in somewhat more detail in the next section.

3.6.2. Fourier Analysis of Detector Response: Spectral Line Shape

If an *N-point* time domain signal can be represented by an *N-point* frequency domain amplitude spectrum [equations (3.14)], then it should seem reasonable to represent a *continuous* time domain signal by a *continuous* frequency domain amplitude spectrum. Mathematically, this progression consists of replacing the sum of equations (3.14) by the integral of equation (3.19)[20]:

$$y(t) = \mathrm{Re}\left(\frac{1}{2\pi}\right)\int_{-\infty}^{\infty} x(\omega)\exp(i\omega t)\,d\omega \tag{3.19}$$

It is worth stopping to note that while $y(t)$ is *real* (in the mathematical sense), $x(\omega)$ appears to be *complex*; moreover there are negative frequencies in equation (3.19). These puzzling features disappear when we explain that the absorption amplitude frequency spectrum $A(\omega)$ corresponds to just the *real* part of $x(\omega)$:

$$x(\omega) = A(\omega) - iB(\omega) \tag{3.20}$$

where $A(\omega)$ is obtained from

$$A(\omega) = \frac{1}{\pi}\int_{-\infty}^{\infty} y(t)\cos\omega t\,dt = \frac{1}{\pi}\int_{0}^{\infty} y(t)\cos\omega t\,dt \tag{3.21}$$

for physically sensible $y(t)$. We will use equation (3.21) to compute $A(\omega)$ only for positive frequencies ($\omega > 0$), for physically sensible time domain signals [i.e., $y(t) = 0$ for $t < 0$, before the detector is turned on]. Thus, while the mathematical expressions of equations (3.19) and (3.21) are conveniently couched in *complex* notation involving *negative* frequencies and *negative* times, the quantities of physical interest, namely $y(t)$ and $A(\omega)$, are both *real* and involve *positive* times and frequencies.

Equation (3.21) provides us with a means for computing the frequency-domain representation of any time domain signal. Now, when a given single oscillator is subjected to irradiation at its resonant frequency (or when a group of oscillators is subjected to coherent irradiation; see Figure 3.6), the amplitude of oscillation will increase. If the irradiating excitation is then removed, the oscillation will persist with an amplitude that decreases (often exponentially) with time, as shown for three convenient limiting situations at the left of Figure 3.9. If the oscillation is *not* appreciably reduced in amplitude during the observation (data acquisition) time T (top trace of Figure 3.9), then the corresponding frequency representation [calculated from (3.21)] has a functional form that resembles the amplitude of (Fraunhofer) diffraction by a slit.[21] If, on the other hand, the oscillation is observed for *several* lifetimes of its decay (middle trace of Figure 3.9), the spectral amplitude representation approaches the familiar Lorentzian line shape encountered in many forms of spectroscopy. Finally, the bottom trace of Figure 3.9 illustrates an intermediate case, for which the data acquisition time T is of the order of the decay lifetime τ. The irreversible decay of the oscillation is due to (i) radiative damping ("spontaneous emission"),[22]

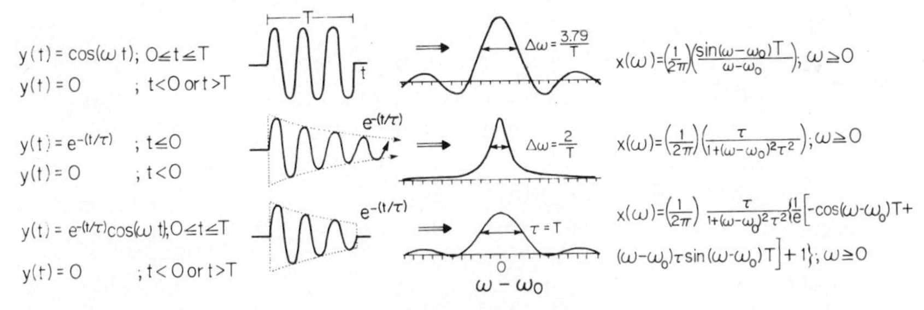

FIGURE 3.9. Fourier analysis of detector response: spectral line shape. Limiting spectral line shape expressions (far right) and frequency domain representation (middle right) for three types of time domain signals (far left) whose time domain representations are shown at middle left. Each line shape expression is obtained from the corresponding time domain signal by use of equation (3.21) (see Section 3.6.2). Top: Fraunhofer line shape, resulting from Fourier transformation of a finite, undamped time domain oscillation. Middle: Lorentzian line shape, resulting from Fourier transformation of an infinitely persisting, exponentially damped time domain signal. Bottom: Intermediate line shape, resulting from Fourier transformation of an exponentially damped time domain signal that is observed for exactly one time constant of the exponential decay.

(ii) interactions of the sample (nucleus, ion, molecule, atom) with its surroundings, where the interaction may be neutral–neutral collisions (microwave, infrared, visible–uv), ion–molecule collisions (ion cyclotron resonance), rotational or translational diffusion (nuclear magnetic resonance, electron spin resonance), or (iii) depletion of the excited species due to chemical reaction. Examples of many of these situations will be found in the applications discussed in subsequent chapters.

3.6.3. Pulsed Monochromatic Coherent Radiation as a Broad-Band Radiation Source

While the top trace of Figure 3.9 was introduced as the frequency representation of a finite pulse of coherent radiation from a *detector*, equation (3.21) applies equally well to the description of the frequency amplitude spectrum of pulsed monochromatic coherent radiation from *any* radiation *source*, such as a laser or radio-frequency oscillator. For example, consider a simple DC (zero-frequency) pulse, which is turned on at time zero and turned off at time T, as shown in the upper-left diagram of Figure 3.10. Equation (3.21) indicates that the frequency representation of such a pulse consists simply of a signal that is spread over a *range* of frequencies near zero. By using a shorter pulse (middle of Figure 3.10), the frequency amplitude spectrum is spread over an even wider range, and in the limit that the DC pulse is made infinitely narrow (bottom of Figure 3.10), the frequency amplitude spectrum is completely flat.

The diagrams in Figure 3.10 suggest that the broad-band frequency *excitation* required for a multichannel (in particular, for a multiplex) spectrometer can be generated by use of a sufficiently narrow pulse of coherent electromagnetic radiation. (If the pulse consists of an AC rather than a DC waveform, then the pictures of Figure 3.10 still apply, except that the frequency amplitude representation is now centered at the AC frequency rather than at zero-frequency—compare the top trace in Figure 3.10 to the top trace in Figure 3.9.)

For nuclear magnetic resonance, for example, Table 3.1 indicates that an excitation bandwidth of about 10 kHz is required—Figure 3.9 (top trace)

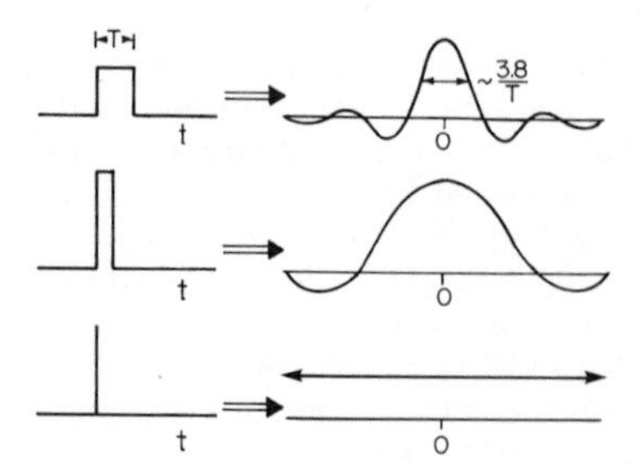

FIGURE 3.10. Time domain (left) and frequency domain (right) representations of DC pulses of three different durations (see Section 3.6.3).

shows that such an excitation may be produced simply by applying a ra frequency electromagnetic pulse whose duration is of the order of 10 μs another example of a short pulse to produce broad-band radiation, electron impact spectroscopy[23] is based on the rapid passage of an electron past a molecule—this passing electron produces a very short, sharp pulse of electric field at the molecule, and thus acts as a very broad-band, nearly flat source of radiation. [In this case, there is no macroscopic coherence in the radiation issuing from excited molecules of the sample, since different incident (source) electrons arrive with random phases.] The electron-impact frequency bandwidth is sufficient to excite the same sorts of energy-level transitions that are more conventionally studied by photoelectron and ESCA spectroscopy.

3.7. SUMMARY: RELATIONS BETWEEN DIFFERENT SPECTROMETERS

The development of the preceding sections is summarized in Figure 3.11, which presents a classification of spectrometers according to whether the whole intensity spectrum is detected at once or one at a time (single-channel or multichannel), and whether the radiation involved is inherently monochromatic or not. Beginning at the upper left of Figure 3.11, the two types of single-channel spectrometer listed are those illustrated in Figures 3.1a and 3.1b. While these spectrometers employ a single-channel (narrow-band) *source*, one can as easily imagine single-channel spectrometers using a broad-band source with a single-channel *detector*. For example, if the disperser and slit of Figure 3.1b were located between the sample and the

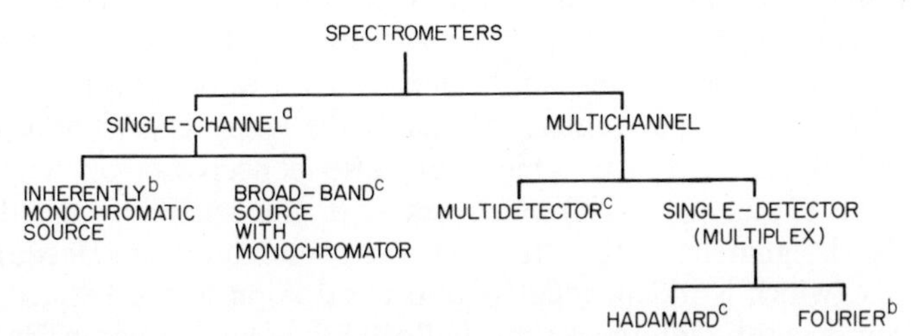

FIGURE 3.11. Relations between different spectrometers (see Section 3.7). (a) While single-channel spectrometers in this diagram are classified according to whether the *source* is inherently monochromatic or inherently broad-band, it is also possible to discriminate on the basis of whether the *detector* is inherently monochromatic or inherently broad-band, as discussed in Section 3.7. (b) In these cases, the source and/or the detector usually operate with coherent radiation. (c) In these cases, the source and detector usually operate with incoherent radiation, which is spatially dispersed in order to resolve spectral absorption or emission at various frequencies.

broad-band detector, the disperser–slit–detector combination could be regarded as a single-channel detector. Similarly, by using an external frequency reference with the mixer–filter detection of Figure 3.6, one would have an inherently monochromatic detector. All four possible combinations of inherently monochromatic or broad-band plus monochromator sources and detectors have been used, although it is most common to combine a coherent source with a coherent detector and a broad-band incoherent source with a broad-band incoherent detector. Finally, it is most usual to find coherent source–detector spectrometers in the longer-wavelength part of the spectrum (microwave and radiofrequency) and incoherent source–detector spectrometers in the shorter-wavelength region (infrared, visible–uv, and X-ray or γ-ray). The reasons for these choices are (i) spatial dispersion of the desired spectral window would require detectors with unreasonably large dimensions for wavelengths much longer than infrared (e.g., microwave or radiofrequency), and (ii) with a few exceptions,[16,17] the mixer–filter scheme of Figure 3.6 fails at short wavelengths (infrared, visible–uv) because of the absence of either a suitable mixer or a suitable conductive medium for routing the detected signal around the circuit of Figure 3.6.

Proceeding to the multichannel portion of Figure 3.11 (upper right of diagram), the multidetector spectrometer of Figure 3.2 is based on spatial dispersion of the spectrum, and is thus practical only for relatively short wavelengths (visible–uv,[24] photoelectron and ESCA).[1] The multiplex methods (Hadamard and Fourier in Figure 3.11) are useful primarily when the noise in the spectrum is detector-limited,* as for radiation of longer wavelength (infrared, microwave, and radio frequency), and the Hadamard scheme is in addition limited to wavelengths short enough that the spectrum may be dispersed over a reasonably small distance. These limits confine Fourier methods primarily to infrared, microwave, and radio-frequency ranges, and Hadamard methods to infrared alone. It should be noted that the discussion in this chapter had been directed toward opening up the *exit* slit of a (dispersive) spectrometer to detect the whole spectrum at once to give the Fellgett advantage. One could also conceive of opening up the *entrance* slit of the (dispersive) spectrometer to give an additional advantage variously designated as the "throughput," "étendue," or "Jacquinot"[25] advantage, which is independent of and in addition to the Fellgett advantage. Thus, even though there is no Fellgett advantage to use of Hadamard or Fourier methods in the visible–uv region (because the noise is source-limited), the Jacquinot advantage may still be exploited by either the Hadamard or Fourier means. For the radio-frequency region, no throughput advantage is required, since the spectral lines are readily saturated by the radiation from either coherent or incoherent sources; besides, the dimensions of dispersive spectrometers would be impractically large for those wavelengths anyway.

* See Appendix for a discussion of various types of noise.

In conclusion, the value of any instrumental improvement must be gauged on the basis of its impact in making possible *new* experiments for experts as well as better *routine* experiments for nonexperts. On this basis, Fourier methods have revolutionized infrared and NMR spectroscopy, by making it possible to obtain spectra of very weak signals, such as infrared spectra of planets and ^{13}C NMR spectra of large organic molecules. Before 1965 (the advent of Fourier data reduction in NMR), for example, ^{13}C NMR spectra were obtainable (by experts) only with great difficulty. As of 1978, most major chemistry departments use ^{13}C NMR spectra routinely in structural and kinetic analysis because it now requires only a few minutes (rather than several hours prior to Fourier methods) to obtain a typical ^{13}C NMR spectrum.

Based on the substantial proven advantages of Fourier data reduction in infrared[26] and NMR[27,28] spectroscopy, the recent application of Fourier techniques to electrochemical,[29] microwave,[15] ion cyclotron resonance,[30] dielectric,[31] optical,[32] and solid-state NMR[33] phenomena promises to make available to practicing chemists a broad new range of experiments not previously instrumentally accessible. Ordinarily, the details of experimental measurement, although crucial to those working in a given field, are relatively uninteresting to chemists in general. In this chapter, we hope to have shown that by taking the time to see how an ordinary double-pan balance should best be used, it is possible to apprehend a wide spectrum of spectroscopic applications of direct chemical interest.

3.8. APPENDIX. NOISE CONSIDERATIONS FOR MULTICHANNEL SPECTROMETERS

All spectroscopic amplitude or intensity measurements are characterized by a degree of imprecision, or noise. While *systematic* noise can usually be minimized by careful experimental design, there still remains *random* noise, which is inherent to the signal and/or to the measurement. As pointed out in Section 3.3, the *ratio* of noise to signal can *usually* (see below) be reduced by repetition of the measurement. The multichannel methods described in this chapter are designed to reduce random noise. However, the advantages of multichannel (multidetector or multiplex) methods depend critically upon the nature of the (random) noise that accompanies the spectroscopic signal.[1,34] For example, when the dominant noise has a magnitude that is proportional to signal strength, there is actually a serious multiplex *dis*advantage in using Hadamard or Fourier multiplexing rather than a scanning (single-channel) spectrometer, as shown in Table 3.5 and the discussion that follows.

It is useful to classify noise according to its relation to the strength of the signal: N_A, noise that is independent of signal strength; N_B, noise that is proportional to the square root of signal strength; and N_C, noise that is proportional to signal strength. When (as is usually the case) these three

TABLE 3.5. Multichannel Advantage, Classified According to the Dominant Type of Experimental Detected Spectral Noise

	Dominant type of noise	Signal-to-noise enhancement[a]			Time factor[b]		
		N-channel multidetector	N-channel multiplex detector		N-channel multidetector	N-channel multiplex detector	
			Fourier[g]	Hadamard		Fourier[g]	Hadamard
	Column No.:	1	2	3	4	5	6
$N_A \gg N_B, N_C$	Noise independent of signal[c]	$N^{1/2}$	$N^{1/2}$	$\frac{N^{1/2}}{2}$	$\frac{1}{N}$	$\frac{1}{N}$	$\frac{4}{N}$
$N_B \gg N_A, N_C$	Noise $\propto \sqrt{\text{signal}}$[d,e]	$N^{1/2}$	1[e]	$\frac{1}{2^{1/2}}$[e]	$\frac{1}{N}$	1[e]	2[e]
$N_C \gg N_A, N_B$	Noise $\propto$ signal[f]	1[h]	$\frac{1}{N^{1/2}}$[e]	$\frac{1}{N^{1/2}}$[e]	[h]	[j,e]	[j,e]

[a] Signal-to-noise ratio for multichannel spectrometer, divided by the signal-to-noise ratio obtained by a scanning (single-channel) spectrometer after the same total observation time.

[b] Time required to obtain a spectrum with a multichannel spectrometer, divided by the time required to obtain a spectrum having the same signal-to-noise ratio by a scanning (single-channel) spectrometer.

[c] Detector-limited noise.

[d] "Photon" or "shot" noise, often described as "source-limited" noise.

[e] In the multiplex case (Fourier or Hadamard), it is assumed that signal strength is approximately uniform over the spectral range.

[f] "Modulation" or "scintillation" noise.

[g] The values in the table are appropriate to Fourier transform methods applied to radio frequency (nuclear magnetic resonance, ion cyclotron resonance, dielectric) spectroscopy. For the special case of Fourier transform–infrared (FT–IR) spectroscopy,[35] half the spectral intensity is lost at the half-silvered mirror of the Michelson interferometer, and the Fourier advantage (either signal-to-noise or time-saving) is then the same as for Hadamard spectroscopy, at the level of the present analysis.

[h] In this case, the signal-to-noise ratio for the multidetector spectrum does not change with time; the multidetector signal-to-noise ratio is the same as for the scanning (single-channel) spectrometer, no matter how long either spectrometer is operated (i.e., time-averaging will *not* increase signal-to-noise).

[j] In the Hadamard and Fourier cases, the signal-to-noise ratio is worse by a factor of $1/N^{1/2}$ compared to the scanning (single-channel) spectrometer, no matter how long the multiplex spectrometer is operated. Time-averaging has no effect on signal-to-noise for this type of noise.

types of noise are statistically independent,[1] the total experimental noise N_{total} is related to the three individual types of noise by

$$N_{\text{total}} = (N_A^2 + N_B^2 + N_C^2)^{1/2} \quad (3.A1)$$

It is clear from equation (3.A1) that a particular type of noise may be dominant in any given experiment, and our further discussion will be restricted to such limiting cases. (For full treatment of all three types of noise present simultaneously, see reference 34.)

3.8.1. $N_B \propto$ (signal)$^{1/2}$: "Source-Limited" Noise

When the photon (electron) energy is greater than kT, where T is the (absolute) temperature of the detector, it becomes possible to detect (count) individual photons (electrons) with approximately unit efficiency. For this case, the imprecision in the measurement is limited by the inherent imprecision in the counting process itself (Poisson statistics), for which the standard deviation (noise) is proportional to the square root of the number of counts (signal). Such noise is sometimes called source-limited noise, since the noise magnitude is related to the strength of the signal. Examples in which source-limited noise is the dominant type of noise include optical (visible–uv) and charged-particle (photoelectron, ESCA, electron impact) spectroscopy.

3.8.2. N_A = constant: "Detector-Limited" Noise

Noise that is independent of signal strength typically dominates in detection of radiation whose photon energy is less than kT, where T is the (absolute) temperature of the detector. Such noise is often called detector-limited noise, because the noise originates in the detector and is unrelated to the strength of the source. Detector-limited noise arises in detection of radio-frequency, microwave, and infrared radiation. [By sufficiently cooling the detector (i.e., by reducing kT) for the near-infrared spectral region, it is possible to reduce the detector-limited noise to a level at which source-limited noise becomes dominant.]

3.8.3. $N_C \propto$ signal: "Fluctuation" Noise

In detection of low-level signals, there is a third type of noise whose magnitude is proportional to signal strength. This type of noise is variously known as "fluctuation," "modulation," or "scintillation" noise, according to its origin.[34] (The corresponding type of noise in electronic circuits is called $1/f$ noise, because its magnitude varies inversely with frequency.) When fluctuation noise is the dominant type of noise, the spectral signal-to-noise ratio *cannot be improved by repeating the measurement* (time-averaging), because the noise accumulates at a rate proportional to the rate at which the signal accumulates after repeated measurements.

The multichannel advantage (stated in terms of signal-to-noise enhancement for constant total observation time, or time factor for fixed signal-to-noise ratio) compared to a scanning (single-channel) spectrometer is given in Table 3.5, for cases in which the noise is dominated by one of the three types listed above. While the signal-to-noise and time factors are clearly related (i.e., it is possible to deduce one from the other, for a given type of noise and given type of spectrometer), it is easiest to account for the *multidetector* advantage using a time-saving argument, and to account for the *multiplex* advantage using a signal-to-noise argument. For example, explanation of the entries in the topmost *row* of Table 3.5 have already been given in prior discussion of multidetector (Sections 3.2 and 3.3), Hadamard (Sections 3.3 and 3.4), and Fourier (Sections 3.3 and 3.6) spectrometers.

The *multidetector* time factors for various types of noise (column 4 of Table 3.5) follow directly from detection of the entire N-channel spectrum in the same time it would take to determine the signal in a single channel using a scanning spectrometer, except as noted in footnote *h* of Table 3.5. Alternatively, for a given total observation time, it would be possible to repeat the multidetector measurement N times, thus increasing the signal by a factor of N and the noise by a factor of $N^{1/2}$, to achieve a signal-to-noise enhancement of a factor of $N^{1/2}$ compared to a scanning (single-channel) spectrometer, as shown in column 1 of Table 3.5 (except as noted in footnote *h*).

In the case of Fourier *multiplex* spectroscopy, the signal strength compared to a scanning (single-channel) spectrometer is increased by a factor of N (for a given total observation time) as for the multidetector spectrometer, but the noise increases so as to give the signal-to-noise enhancements shown in column 2 of Table 3.5. Alternatively, the Fourier multiplex advantage may be expressed in terms of the time factors shown in column 5 of Table 3.5. Similarly, for Hadamard multiplex spectroscopy, the signal strength (for a given total observation time) is increased by a factor of $N/2$ compared to a scanning (single-channel) spectrometer, since the signals from only half the channels are detected at any given instant, while the noise in Hadamard multiplexing increases as in Fourier multiplexing to give the signal-to-noise enhancements shown in column 3 of Table 3.5. Finally, the Hadamard multiplex advantage may be expressed in terms of the corresponding time factors shown in column 6 of Table 3.5.

REFERENCES AND NOTES

1. J. D. Winefordner, J. J. Fitzgerald, and N. Omenetto, *Appl. Spectrosc.* **29**, 369 (1975). This paper gives an excellent review of multidetector, scanning, and multiplex spectrometers as applied to atomic absorption and emission spectra.
2. K. Siegbahn, C. Nordling, A. Fahlman, R. Norderg, K. Hamrin, J. Hedman, G. Johans-

son, T. Bergmark, S. Karlson, I. Lindgren, and B. Lindberg, *ESCA: Atomic, Molecular, and Solid State Structure Studied by Means of Electron Spectroscopy*, Almqvist and Wiksell, Uppsala, Sweden, 1967; K. Siegbahn, C. Nordling, G. Johansson, J. Hedman, P. F. Heden, K. Hamrin, U. Gelius, T. Bergmark, L. O. Werme, R. Manne, and Y. Baer, *ESCA: Applied to Free Molecules*, North Holland, Amsterdam, The Netherlands, 1969.
3. D. W. Turner, A. D. Baker, C. Baker, and C. R. Brundle, *High Resolution Molecular Photoelectron Spectroscopy*, Wiley, New York, 1970.
4. H. Fellner-Feldegg, U. Gelius, B. Wannberg, A. G. Nilsson, E. Basilier, and K. Siegbahn, *J. Electron Spectrosc.* **5**, 643 (1974).
5. R. Kaiser, NMR Spectroscopy with Pseudo-Noise Excitation and Hadamard–Fourier Transform Processing, *NMR Symp. Univ. of Western Ontario*, London, Ontario, 1974. The first author to point out the advantages of weighing several objects together appears to be F. Yates, *J. Roy. Stat. Soc. Suppl.* **2**, 181 (1935).
6. P. Fellgett, *J. Phys. Radium* **19**, 187 (1958).
7. N. J. A. Sloane, T. Fine, P. G. Phillips, and M. Harwit, *Appl. Opt.* **8**, 2103 (1969).
8. N. J. A. Sloane, T. Fine, and P. G. Phillips, *Opt. Spectra* **4**, 50 (1970); J. A. Decker, Jr., *Appl. Opt.* **10**, 510 (1971).
9. G. Forsythe and C. B. Moler, *Computer Solution of Linear Algebraic Systems*. Prentice–Hall, Englewood Cliffs, New Jersey, 1967.
10. R. N. Ibbett, D. Aspinall, and J. F. Grainger, *Appl. Opt.* **7**, 1089 (1968).
11. E. D. Nelson and M. L. Fredman, *J. Opt. Soc. Am.* **60**, 1664 (1970).
12. P. G. Phillips, M. Harwit, and N. J. A. Sloane, *Proc. Aspen Internat. Conf. Fourier Spectroscopy, 1970*, Paper # 48 (G. A. Vanasse, A. T. Stair, and D. J. Baker, eds.), AFCRL-71-0019, Air Force Systems Command, 1971.
13. J. A. Pople, W. G. Schneider, and H. J. Bernstein, *High-Resolution Nuclear Magnetic Resonance*, Chapter 4, McGraw-Hill, New York, 1959.
14. M. B. Comisarow, *J. Chem. Phys.* **55**, 205 (1971).
15. J. Ekkers and W. H. Flygare, *Rev. Sci. Instrum.* **47**, 448 (1976); J. C. McGurk, T. G. Schmalz, and W. H. Flygare, *J. Chem. Phys.* **60**, 4181 (1974); *Adv. Chem. Phys.* **25**, 1 (1974).
16. M. Mumma, T. Kostiuk, S. Cohen, D. Buhl, and P. C. von Thuna, *Nature* **253**, 514 (1975).
17. R. G. Brewer, *Science* **178**, 247 (1972).
18. H. C. Andrews and K. L. Caspari, *IEEE Trans. Computers* **C-19**, 16 (1970).
19. J. W. Cooley and J. W. Tukey, *Math. Comp.* **19**, 297 (1965).
20. R. Bracewell, *The Fourier Transform and Its Applications*, McGraw-Hill, New York, 1965.
21. W. H. Furry, E. M. Purcell, and J. C. Street, *Physics for Science and Engineering Students*, McGraw-Hill, New York, 1960.
22. J. C. Davis, Jr., *Advanced Physical Chemistry: Molecules, Structure, and Spectra*, pp. 252–254, Ronald, New York, 1965; L. Pauling and E. B. Wilson, Jr., *Introduction to Quantum Mechanics*, pp. 299–301, McGraw-Hill, New York, 1935.
23. C. E. Brion, *in: MTP International Review of Science, Mass Spectroscopy, Physical Chemistry*, Series One (A. D. Buckingham and A. Maccoll, eds.), Vol. 5, Chapter 3, Butterworths, London, 1972.
24. G. Horlick and E. G. Codding, *Anal. Chem.* **46**, 133 (1974).
25. L. Mertz, *Transformations in Optics*, Wiley, New York, 1965.
26. P. Connes, *Annu. Rev. Astron. Astrophys.* **8**, 209 (1970); J. P. Maillard, *in*: *I.A.U. Highlights of Astronomy 1973* (Contopoulos *et al.*, eds.), Reidel, Dordrecht, Holland, 1974; E. G. Codding and G. Horlick, *Appl. Spectrosc.* **27**, 85 (1973).
27. R. R. Ernst, *Adv. Mag. Res.* **2**, 1 (1968).
28. J. B. Stothers, *Carbon-13 NMR Spectroscopy*, Academic Press, New York, 1972; G. C. Levy and G. L. Nelson, *Carbon-13 Nuclear Magnetic Resonance for Organic Chemists*, Wiley, New York, 1972; L. F. Johnson and W. C. Jankowski, *Carbon-13 NMR Spectra*, Wiley-Interscience, New York, 1972.
29. S. C. Creason, J. W. Hayes, and D. E. Smith, *Electroanal. Chem. Interfacial Electrochem.*

47, 9 (1973); S. C. Creason and D. E. Smith, *Anal. Chem.* **45**, 2401 (1973), and references therein.

30. M. B. Comisarow and A. G. Marshall, *Chem. Phys. Lett.* **25**, 282 (1974); **26**, 489 (1974); *Can. J. Chem.* **52**, 1997 (1974); *J. Chem. Phys.* **62**, 293 (1975); **64**, 110 (1976).
31. G. A. Brehm and W. H. Stockmayer, *J. Phys. Chem.* **77**, 1348 (1973); R. H. Cole, *J. Phys. Chem.* **78**, 1440 (1974).
32. G. Horlick and W. K. Yuen, *Anal. Chem.* **47**, 775A (1975).
33. A. Pines, J. J. Chang, and R. G. Griffin, *J. Chem. Phys.* **61**, 1021 (1974).
34. J. D. Winefordner, R. Avni, T. L. Chester, J. J. Fitzgerald, L. P. Hart, D. J. Johnson, and F. W. Plankey, *Spectrochim. Acta* **31B,** 1 (1976).
35. M. J. D. Low, *J. Chem. Educ.* **47**, A163, A255, A349, A415 (1970).

Chapter 4

Data Handling in Fourier Transform Spectroscopy

James W. Cooper

4.1. THE COMPUTER SYSTEM

4.1.1. Introduction to Computers

The growing use of the Fourier transform in various forms of spectroscopy has necessitated the introduction of the minicomputer as an integral part of the spectrometer system. It is used both for data collection and processing and cannot be used effectively without some understanding of its potential limitations.

Briefly, the minicomputer consists of an arithmetic unit for performing calculations, some data acquisition hardware, and some memory. Memory is divided into *words*, each consisting of a number of *bits*. The larger the number of bits in a word, the larger or more precise the number that can be stored in the computer's memory, just as the more digits in a number indicates more significant figures. Now, when we write an ordinary decimal number such as 28994_{10}, where the subscript 10 indicates base ten, we mean

$$\begin{aligned} 2 \times 10^4 &= 20{,}000 \\ +8 \times 10^3 &= 8{,}000 \\ +9 \times 10^2 &= 900 \\ +9 \times 10^1 &= 90 \\ +4 \times 10^0 &= 4 \\ \hline & \quad 28{,}994 \end{aligned}$$

James W. Cooper • Department of Chemistry, Tufts University, Medford, Massachusetts 02155

We will consider the same number as a binary number. Our examples in this chapter will be drawn from the 16-bit computer, although it should be recognized that other word length computers, in particular the 20-bit word length, are widely used in Fourier transform data systems.

Since each bit in a computer word can be set to one or zero, numbers stored in a computer word are represented as ones and zeros in base-2 or *binary* notation. Thus the binary number 0111000101000010 equals

$$
\begin{aligned}
1 \times 2^{14} &= 16{,}384 \\
+1 \times 2^{13} &= 8{,}192 \\
+1 \times 2^{12} &= 4{,}096 \\
+1 \times 2^{8} &= 256 \\
+1 \times 2^{6} &= 64 \\
+1 \times 2^{1} &= 2 \\
\hline
& \quad 28{,}994
\end{aligned}
$$

Now the computer can be used to represent integers between 0000000000000000 and 1111111111111111 or between 0 and 65,535. In general, a w-bit word will allow the representations of 2^w numbers from 0 to $2^w - 1$. This does not allow for sign, however, and it is often found that it is more convenient to adopt a format where half of all possible numbers are negative and half positive. In most minicomputers, negative numbers are represented by their *two's complement*, which is obtained by interchanging all the ones and zeros and then adding one. Thus, the two's complement of 5 or 0000000000000101 is 1111111111111011. The continuum of all numbers in a 16-bit computer can be represented by

1000000000000000	1111111111111111	0	0111111111111111
−32768	−1	0	+32767

Since the leftmost bit of any negative number will be 1, this bit is often called the *sign bit*. For convenience, binary numbers are often converted to their *octal* or base-8 representation. This is easily done by dividing any binary number into groups of three from the right and writing down the octal equivalent of each group. Thus we can write 28994_{10} as $0\,111\,000\,101\,000\,010_2$ or 070502_8, and -28994_{10} as 1 000 111 010 111 110 or 107276_8.

Finally, it is quite possible to adopt some sort of scientific notation where we would represent 28994 as 2.8994×10^4 or analogously in binary. These representations vary in detail with the computer but are generally referred to as *floating-point* representations and usually take at least two minicomputer words per number. Binary, octal, and floating-point arithmetic have been thoroughly discussed in a number of places.[1-3]

4.1.2. Data Acquisition

Data are acquired in the minicomputer attached to the laboratory instrument through the use of an analog-to-digital converter or ADC (sometimes also known as a "digitizer") and a clock to cause data points to be taken at equally spaced time intervals. The analog-to-digital converter converts the input voltage at the instant the clock "ticks" into a binary number. Typically, the ADC has a somewhat smaller number of bits or *resolution* than the computer word, generally 12 or 13 bits maximum, although ADCs having up to 16 bits of resolution are commercially available.

If a given ADC has an input range of −1 to +1 V, and 6 bits of resolution, it may convert −1 V to 000000 and +1 V to 111111. This conversion scheme is called *unipolar*, since the most negative input value is converted to a zero and the most positive input voltage to all ones. A more useful conversion scheme is the *bipolar* one, where zero volts is converted to zero and more negative voltages to a negative number (two's complement) representation. Our bipolar 6-bit ADC might convert −1 V as 100000, 0 V as 000000, and +1 V as 011111. This last method is preferable for computers involved in signal *averaging*, since in this scheme zero volts is indeed zero and will not add a bias to a running sum.

4.1.3. Timing in Data Acquisition

The acquisition of data must be at precisely known intervals so that the time of each point can be calculated in processing the data. In Fourier transform spectroscopy, the data points must not only be accurately timed but equally spaced in time if a discrete transform is to be applied to them. To illustrate the consequences of ignoring this requirement, Figure 4.1 shows the Fourier transform of a 2500-Hz sine wave sampled with varying amounts of timing error or jitter. The errors that this jitter introduces are quite severe even with a small amount of variation.

This requirement virtually eliminates the possibility of using a computer's memory cycle timing instead of a clock to generate the timing for high-speed data acquisition, since this timing is a function of the minimum instruction time in the computer and of other somewhat random computer parameters that may not be controllable, such as memory refresh cycles and interrupts. Lower-speed acquisition such as in gas chromatography can be timed by counting memory cycles without any trouble.

The case of Fourier transform IR is somewhat different, since data points are not taken as a function of time but as a function of mirror position. The mirror travel is converted to pulses through the use of a sinusoidal interferogram generated from a laser source, which triggers the computer at regularly spaced intervals along the mirror's path. It is this travel distance rather than the time interval that is crucial in acquiring accurate data from an infrared interferometer.

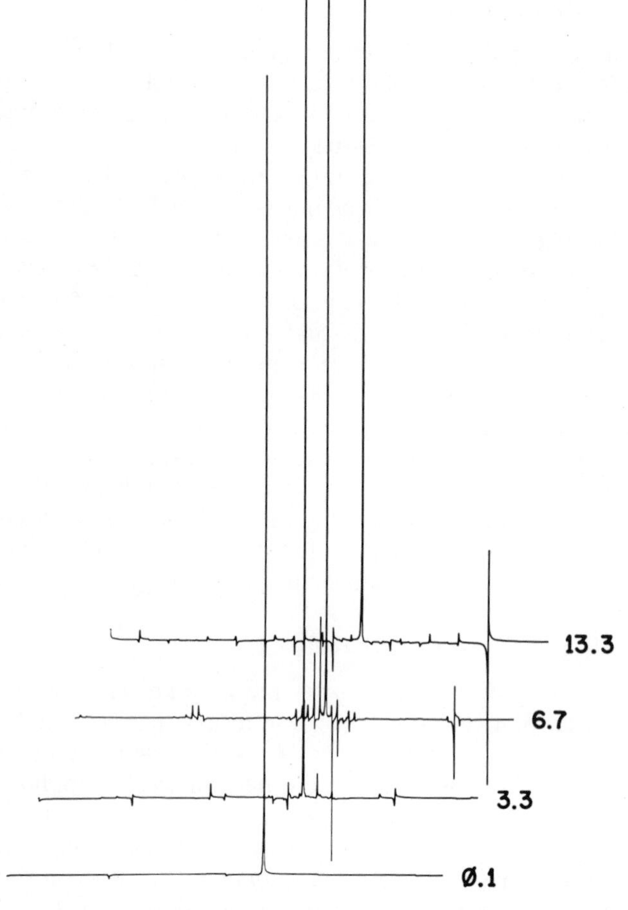

FIGURE 4.1. Fourier transforms of a 2500-Hz sine wave sampled with the indicated amounts of random timing jitter per point. Courtesy of John Wiley and Sons, Inc., New York.

4.1.4. The Sampling Theorem

Briefly stated, the sampling theorem states that the highest-frequency sine wave in a given spectrum, from whatever source, must be sampled at a rate of at least two points per cycle in order to be represented accurately. If less than two points per cycle are used, the computer will "see" this frequency as one less than the actual frequency. Thus, if we wish to examine a spectrum having a 5000-Hz bandwidth, we must sample at a rate of 10,000 Hz or once every 100 μs. The highest frequency that we can accurately represent is 5000 Hz here and this frequency is known as the *Nyquist*[4] frequency.

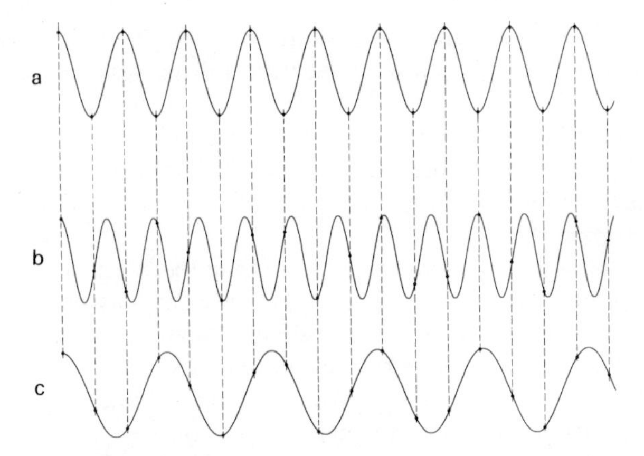

FIGURE 4.2. Sampling of a sine wave at (a) the Nyquist frequency N, (b) $N + \Delta f$, (c) $N - \Delta f$.

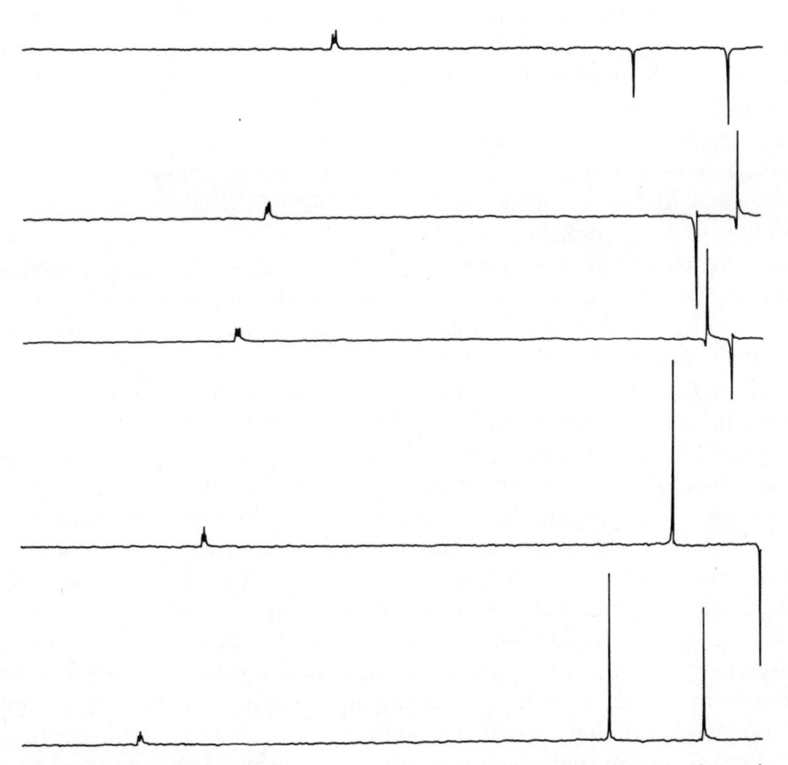

FIGURE 4.3. Illustration of fold-back when lines occur at frequencies above the Nyquist frequency. The sample is ethyl iodide in chloroform-*d*.

Figure 4.2a shows a sine wave sampled at the Nyquist frequency, or twice per cycle. Figure 4.2b shows a sine wave having a frequency of $N + \Delta f$ and Figure 4.2c shows a sine wave having a frequency of $N - \Delta f$. All three waves are sampled at the same rate. Note that the dots representing the data points that the computer sees are in the same places in the waves having frequencies of $N + \Delta f$ and $N - \Delta f$, and that the computer will "see" them as the same frequency.

Thus a sine wave of a higher frequency than the Nyquist frequency will be *folded back* into the spectrum at a frequency exactly as much lower than the Nyquist frequency as the sine wave is higher than the Nyquist frequency. This phenomenon is also referred to as *aliasing.* Figure 4.3 shows just such a case, where a three-line FT–NMR spectrum is gradually shifted by changing the carrier frequency so that the lines one by one fold back into the spectrum. Note that the phase of lines folded back changes; this is often a way of detecting such a line folded back.

The sampling rate in FT spectroscopy thus is determined not by the desired resolution, but by the frequency width to be observed. A 1000-Hz spectrum will be obtained only if the data system samples at 2000 Hz, or once every 500 μs. Sampling faster will place the desired peaks in a cluster at one end of the spectrum: it will not increase the number of data points per peak or the resolution. The resolution can only be increased by sampling for longer times at the same rate so that more points make up the block to be transformed. Furthermore, it is not possible to examine only part of a spectrum in FT spectroscopy, since a smaller sampling frequency will only cause the data outside the "window" to fold back into the spectrum, leading to very confusing results.

One cannot always sample for a longer period of time, however, and obtain real information, since in NMR data the nuclei have relaxed completely after a multiple of the relaxation time T_2^* and no further information is being presented to the ADC. Similarly, in FT–IR data, the resolution need only be about 1/5 the frequency width at half-height of the narrowed lines. In this case the resolution of the spectrum may actually be *degraded* by taking more data points, since more noise is being included in the spectrum than would have been present in a smaller number of data points. Therefore, it is advisable to simply add a block of zero data points to the acquired data and transform the data with this block of zeros included. This process, called *zero filling,* has been shown by Ernst[5] to be entirely mathematically justifiable, as long as the block of zeros does not exceed the length of the original sampled data block. In practice, data are acquired in blocks of powers-of-2 numbers of data points since these are easiest to transform, and the zero filling is usually up to the next power of two number of points. Zero filling beyond this point does not lead to any further resolution enhancement, but is equivalent to the interpolation according to the instrument's line shape function.

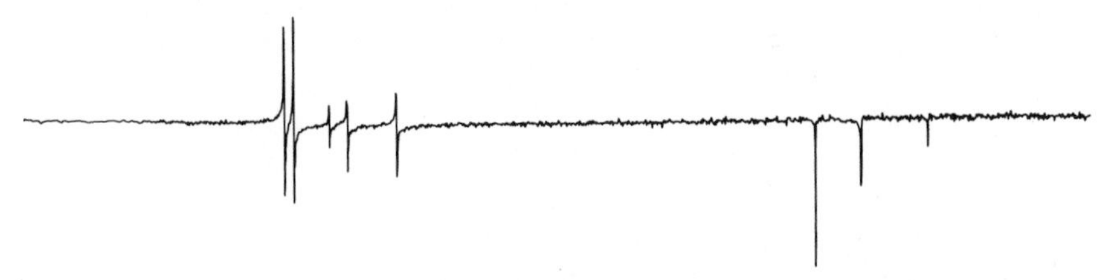

FIGURE 4.4. Un-phase-corrected spectrum of 3-ethylpyridine.

4.1.5. Digital Phase Correction

After a spectrum is acquired and Fourier transformed, it may look more like Figure 4.4 than like a normal spectrum because of phase errors. These errors are of two kinds, zero order and first order, or frequency dependent. Their causes are instrumental but the phase correction may be most conveniently performed digitally following the Fourier transform, since the eye cannot usually detect instrumental phase adjustments in the time domain spectrum.

Zero-order phase errors in NMR spectra are usually caused by improper adjustment of the instrument's phase-sensitive detector during the data acquisition process. The phase detector can be adjusted by successive acquisition and retransformation, but since all of the phase information is in either the real or the imaginary portion of the data, a simple phase correction algorithm can be devised to make this correction digitally.

First-order phase errors are caused by two parameters in NMR spectra: the predelay time and the filters used before the ADC. To see how the delay time can affect the phase of the spectrum, let us consider the sine waves in Figure 4.5, being sampled at some low frequency. Sampling starts in Figure 4.5a exactly at the start of the data, at the top of the cosine wave. However, in Figure 4.5b, a one-sampling interval delay has been introduced between the start of the excitation pulse and the first acquired data point. Clearly the Fourier transform of this wave should produce a peak that is about 70° out of phase.

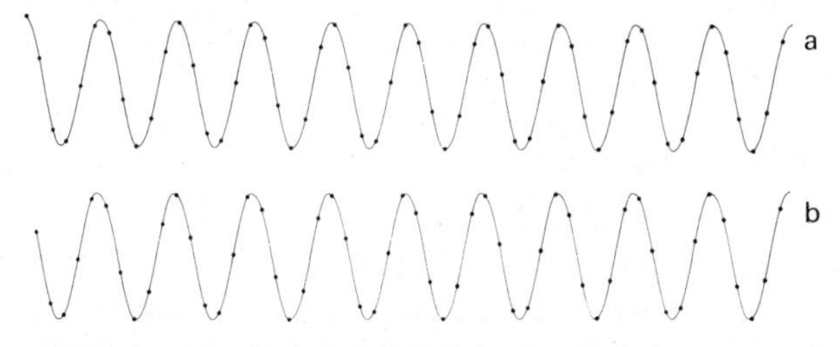

FIGURE 4.5. Sampling of low-frequency sine wave (a) without predelay, (b) with one address of predelay.

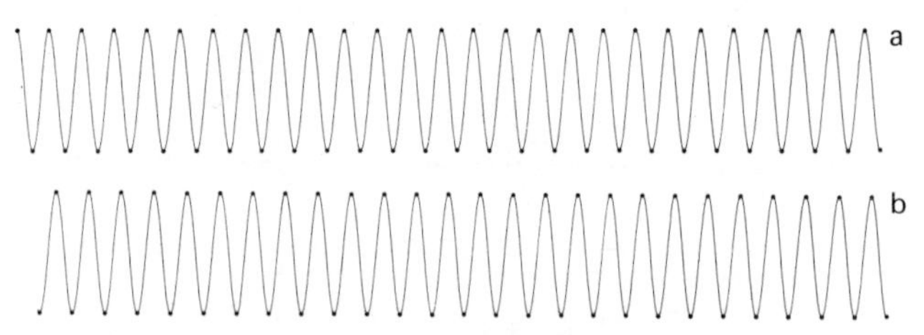

FIGURE 4.6. Sampling of Nyquist frequency sine wave (a) without predelay, (b) with one address of predelay.

Now, let us consider the effect of this one-address delay on the line sampled at the Nyquist frequency in Figure 4.6. If a one-sampling-interval delay is introduced as shown in Figure 4.6b, the transformed line will be 180° out of phase rather than just 70°. This is called a frequency-dependent phase error or a first-order phase dependence and it can be stated that there will be a linearly increasing phase error from 0 to 180° across the spectrum for each one-sampling-interval delay introduced before the onset of data acquisition. This, again, can be corrected digitally if the delay is small. but data with very large delays can cause such a "wrapping around" of the phase information into a spiral that phase correction becomes very difficult, and impossible if the amount of delay is not accurately known. Since such delays are introduced only to minimize switching transients, they are usually small and well controlled.

The data system itself can introduce a negative frequency-dependent phase error through its input filters. Filters are generally used to cut out any spurious lines outside the frequency range of interest, but they also introduce a phase error. Since this error is not always linear with frequency, these filters should be carefully chosen, so that the phase error is as linear as possible or it will not be easily correctable. One set of filters that satisfies this criterion are four-pole Butterworth filters, available from a number of sources. A thorough discussion of phase correction algorithms has been given in several places.[3,6]

Phase errors in FT–IR data are caused by instrumental rather than computer parameters, such as dispersion due to the beamsplitter and alignment of the mirrors. These problems are covered in detail in Chapter 2. The mathematics of FT–IR phase correction is discussed in reference 3.

4.1.6. Signal Averaging

Signal averaging[5,6] is a process in which successive scans through the data are summed in memory. Using this technique, weak signals can be enhanced since the signal is assumed to be coherent and the noise random. As additional scans through a spectrum are co-added the signal grows linearly with the number of scans,

$$\text{signal} = k_1 N \tag{4.1}$$

while the noise grows at a rate proportional to the square root of the number of scans,

$$\text{noise} = k_2 N^{1/2} \tag{4.2}$$

Dividing equation (4.1) by (4.2) we have

$$\frac{\text{signal}}{\text{noise}} = \frac{k_1 N}{k_2 N^{1/2}} = K N^{1/2} \tag{4.3}$$

or the fact that the signal to noise grows at a rate proportional to the square root of the number of scans.

Signal averaging depends on the fact that each scan must start at the same place in the spectrum, either by controlling an excitation pulse, as in NMR, or by exact triggering from the scanning apparatus, as in infrared. Then each scan is added to the rest. In order for this to be true averaging, this sum must at some point be divided down by the total number of scans. This is seldom done, however, since in practice the output of the spectrum after the Fourier transform can be appropriately scaled.

Since no division by the number of scans can occur during the accumulation process under normal signal-averaging conditions, the computer's memory may overflow under some conditions. While memory overflow is allowable in slow scan-averaging processes such as CW NMR, since perhaps only one large solvent peak may be affected, overflow cannot be permitted in Fourier transform spectroscopy, since this overflow will cause a signal that was formerly a sine wave to appear more and more like a square wave. The results of allowing memory overflow in an FT–NMR spectrum are shown in Figure 4.7, where the number of points of the free-induction decay that overflowed are shown in the margin of each transformed spectrum.

It has been assumed by many workers that the total number of scans TS that can be obtained before a computer memory word overflows is related only to the word length w and the ADC resolution d,

$$TS = 2^{w-d} \tag{4.4}$$

In other words, a 16-bit computer having a 12-bit ADC would allow only $2^{16-12} = 2^4 = 16$ scans before memory overflow occurs. In actual fact, the number of scans is this small only when the spectrum being averaged has an infinitely large signal-to-noise ratio (S/N). In the cases where most signal averaging is needed, where the initial signal-to-noise ratio is very low, the total number of scans is much larger. In FT–IR, the zero frequency peak always leads to a large time domain dynamic range.

Suppose we have a spectrum where the initial signal-to-noise ratio of a single scan is 1:1. If we add together 16 scans using a 12-bit digitizer, we ought to fill memory if equation (4.4) holds. Consider, however, what the ADC sees in a single scan. In those parts of the spectrum containing the most

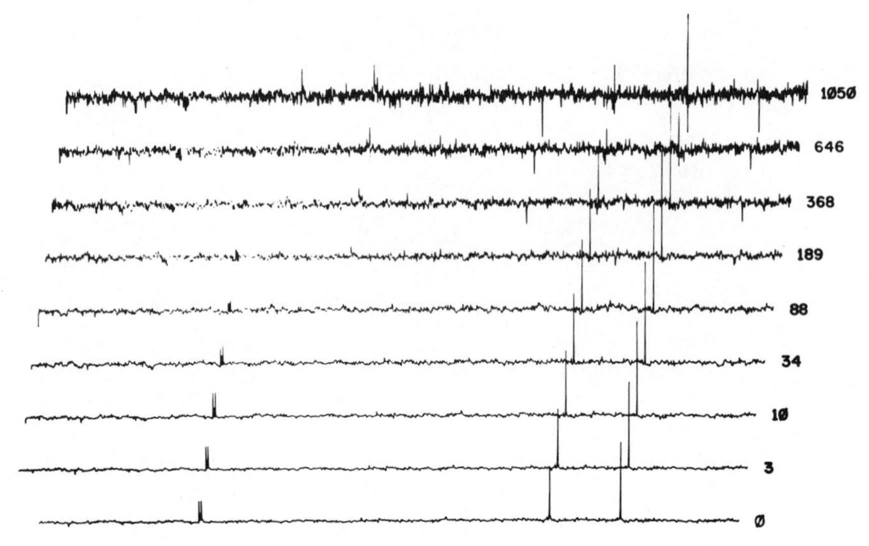

FIGURE 4.7. Fourier transform of ^{13}C spectrum of ethyl iodide after the indicated number of memory channels have been allowed to overflow. Courtesy of John Wiley and Sons, Inc., New York.

intense information the ADC will contain half signal and half noise. A 12-bit ADC would contain 2048 counts of signal and 2048 counts of noise. After 16 scans the total number of counts can be calculated from equations (4.1) and (4.2):

Counts due to signal	= 16(2048) =	32,768
Counts due to noise	= 4(2048) =	8,192
Total counts in memory		40,960

Comparing the total number of counts in memory 40,960 with the total number in a 16-bit word 2^{16} or 65,535 counts, it is clear that there is room for considerably more averaging before memory overflow occurs.

We can calculate the total number of scans possible for any initial signal-to-noise by simply generalizing the above calculation. The ADC contains 2^d counts divided between signal and noise. The total amount taken up by signal is $2^d S/(S + N)$ and the amount taken up by noise is $2^d N/(S + N)$. When we have taken TS scans, that total number of counts due to signal will be $TS \cdot 2^d \cdot S/(S + N)$ and the total counts due to noise will be $TS^{1/2} \cdot 2^d \cdot N/(S + N)$. Memory will be full when the sum of these two numbers equals the total number of counts in the memory word 2^w,

$$2^w = \frac{TS \cdot S + TS^{1/2}(1)}{S + 1} \cdot 2^d \tag{4.5}$$

where we have set $N = 1$, since signal to noise is always referred to as S:1. This equation is most easily solved for $TS^{1/2}$, giving

$$TS^{1/2} = \frac{-1 + [1 + 4S(S+1) \cdot 2^{w-d}]^{1/2}}{2S} \tag{4.6}$$

Note that this does reduce to equation (4.4) for $S \gg 1$. A tabulation of the total number of scans possible for various values of S/N and $w - d$ has been reported.[7] Note, however, that the maximum number of scans is dependent only on $w - d$ and not on either separately.

Now, suppose we want to take 262,000 scans of a given spectrum in order to enhance the signal to noise. This represents about three days of averaging at one second per scan. We know from equation (4.3) that the initial signal to noise will be improved by $262{,}000^{1/2}$ or a factor of 512. If we accept 4:1 as the minimum recognizable signal to noise, we can resolve an initial signal to noise of 4/512 or 0.00781 in this time. Rearranging equation (4.6) we have

$$2^{w-d} = \frac{(2S \cdot TS^{1/2} + 1)^2 - 1}{4S(S+1)} \tag{4.7}$$

Solving, we find that $2^{w-d} = 1454$. Thus $w - d$ must be 11 for us to complete 262,000 scans without memory overflow occurring. Note, however, that we have no way of measuring an initial S/N of 0.00781 or of any value much less than about 2, and so we must instruct our computer to keep careful track of potential memory overflow and prevent it before it occurs. This can be done by dividing down the running average by 2 as memory overflow becomes imminent and then dividing all further ADC readings by 2 as well. This process can be continued until there is only one bit of resolution left in the ADC. In fact, if the initial S/N is less than 1 to start with, a one-bit ADC is quite sufficient for signal averaging. While no commercial data systems boast of their 1-bit ADCs for psychological reasons, most of them have software that divides down the ADC reading as needed to prevent overflow and this division can result in just that.

4.1.7. Signals Having High Dynamic Range

When it is necessary to measure small, weak peaks in the presence of large ones, the problems become quite great in Fourier transform spectroscopy. Unlike swept frequency methods, such as CW NMR, or dispersive IR, at no point in the scan can the signal be allowed to overflow memory even though it is not of interest. Further, since both the large and the small peaks must be detected accurately, the ADC cannot be divided down when memory overflow is imminent. Therefore, the total number of scans that can be obtained before memory overflow occurs is given by equation (4.4). Further, the dynamic range that can be measured is directly related to the

ADC resolution. For example, it would take a 12-bit ADC to measure an 8000:1 dynamic range, since the smallest signal that can be detected is about 1/2 of one bit. This small signal occasionally sets the lowest bit during successive averaging scans when riding on top of some random noise. Larger dynamic range signals can only be measured by higher resolution ADCs and, as we shall see later, these larger dynamic ranges often do not survive the noise of the Fourier transform itself.

Rather than building more and more expensive long ADCs, experimenters have devised a number of spectroscopic tricks to deal with dynamic range problems. These include chemically or physically removing the large peak or adjusting its persistence in the spectrum so that there is a larger part of the spectrum having no large peak, and amplifying this portion over the high dynamic range part.

Computer techniques for dealing with dynamic range include double precision averaging, block averaging, and adjustment of the predelay time and spectrum position so that the large peak is sampled at the Nyquist frequency and at its zero crossing point.(8) Block averaging is only applicable to cases such as ^{1}H NMR, where large solvent peaks obscure the measurement of small signals. Freeman(9) and Anderson(10) have suggested another complex method of dealing with dynamic range during signal averaging using a ramp function, but this method has not been thoroughly accepted.

Double-precision averaging simply amounts to using two computer words for each data point acquired. This obviously allows many more scans before memory overflows but has the obvious disadvantage of halving the available memory for averaging. This reduced memory size may well affect the resolution of the final spectrum. Further, double-precision acquisition requires fairly tricky software intervention and usually slows down the computer from its maximum data acquisition rate. Since acquisition rate is related only to the spectral width sampled, this is a limitation only when examining large frequency ranges. One variation of the double-precision averaging technique allows data to be acquired in single precision, storing the full memory blocks on magnetic disk. The double-precision summation need only be performed when the experiment is complete or when no more room for the storage of these subblocks exists on disk. This technique can be carried out on a more limited scale in computers without disks but large amounts of memory are then required.

Block averaging is an increasingly popular way of avoiding the single-precision memory overflow, although it too requires double the memory of a usual single-precision average. In this technique, time domain data are acquired until memory overflow is imminent and then the data are transformed and placed in another memory block.

Then the first block is zeroed and more scans accumulated until memory overflow is again imminent. The data are again transformed and added to the already transformed block. This additional averaging occurs in the

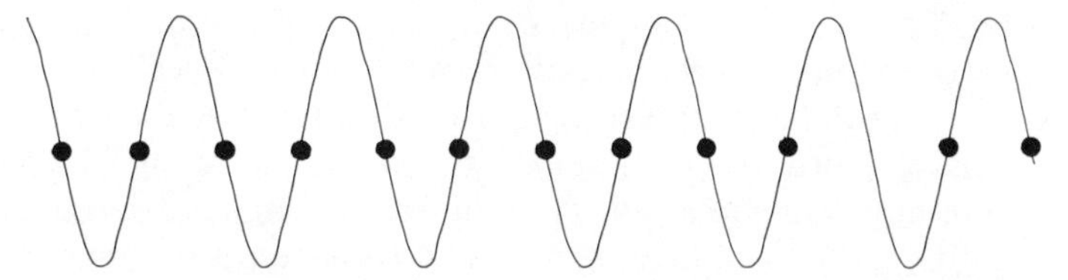

FIGURE 4.8. Sampling at zero-crossing point at Nyquist frequency.

frequency domain, where the large solvent peak is allowed to overflow memory without affecting the rest of the spectrum. The principal advantage of block averaging is that this second block can be stored on a magnetic disk or tape since it is accessed infrequently and the main high-speed memory reserved for data acquisition. If no secondary storage device is available block averaging is still more attractive than double-precision averaging, since it eliminates the potential necessity of a double-precision Fourier transform. Block averaging does require that the experiment be interruptible for as much as a minute every few scans, so that the processing can take place and that the data of interest not be on the shoulder of the large peak, which is also likely to overflow memory during averaging. Since there is seldom a large solvent peak in FT–IR, but always a large dynamic range, double-precision averaging and transforms are the method of choice.

Recently, the possibility of careful spectral placement before acquisition so that the large peak in FT–NMR spectra lies at or near the Nyquist frequency has been discussed.[8] This method is illustrated in Figure 4.8. Here the predelay is adjusted so that the sampling of the first point of the free induction decay occurs at the zero-crossing point of the sine wave. This leads to a much reduced intensity, approaching a zero intensity in the ideal case if the carrier position can be quite accurately controlled. Experimentally, the carrier must be movable in very small, accurate, increments to achieve this sampling condition.

Several other techniques for decreasing dynamic range are in use and will be discussed in Chapter 9.

4.1.8. Other Computer Requirements

As we have seen, one principal requirement for a good laboratory data system is an ADC having 12 to 15 bits of resolution, which can be conveniently stepped down to smaller resolutions for long-term low-dynamic-range averaging. Second, an accurate method of timing data acquisition must be available, usually a programmable clock that will generate highly accurate time intervals between data points with minimal software intervention, which might cause timing inaccuracies. These two requirements can be combined in some hardwired devices called *signal averagers*, which do nothing but acquire points at various rates, add them into memory, and display the results. They are not computers, athough they contain many of the same components as general-purpose computers. One ideal mix between the averager and the computer is one in which a signal averager-type subprocessor is included preprogrammed to do the entire signal-averaging task without any software at all. This allows accurate high-speed data acquisition and frees the computer segment to do other types of data processing and displaying simultaneously. With the advent of low-cost microcomputers, this is an increasingly attractive route for home-built as well as for commercial systems.

While the signal averager itself cannot be used in Fourier transform spectroscopy, it can still be an integral component by being attached to a minicomputer for the actual Fourier transform processing. The criteria for the choice in such systems have been described previously.[11]

4.1.9. Disk-Based Data Acquisition

Signal acquisition need not be into a computer's main memory in all cases. It is sometimes quite feasible to store most of the data being acquired on a rotating magnetic disk, rather like a phonograph record. Transfer to a disk from a computer system is slower than to core memory, but if the data are acquired into memory in small blocks and then written onto disk while more are being acquired, this can be accomplished fairly efficiently. In order for this to happen, either the ADC or the disk or both must have some subprocessor that directs their data transfer so that two operations can occur essentially simultaneously. Most computer systems have a special channel for transfer to and from the disk system, which operates independently of other devices such as the ADC once the computer's central processor has started it. A few also have a peripheral processor for the ADC. This transfer processor is commonly referred to as direct memory access (DMA).

The general procedure for disk-based data acquisition has been described[6] and requires far less main memory. It requires two blocks of memory into which data are acquired alternately. The size of these blocks determines how closely the data acquisition rate approaches the disk transfer rate.

Data points are acquired into memory block 1 until all words have been used. Then a software flag is set so that future data are routed into block 2. Meanwhile, the disk controller is activated to begin writing the contents of data block 1 onto disk. This writing process must be finished before all words of block 2 have been acquired. When block 2 has been completed, data are again routed to block 1 while block 2 is written onto another disk area. This process continues until all desired data points have been collected. The main advantage of this process is that disk memory is much cheaper per bit than core or other high-speed memory and a much larger amount of data can be stored on disk. Indeed 64,000 or even 512,000 point arrays are possible and not uncommon.

The principal disadvantages of disk-based acquisition are that the computer is kept extremely busy switching from block 1 to block 2 and telling the disk what to do next, and that the *averaging* must take place during a pause between scans. This may preclude a data acquisition rate of greater than 40 kHz, meaning that the spectrum bandwidth to be observed can be no more than 20 kHz. Alternatively, averaging can take place during the data acquisition process by continually reading the data from previous scans into memory from the disks before acquisition occurs into that block and then adding the data into the block rather than depositing them. This decreases the sampling rate further, but decreases the intersweep interval. This is a problem principally in obtaining the FT–NMR spectra of "other" nuclei such as ^{19}F and ^{31}P, or in superconducting magnet systems. Another disadvantage to disk-based data acquisition is that it is difficult to maintain a display of the data on a CRT during acquisition, since the processor that would normally be generating the display is so busy. Unless an additional processor for a display is provided or the computer is extremely fast, this feature is often discarded as the cost of large data blocks with concomitant high resolution. The most telling disadvantage of disk-based acquisition, however, is in the disk-based Fourier transform, which is necessarily somewhat slower than the memory-based one, since so little of the data can be contained in memory at one time.

4.1.10. Comparison of Data System Requirements in NMR and IR

The principal requirements for NMR data systems include a moderate resolution ADC of 12 to 13 bits, and a sampling rate of at least 50 kHz, leading to a bandwidth of 25 kHz. For some nuclei observed at superconducting fields, the requirements may be more stringent, requiring sampling rates of up to 100 kHz. NMR spectroscopists also expect a rapid Fourier transform, interactive phase correction software, and a good display while averaging, which usually means that data must be kept in memory rather than on disk. Longer computer word lengths are desirable to measure high dynamic range signals on occasion.

FT infrared spectroscopists, on the other hand, expect high dynamic range at all times, since the nature of the interferometer always provides a large zero frequency peak and very small signals in the tail of the interferogram. Sampling rate is not as important except for kinetics studies since it is a function of mirror velocity rate as well as bandwidth. However, the high dynamic range requires a longer word length computer and/or double-precision averaging and Fourier transform. Disk-based acquisition and transformation are common in single precision for highest resolution, but core-based averaging may be necessary to make a true double-precision transform tractable.

4.2. THE FOURIER TRANSFORM

4.2.1. Introduction

The Fourier transform process has already been discussed in detail in Chapter 2, and we will content ourselves here with the simple definition that the Fourier transform is a method of converting time domain data to frequency domain data. Since the data we will have available will always be discrete data points in a digital computer, we will give the transform in its discrete form as

$$A_r = \sum_{k=0}^{N-1} X_k W^{rk}, \qquad r = 0, 1, \ldots, N-1 \tag{4.8}$$

where

$$W = e^{-2\pi i/N} \tag{4.9}$$

X_k is the kth time domain point and A_r the rth frequency domain point. The X's may be complex numbers and the A's are always complex. Since it clearly takes N multiplications to calculate one A_r, it must take N^2 multiplications to calculate all N A_r's. Since these are all complex multiplications and multiplication is one of the slower computer operations, this can be a very time-consuming process indeed. In fact, Fourier transforms were once the bottleneck of any procedure that required them, leading to endless computer time with its associated costs.

4.2.2. The Cooley–Tukey Algorithm

However, in 1965, Cooley and Tukey[12] proposed a method for simplifying the Fourier transform process that relied on the ability to factor the data into sparse matrices containing mostly zeros. A good matrix algebra proof of the Cooley–Tukey method or *algorithm* has been given by Brigham.[13] We will briefly show how the algorithm works here by giving a simplified proof of the method following one originally given by the G-AE Subcommittee on the Fourier transform.[14]

Let us consider a time series X_k consisting of N complex points. We first divide this series into two subseries Y_k and Z_k, consisting of the odd and even points.

Then

$$Y_k = X_0, X_2, X_4, \ldots = X_{2k}, \qquad k = 0, 1, 2, \ldots, N/2 - 1 \tag{4.10}$$

$$Z_k = X_1, X_3, X_5, \ldots = X_{2k+1} \tag{4.11}$$

Now both Y_k and Z_k have discrete Fourier transforms according to equation (4.8). We shall call them B_r and C_r. Then we can write

$$B_r = \sum_{k=0}^{N/2-1} Y_k \exp(-4\pi irk/N), \qquad r = 0, 1, \ldots, (N-1)/2 \quad (4.12)$$

$$C_r = \sum_{k=0}^{N/2-1} Z_k \exp[-2\pi ir(2k+1)/N] \quad (4.13)$$

Now let us rewrite the transform A_r in terms of its odd and even numbered points and relate them to B_r and C_r:

$$A_r = \sum_{k=0}^{N/2-1} Y_k \exp(-4\pi irk/N) + Z_k \exp[-2\pi i(2k+1)/N],$$

$$r = 0, 1, \ldots, N-1 \quad (4.14)$$

Expanding the second term we find that we can remove $\exp(-2\pi ir/N)$ so that

$$A_r = \sum_{k=0}^{N/2-1} Y_k \exp(-4\pi irk/N) + \exp(-2\pi ir/N) \sum_{k=0}^{N/2-1} Z_k \exp(-4\pi irk/N) \quad (4.15)$$

Now using the definitions of B_r and C_r from (4.12) and (4.13), we find that we can simplify (4.15) to

$$A_r = B_r + \exp(-2\pi ir/N)\, C_r, \qquad r = 0, 1, \ldots, (N/2) - 1 \quad (4.16)$$

For values of r above $(N/2) - 1$, we find that B_r and C_r repeat periodically. If we substitute $(r + N/2)$ for r, we have

$$\begin{aligned} A_{r+N/2} &= B_r + \exp[-2\pi i(r + N/2)/N]\, C_r \\ &= B_r + \exp(-2\pi ir - 2\pi iN/2N)\, C_r \\ &= B_r + \exp(-2\pi ir/N - \pi i)\, C_r \\ &= B_r + \exp(-2\pi ir/N)\exp(\pi i)\, C_r = B_r + \exp(-2\pi ir/N)(-1)\, C_r \\ &= B_r - \exp(-2\pi ir/N)\, C_r \end{aligned} \quad (4.17)$$

since

$$e^{i\pi} = -1 \quad (4.18)$$

by Euler's formula. Summarizing and using (4.9), we have

$$A_r = B_r + W^r C_r, \qquad \text{for } 0 < r < N/2 \quad (4.19)$$

$$A_{r+N/2} = B_r - W^r C_r, \qquad \text{for } 0 < r < N/2 \quad (4.20)$$

These equations show that the first and last $N/2$ points of the transform A_r can be obtained from the transforms of two $N/2$-point arrays Y_k and Z_k. Now, we found that the definition of the discrete transform [equation (4.8)] predicted that N^2 multiplications would be required to perform the Fourier transform for A_r. Here we have found that we can perform two $N/2$-point transforms, each of which requires $(N/2)^2$ multiplications, and the total number of multiplications to obtain A_r is reduced to

$$2(N/2)^2 = N^2/2$$

Thus, this shuffle of points has enabled us to obtain A_r in only half as many complex multiplications as we expected; thus we have doubled our efficiency.

Needless to say, we can continue this halving process by dividing Y_k into T_k and V_k, having transforms D_r and E_r, and Z_k into V_k and W_k, each having transforms F_r and G_r, and so forth, until we have divided each array down into one-point arrays. Now the Fourier transform of a single point is the point itself, as can readily be seen from equation (4.8), and we can thus reduce the transform process to a series of point shuffles and a recombination process such as is shown in (4.19) and (4.20). If we do this, we have reduced the number of multiplications from N^2 to $N \log_2 N$, a very substantial time saving.

For example, if we had a 4096-point array to Fourier transform, equation (4.8) predicts that it would require $(4096)^2$, or 16.7 million multiplications. The Cooley–Tukey algorithm allows us to reduce this to (4096) $\times \log_2 4096$ or $4096(12) = 49{,}152$ multiplications, a saving of a factor of 341 in time.

4.2.3. The Signal Flow Graph

The easiest way to regard the point shufflings and recombinations necessary to realize the simplifications of the Cooley–Tukey method is to view the transform as a series of $\log_2 N$ steps or *passes*, which are performed on all data points. The necessary steps are shown schematically in a signal flow graph in Figure 4.9. This graph shows the array X_k at the left side and

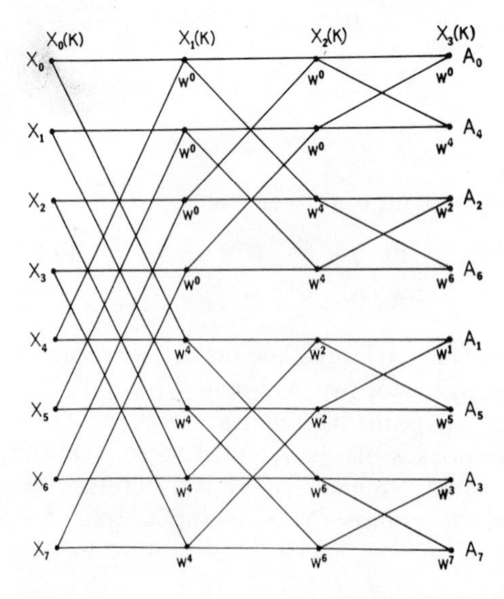

FIGURE 4.9. Signal flow graph of fast Fourier transform with bit-inverting last. Courtesy of John Wiley and Sons, Inc., New York.

the array A_r at the right side. Each arrow leads from an X_k in one column to a new point in the next column. The data point in the new column is calculated by adding together the data points from which the arrows are drawn. If a number W^y is written alongside the arrow, that value of W is multiplied by that data point before addition. Thus each new data point is calculated from two points in the previous column by

$$X_i' = X_i + W^y X_k \tag{4.21}$$

where i is the index of the new point and i and k the indices of the old points. The operations in equation (4.21) are a complex multiplication and a complex addition, since the X are assumed to be complex. Since most physical data are not in themselves complex, a scheme for transforming real data will be given following this fundamental discussion.

Examining Figure 4.9 in detail, we find that while the X's along the left side start in their usual order, the A's along the right side end up in a scrambled order, A_0, A_4, A_2, A_6, A_1, A_5, A_3, A_7. The generality of this order may not be readily apparent but it can be best described by looking at the subscripts as binary numbers:

0	000	4	100	2	010	6	110
1	001	5	101	3	011	7	111

If these subscripts are bit-reversed in binary, or read from right to left, we will have 000, 001, 010, 011, 100, 101, 110, 111, or the numbers from 0 to 7 in their natural order. Thus, the Cooley–Tukey algorithm produces a transform whose points are shuffled into *bit-inverted* order. Fortunately, unshuffling such an array is much faster than reverting to the classical transform method given in equation (4.8). The coefficients of W, further, as

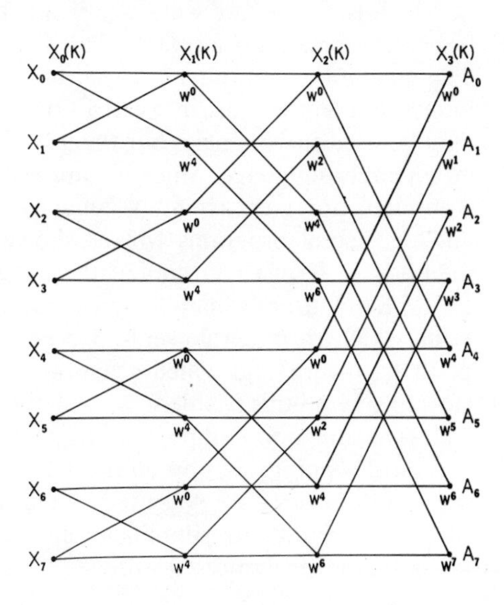

FIGURE 4.10. Signal flow graph of fast Fourier transform with bit-inversion first. Courtesy of John Wiley and Sons, New York.

shown, are also advanced down each column in a bit-inverted order, so that each new W calculation must be preceded by a bit inversion and a division related to the number of the column. However, let us consider the Fourier transform shown in the flow-graph in Figure 4.10. Here the X's are shuffled into bit-inverted order first, and then the A's end up in their natural order at the end of the transform. Further, the coefficients of the W now also appear in their natural order and their calculation is much simplified. This second method is called *decimation in frequency* and is the method of preference for minicomputer programming.

4.2.4. In-Place Transforms

Examining the signal flow graphs of Figures 4.9 and 4.10, one other major advantage becomes clear: the transforms can be done *in place*. In other words, the X_k's can be converted to A_r's with no additional storage. This is of great importance in minicomputers, since the amount of available memory is always limited. Examining either of the flow charts, the source data for any two points are always just two other points and these two old points are used to form only one other new point. Thus data points can be calculated in pairs from old pairs, leading to even more efficiency, and only two spare storage locations are needed regardless of the size of the transform.

4.3. WRITING A FOURIER TRANSFORM FOR A MINICOMPUTER

4.3.1. Introduction

In this section, we will discuss in detail a method for writing a Fourier transform for a minicomputer. This is almost always done in the assembly language of the computer in order to obtain the highest possible speed. As minicomputers increase in speed, FORTRAN and other high-level languages may be used increasingly, but the greatest speed will always be obtainable in assembly language. Since we are not speaking of any particular minicomputer, however, assembly language examples would be inappropriate, and the actual programs will be shown in FORTRAN as models of the methods to be used in coding the program in assembly language. Now, FORTRAN usually handles most numbers in floating-point representations, which are much slower to operate on unless the computer has floating-point hardware, and which generally require two minicomputer words to store. To avoid this problem, we will operate entirely in integer mode, which will save time and computer space.

At this point we know all the details of the transform process, including the form of the W we will need to march across the flow chart. However, there are a number of simplifications that can be pointed out here to speed up both the programming process and the resulting transform.

4.3.2. The Form of W

We defined the number W in equation (4.9) as

$$W = \exp(-2\pi i/N) = e^{-2\pi i/N} \tag{4.9}$$

We proceed through the transform multiplying points by W^y, where y varies in each pass through the flow graph. If we look at the first column or pass in the flow graph, we see that the only terms are W^0 and W^4. We know that $W^0 = 1$ and further

$$W^4 = \exp[-(2\pi i/8)(4)] = -1$$

All of the other powers of W can be quickly evaluated as well by remembering that Euler's formula also tells us that

$$e^{iy} = \cos y + i \sin y \tag{4.22}$$

and some odd power of W thus becomes

$$\begin{aligned} W^3 &= \exp(-6\pi i/8) = \cos(-0.75\pi) + i\sin(-0.75\pi) \\ &= -0.707 - 0.707i \end{aligned} \tag{4.23}$$

Since all of the X's are assumed to be complex, we now have a complex multiplication between a complex number derived from W and the complex number X to carry out for each new point, followed by a complex addition.

Now let us consider any two complex points $X1'$ and $X2'$, which are calculated from $X1$ and $X2$ in the previous column of the graph along with a power of W. We find by examining the chart that we can always write

$$X1' = X1 + X2W^y \tag{4.24}$$

$$X2' = X1 + X2W^z \tag{4.25}$$

where y and z are two different exponents of W. These exponents are related, however, in that they always differ by $N/2$:

$$z = y + N/2 \tag{4.26}$$

so we can write

$$W^z = W^{y+n/2} = W^y[\exp(-2\pi i N/2N)] = W^y(-1)$$

$$W^z = -W^y \tag{4.27}$$

Thus, the two equations that we calculate for any two new points require calculation of only one W and we can write

$$\begin{aligned} X1' &= X1 + X2W^y \\ X2' &= X1 - X2W^y \end{aligned} \tag{4.28}$$

4.3.3. The Fundamental Operations

These equations are still in the form of complex numbers and they must be reduced to simple operations. We can do that if we let

$$X1 = R1 + iI1 \tag{4.29}$$

$$X2 = R2 + iI2 \tag{4.30}$$

and expand (4.28) and (4.29) to

$$X1' = R1' + iI1' = (R1 + I1) + (R2 + I2)(\cos y + i \sin y) \tag{4.31}$$

$$X2' = R2' + iI2' = (R1 + I1) - (R2 + I2)(\cos y + i \sin y) \tag{4.32}$$

Then, collecting terms, we have the fundamental equations that are carried out in any Fourier transform:

$$R1' = R1 + R2 \cos y - I2 \sin y \tag{4.33}$$

$$R2' = R1 - R2 \cos y + I2 \sin y \tag{4.34}$$

$$I1' = I1 + R2 \sin y + I2 \cos y \tag{4.35}$$

$$I2' = I1 - R2 \sin y - I2 \cos y \tag{4.36}$$

Note that only the $R2$ and $I2$ terms are multiplied by the sine and cosine and that $R1$ and $I1$ are always added in. This simplifies the programming since we need only calculate the sine and cosine of y and then multiply by $R2$ and $I2$ to get the four terms needed to evaluate (4.33)–(4.36).

One last programming trick that we will use is the fact that we need no multiplications in the first pass through the transform because only W^0 and $W^{N/2}$ are needed. For the first pass we can write a separate portion of the program, which reduces (4.33)–(4.36) to

$$R1' = R1 + R2 \tag{4.37}$$

$$R2' = R1 - R2 \tag{4.38}$$

$$I1' = I1 + I2 \tag{4.39}$$

$$I2' = I1 - I2 \tag{4.40}$$

4.3.4. The Sine Look-Up Table

The evaluation of (4.33)–(4.36) continuously requires the calculation of sine and cosine functions. In nearly all computers, these are calculated by the evaluation of an appropriate Taylor series.[15] While this is all right when only a few sines and cosines are to be calculated in a program, the evaluation of $N \log_2 N$ sines and cosines will result in a tremendous slow-down in the calculation even in a large-scale computer. In a minicomputer where high-speed floating-point hardware is not generally available, this can be disastrous. Therefore, most efficient Fourier transforms utilize a table of sines along with a look-up routine to calculate each sine as fast as possible. Typically 513 or 1,025 memory locations are used to generate a table of numbers representing the values in the first quadrant of the sine function, and all sines and cosines are calculated by looking up values in this table.

Generation of this table requires 1,025 sine function calculations and prevents the necessity of, say, 49,000 such calculations for a 4,096 point Fourier transform. The table generated by calculating the sine of $\pi/2$ times 0/1,024, 1/1,024, 2/1,024, . . . , 1,024/1,024, and storing it in 1025 consecutive addresses. These numbers could be stored as floating-point values, but in most minicomputers they are converted to a fixed binary fraction.

4.3.5. Binary Fractions

When we multiply together two numbers in a computer, the multiplication hardware or software usually produces a double-length result. Thus, a multiplication of two 16-bit numbers produces a 32-bit product. If we consider for a moment a simple 4-bit computer then multiplying 2 times 2 is the multiplication of 0010 and 0010 to produce 0000 0100.

Similarly, the multiplication of 4 times 4 occurs as follows:

$$\begin{array}{ccccc} 0100 & \times & 0100 & = & 0001\ 0000 \\ 4 & \times & 4 & = & 16 \end{array}$$

All of the numbers described above are integers, and we know that in our Fourier transform we will have to utilize some fractional notation to represent all of the sines and cosines. Since all of these numbers are less than 1, however, we might as well let the entire computer word represent a fraction with the binary point lying just to the right of the sign bit:

$$0\,.000\ 000\ 000\ 000\ 000$$

for a 16-bit computer word. The leftmost data bit then represents 2^{-1}, the next 2^{-2}, and so forth. The number 1.0000 can be approximated by turning on all of the bits in the fraction

$$0\,.111\ 111\ 111\ 111\ 111 = 0.9999694825 \cong 1.0$$

Then we simply arrange our sine look-up table values to vary from 0 to 0 111 111 111 111 111.

Fractional multiplication is somewhat different from integer multiplication, however. If we again consider our 4-bit computer using this same fractional representation and we wish to multiply 0.5 times 0.5, we would write

$$0.\ 100 \times 0\ .100$$

We expect the answer to be 0.25, of course, but we have just shown that 0100 × 0100 will give us 0001 0000, while what we want is 0010 0000. Therefore, when performing fractional multiplications we must *multiply the result by* 2 or shift it one place to the left and then take the *upper* word of the double-length product as our answer. This process must be carefully allowed for in any computer system which does not have a fractional multiply instruction.

4.3.6. The Sine Look-Up Routine

The evaluation of the various W's in the Fourier transform requires continuous calculation of sines and cosines that always contain the constant π, since

$$W^M = \exp(-2M\pi i/N) = \cos(-2M\pi/N) + i\sin(-2M\pi/N)$$

Thus, the values of M/N simply become the distance down a look-up table whose length is 2π. Therefore, the coefficients of W can be directly converted to the address of the correct sine by a simple multiplication process. Since we need to vary the sine over 2π, we include in our look-up routine tests for which quadrant the sine will lie in and return the adjusted value for the sine. We return values as in Table 4.1.

For our 16-bit computer, we can allow the two most significant bits to represent the quadrants so that sines in the first quadrant are returned only if these two bits are 0. This is shown in Table 4.2. Our look-up routine simply complements y if bit 14 is set and negates the sine if bit 15 was set.

TABLE 4.1

Quadrant	Input value	Sine
1	0–$\pi/2$	$\sin y$
2	$\pi/2$–π	$\sin(-y)$
3	π–$3\pi/2$	$-\sin y$
4	$3\pi/2$–2π	$-\sin(-y)$

TABLE 4.2

Input value (octal)	Bit 15	Bit 14	Return sine
0– 37777	0	0	$\sin y$
40000– 77777	0	1	$\sin(-y)$
100000–137777	1	0	$-\sin y$
140000–177777	1	1	$-\sin(-y)$

```
      INTEGER FUNCTION ISIN(L)
      COMMON X(8192),IS(1025),N
      WRDLEN=36
      IMAX=2**(WRDLEN-2)
C DEFINE LOG2(1024) AS LOG OF TABLE LENGTH
      LENTAB=10
C DEFINE INTERPOLATION MASK FOR SINE TABLE
C ONLY THE LEFTMOST LENTAB BITS PER QUADRANT ARE SIGNIFICANT
C THE REMAINDER ARE USED IN LINEAR INTERPOLATION
C BETWEEN LOOK-UP VALUES
      IMASK=2**(WRDLEN-LENTAB-2)-1
C DEFINE MASK TO MASK OUT ALL BUT THE 10 MOST SIG. BITS
C PER QUADRANT, EXCLUDING SIGN BIT AND QUADRANT BIT
      LMASK=((2**LENTAB)-1)*2**(WRDLEN-LENTAB-2)
C IN 16 BIT WORD:
C     0 000 000 000 000 000
C     0 011 111 111 110 000 THIS WOULD MASK AS SHOWN
C     ^ - SIGN BIT
C       ^  QUADRANT BIT
      M=L
      ISIGN=0
      IF (M.LT.0) ISIGN=1
10    IF ((M.AND.IMAX).EQ.0) GO TO 20
      M=.NOT.M
20    INTERP=M.AND.IMASK
      M=M.AND.LMASK

      M=M/2**(WRDLEN-LENTAB-2)
      LMAX=2**(WRDLEN-LENTAB-2)
C GET LOWER SINE
      ISUM=0
      ISIN=IS(M+1)
C PREVENT ADDRESSING OUTSIDE ARRAY
      IF (INTERP.EQ.0)GO TO 200
      IF ((M+2).GT.1025) GO TO 200
C GET DIFFERENCE FROM NEXT HIGHER VALUE
      IDIF=IS(M+2)-ISIN
C NOW ADD ON FRACTION OF DIFFERENCE FOR INTERPOLATION
      B=FLOAT(INTERP)*FLOAT(IDIF)/2**(WRDLEN-LENTAB-2)
      ISUM=IFIXIT(B)
200   ISIN=(ISIN+ISUM)
      IF(ISIGN.EQ.1)ISIN=-ISIN
      RETURN
      END


      INTEGER FUNCTION ICOS(M)
      COMMON X(8192),IS(1025),N,MASK
C CONVERT TO COSINE POSITION BY ADDING 90 DEGREES
C BUT AVOID SETTING INTEGER OVERFLOW FLAG
      IO=M.AND.3
C SAVE BIT 0
      L=(M/4).AND."177777777777
C NOW ADD 90 DEGRESS/2
      L=(L+"40000000000).AND."177777777777
      IF((L.AND."100000000000).NE.0)L=L.OR."600000000000
C EXTEND SIGN W/O SETTING OVERFLOW
C THEN RESTORE BIT 0 AFTER DOUBLNG
      L=L*4 + IO
      ICOS=ISIN(L)
      RETURN
      END
```

FIGURE 4.11. Sine and cosine look-up routines.

Because of the fact that the two's complement of 0 100 000 000 000 is also 0 100 000 000 000, however, the one's complement is taken if bit 14 is set, rather than the negative or two's complement. This along with the extra point in the sine table ensures that both the sine of 0 100 000 000 000 and the cosine of 0 will be 0 111 111 111 111 111. The actual address of the correct sine value is then found by dividing the input angle value by 2^{W-2} over the table length. In the 16-bit case with a 1025-point look-up table, the input value is converted to the address of the correct sine by dividing by 16384/1024 = 16. A complete sine look-up routine is shown in Figure 4.11.

The cosine of the input value is calculated by simply adding $\pi/2$ to the input value before the look-up as is shown in Figure 4.11. Care must be taken in many computers that adding the large positive number that represents $\pi/2$ to the input value does not cause an error condition when the sign of the input value is changed by the addition, since this is an integer overflow condition.

4.3.7. Scaling during the Transform

Evaluation of terms like $I2 \sin y$ or $R2 \cos y$ will not cause any problems with memory overflow since the multiplication of any number by a sine will always produce a product less than or equal to the original value in magnitude. However, the two *additions* in (4.33)–(4.36) can cause memory

```
      SUBROUTINE SCALE
      COMMON X(8192),IS(1025),N
      INTEGER X,O1,O2,WRDLEN
      WRDLEN=36
      O1=2**(WRDLEN-3)
      O2=2**(WRDLEN-2)
C DIVIDE ARRAY BY 2 IF ANY WORD IS MORE THAN 1/4 FULL
C DIVIDE ARRAY BY 4 IF ANY WORD IS MORE THAN 1/2 FULL
      DO 100 I=1,N
      IT=X(I)
C TAKE INTEGER ABSOLUTE VALUE
      IF(IT.LT.0)IT=-IT
      IF(IT.GE.O2) GO TO 200
      IF(IT.GE.O1) GO TO 300
      GO TO 100
200   CALL DIV2
300   CALL DIV2
100   CONTINUE
      RETURN
      END

      SUBROUTINE DIV2
C DIVIDES X ARRAY BY 2
      COMMON X(8192),IS(1025),N
      INTEGER X,XI
      DO 200 I=1,N
      X(I)=(X(I)/2)
200   CONTINUE
      RETURN
      END
```

FIGURE 4.12. Scaling routine.

overflow, and in fact, the numbers being transformed must be continually tested to prevent them from becoming too large during the transform. Since a number may become more than twice as large during a particular pass through the transform, one simple way to prevent overflow is to scale all numbers to be less than $\frac{1}{4}$ full scale before each pass of the transform. A simple scaling routine is shown in Figure 4.12.

4.3.8. Forward and Inverse Transforms

We define a *forward* transform as one from the time domain to the frequency domain:

$$A_r = \sum_{k=0}^{N-1} X_k W^{rk}, \qquad r = 0, 1, \ldots, N-1 \tag{4.8}$$

The *inverse* transform, or the conversion from the frequency domain to the time domain can be calculated by simply converting equation (4.8) to its inverse:

$$X_k = \sum_{k=0}^{N-1} A_r W^{-rk}, \qquad k = 0, 1, \ldots, N-1 \tag{4.41}$$

This simply amounts to changing the sign of the terms during the calculation, and this can be done in several ways depending on the program that is developed.

In fact, since we define W as $\exp(-2\pi i/N)$ we can use (4.33)–(4.36) with positive angles for the inverse transform and with negative angles for the positive transform. For systems where only forward transforms will be done, we can rewrite these equations by recalling that $\sin(-y) = -\sin y$ and $\cos(-y) = \cos y$:

$$R1' = R1 + R2\cos y + I2\sin y \tag{4.42}$$
$$R2' = R1 - R2\cos y - I2\sin y \tag{4.43}$$
$$I1' = I1 - R2\sin y + I2\cos y \tag{4.44}$$
$$I2' = I1 + R2\sin y - I2\cos y \tag{4.45}$$

In other words, the difference between the forward and the inverse transform is simply the sign of the sine terms.

4.3.9. Fourier Transforms of Real Data

We have so far discussed only the Fourier transforms of complex data, where two arrays of numbers are involved, containing the real and the imaginary coefficients, respectively. This technique is useful in quadrature detection methods,[16] where there are indeed two data inputs 90° apart in phase, but in many cases there are only real data produced by the experiment.

The Fourier transform of real data can be derived from the $2N$ point transform method described by Cooley *et al.*[17] and Brigham.[13]

We simply divide the real input data into two functions consisting of the even and odd points,

$$h_k = X_{2k} \tag{4.46}$$

$$g_k = X_{2k+1} \tag{4.47}$$

and then treat the real data as if alternate points were imaginary and the function were really the complex function Y_k:

$$Y_k = h_k + ig_k, \qquad k = 0, 1, \ldots, (N-1)/2 \tag{4.48}$$

Then we compute the FFT of Y_k:

$$B_r = \sum_{k=0}^{N-1/2} Y_k W^{rk} \tag{4.49}$$

We then have the transformed function B_r consisting of a real and an imaginary part;

$$B_r = R_r + iI_r \tag{4.50}$$

To convert the function B_r to the desired function A_r, the transform of X_k, we then perform a simple one-pass multiplication of the final points after first scaling the data so that no overflow can occur:

$$A_n = R_p + \cos(\pi n/M)\, I_p - \sin(\pi n/M)\, R_m + i(I_m - \sin(\pi n/M)\, I_p - \cos(\pi n/M)\, R_m) \qquad n = 0, 1, \ldots, M \tag{4.51}$$

where

$$R_p = R_n + R_{M-n} \tag{4.52}$$
$$I_p = I_n + I_{M-n} \tag{4.53}$$
$$R_m = R_n - R_{N-n} \tag{4.54}$$
$$I_m = I_n - I_{M-n} \tag{4.55}$$

and $M = N/2$.

Note that the evaluation of (4.51) generates $N/2$ complex points but requires $N/2 + 1$ input points. Clearly there will be only $N/2$ points to start with, but the symmetry of the transform given by equation (4.20) allows us to calculate the first and last points separately, since R_N and I_N are required to calculate A_0. For the first and last points the following equations hold:

$$R_p = R_0 \tag{4.56}$$
$$I_p = 0 \tag{4.57}$$
$$R_m = 0 \tag{4.58}$$
$$I_m = -R_0 \tag{4.59}$$

```
      SUBROUTINE POST(INV)
C POST PROCESSING FOR REAL DATA
C FORWARD TRANSFORM POST-PROC IF INV=0
C INVERSE TRANSFORM PRE-PROC IF INV=1
      COMMON X(8192),IS(1025),N
      INTEGER X, WRDLEN,RP,RM,RMSIN,RMCOS
      WRDLEN=36
C POST-PROCESSING SHUFFLE
      CALL SCALE
      N2=N/2
      N4=N/4
C IMAX REPRESENTS PI/2
      IMAX=(2**(WRDLEN-2))
      DO 500 L=1,N4
C START AT ENDS AND WORK TOWARD MIDDLE OF REAL AND IMAGINARY PARTS
      I=L+1
      M=N2-I+2
      RP=X(I)+X(M)
      RM=X(I)-X(M)
      IP=X(I+N2)+X(M+N2)
      IM=X(I+N2)-X(M+N2)
C SINEOF PI/2N
      IARG=(IMAX/N4)*(I-1)
      IC=ICOS(IARG)
C COSINE TERM IS - IF INVERSE REAL TRANSFORM
      IF(INV.EQ.1)IC=-IC
      IS1=ISIN(IARG)
      IPCOS=IMULT(IP,IC)
      IPSIN=IMULT(IP,IS1)
      RMSIN=IMULT(RM,IS1)
      RMCOS=IMULT(RM,IC)
      X(I)=RP+IPCOS-RMSIN
      X(I+N2)=IM-IPSIN-RMCOS
      X(M)=RP-IPCOS+RMSIN
      X(M+N2)=-IM-IPSIN-RMCOS
500   CONTINUE
      RETURN
      END
```

FIGURE 4.13. Postprocessing routine.

In actual programming, the points are done a pair at a time starting at the two ends of the real and imaginary arrays and moving toward the center. An example of a postprocessing routine such as this is shown in Figure 4.13.

4.3.10. Inverse Real Transforms

Inverse real transforms can be performed by reversing the effect of equation (4.51) and then taking an inverse transform such as is given in equation (4.41). Reversing the effect of equation (4.51) can be accomplished by preprocessing the data with the signs of the cosine terms reversed:

$$\begin{aligned} B_n = R_p - \cos(\pi n/M)\, I_p - \sin(\pi n/M)\, R_m \\ + i[I_m - \sin(\pi n/M)\, I_p + \cos(\pi n/M)\, R_m] \end{aligned} \tag{4.60}$$

4.3.11. Baseline Correction

Before transforming real data, acquired from some scientific instrument, it is common to perform baseline correction. This simply makes the integral of the entire array equal to zero, and is most commonly performed by adding the entire array together, dividing by N, and then subtracting that value

```
C CALLS FOR VARIOUS TYPES OF FOURIER TRANSFORMS

C COMPLEX TRANSFORM WHERE POINTS ARE 1ST HALF REAL, 2ND HALF IMAG
      CALL FFT(0)

C COMPLEX INVERSE
      CALL FFT(1)

C COMPLEX FORWARD TRANSFORM WHERE ALTERNATE POINTS ARE REAL, IMAG, REAL...
      CALL SHUFFL(0)
      CALL FFT(0)

C COMPLEX INVERSE WHERE
C FREQ DOMAIN POINTS ARE REAL-1ST HALT, IMAG-2ND HALF
      CALL FFT(1)
      CALL SHUFFL(1)

C REAL FOWARD TRANSFORM
C FIRST PUT IMAGINARY POINTS IN SECOND HALF
      CALL SHUFFL(0)
      CALL FFT(0)
C THEN DO POST PROCESSING
      CALL POST(0)

C REAL INVERSE TRANSFORM
C PRE-PROCESS
      CALL POST(1)
      CALL FFT(1)
C THEN SHUFFLE BACK INTO ALTERNATE ORDER
      CALL SHUFFL(1)
```

FIGURE 4.14. Calling routines for various transforms.

TABLE 4.3. Intermediate Results during Forward Real Fourier Transform

I	Input	Shuffle[a]	Bit invert	Pass 1[b]	Pass 2	Pass 3	Post
1	8	8	8	80	223	509	509
2	16	24	72	−64	−127	−216	−130
3	24	40	40	144	−63	−125	−128
4	32	56	104	−64	−1	−90	−128
5	40	72	24	112	287	−63	−63
6	48	88	88	−64	−127	−38	−128
7	56	104	56	176	−63	−1	−124
8	64	120	120	−64	−1	88	−126
9	72	16	16	96	255	573	573
10	80	32	80	−64	−1	88	632
11	88	48	48	160	−63	−1	300
12	96	64	112	−64	−127	−38	189
13	104	80	32	128	319	−63	125
14	112	96	96	−64	−1	−90	85
15	120	112	64	192	−63	−125	52
16	128	128	128	−64	−127	−216	24

[a] Shuffled from alternate points to first-half real, second-half imaginary.
[b] No baseline correction used.

from each data point. Occasionally, if the data are not symmetric about zero, this process will be omitted or modified to include only that part of the data that is symmetric about zero. Baseline correction is generally *not* performed on frequency domain data that are to be inverse transformed to the time domain. The consequences of omitting baseline correction in the time domain are observed as a spike in address zero of the transformed data, which is the zero frequency or DC term in the time domain. When the number of cells to do in the pass becomes zero, the transform is complete and exit occurs. The FFT subroutine as written is an entirely complex one. It can be an inverse by changing the sign of the sine terms. It can be a real transform by appropriate preshuffling and postprocessing. The calling routines to accomplish these things are shown in Figure 4.14.

Since most of the difficulty in preparing Fourier transform program lies in its debugging, Table 4.3 shows the results of a 16-point real transform at various stages. These are after shuffling, after each pass, and after postprocessing. Baseline correction is not performed in this table.

4.3.12. A Fourier Transform Routine

The Fourier transform routine shown in Figure 4.15 is a model for the preparation of an assembly language routine using the fractional integer representation discussed above along with a sine look-up table. The integer multiplication routine IMULT in the figure is assumed to be one that produces a double-length product multiplied by two and returns the upper word of that product as the single-precision integer value. There is no easy way to do this in FORTRAN, which does not provide for double-precision integers, and the IMULT routine that was actually used to test this program amounted to converting each integer to a double-precision floating-point number, multiplying, and then fixing the product to single precision using a specially written fixing routine. The FFT subroutine itself assumes that there are two separate arrays having the real and imaginary parts of x, which were originally the even and odd points of X_k. The routine to shuffle them into this somewhat more convenient order also requires $N \log_2 N$ operations, although no multiplications are involved, and is shown in Figure 4.16.

The important simplification of the decimation in frequency method of this FFT subroutine as diagrammed in Figure 4.10 is that the exponents of W are not calculated by any complex procedure. They are produced by simply starting an angle Y at 0 in each cell and then advancing it by an increment DELTAY, which is $\pi/2$ in the first pass and $\pi/4$ in the second pass. DELTAY is divided by two at the end of each pass. Each new pass has half as many cells, twice as many pairs per cell, the cells are farther apart, and DELTAY is half as great.

4.3.13. Correlation

Classically, correlation can be regarded as the multiplication of each point in a spectrum by all points in another spectrum. When the two spectra are different, this is known as *cross-correlation* and when they are the same, this is known as *autocorrelation*. Correlation is usually used to uncover

```
      SUBROUTINE FFT(INV)
C COMPLEX TRANSFORM OF N-POINT ARRAY
C FIRST HALF IS REAL, SECOND HALF IMAGINARY
C FORWARD TRANSFORM IF INV=0, INVERSE IF INV=1
      COMMON X(8192),IS(1025),N
      INTEGER X,TR,TI,Y,WRDLEN,DELTAY
      INTEGER CELNUM,PARNUM,COSY,SINY,R2COSY,R2SINY
      N2=N/2
C DEFINE WORD LENGTH FOR THIS COMPUTER SYSTEM
C WOULD NORMALLY BE 16 OR 20 FOR MOST MINICOMPUTERS
      WRDLEN=36
C BASELINE CORRECT DATA
C CALCULATE NU=LOG2(N).
      NU=0
      N1=N/2
10    NU=NU+1
      N1=N1/2
      IF(N1.GE.2)GO TO 10
C SHUFFLE. INTO BIT INVERTED ORDER
      DO 20 I=1,N2
      K=IBITR(I-1,NU) +1
      IF(I.LE.K) GO TO 20
      TR=X(K)
      TI=X(K+N2)
      X(K)=X(I)
      X(K+N2)=X(I+N2)
      X(I)=TR
      X(I+N2)=TI
20    CONTINUE
      CALL SCALE
C FIRST PASS HAS NO MULTIPLICATIONS
      DO 30 I=1,N2,2
      K=I+1
      K1=X(I)+X(K)
      X(K)=X(I)-X(K)
      X(I)=K1
      K1=X(I+N2)+X(K+N2)
      X(K+N2)=X(I+N2)-X(K+N2)
      X(I+N2)=K1
30    CONTINUE
C SET UP SECOND PASS WITH DELTAY= PI/4
      DELTAY=(2**(WRDLEN-2))
      CELNUM=N2/4
      PARNUM=2
      CELDIS=2
C EACH NEW PASS STARTS HERE
100   CALL SCALE
      I=1
C NEW CELLS START HERE
      DO 70 L=1,CELNUM
      Y=0
C DO THE NUMBER OF PAIRS IN EACH CELL
      DO 50 IN=1, PARNUM
      J=I+CELDIS
      COSY=ICOS(Y)
      SINY=ISIN(Y)
C NEGATE SINE TERMS IF INVERSE TRANSFORM
      IF(INV.EQ.1)SINY=-SINY
      R2COSY=IMULT(X(J),COSY)
      R2SINY=IMULT(X(J),SINY)
      I2COSY=IMULT(X(J+N2),COSY)
      I2SINY=IMULT(X(J+N2),SINY)
```

FIGURE 4.15. Complex fast Fourier transform.

```
C NOW COMBINE TERMS TO MAKE FOUR POINTS
        K1=X(I)+R2COSY+I2SINY
        K2=X(I+N2)-R2SINY+I2COSY
        X(J)=X(I)-R2COSY-I2SINY
        X(J+N2)=X(I+N2)+R2SINY-I2COSY
C REPLACE I TERMS
        X(I)=K1
        X(I+N2)=K2
C ADVANCE ANGLE FOR NEXT PAIR
        Y=Y+DELTAY
        I=I+1
50      CONTINUE

C END OF CELL, RESET YN AND PAIR COUNTER
        I=I+CELDIS
70      CONTINUE

C PASS DONE - CHANGE CELL DISTANCE AND NUMBER OF CELLS
        CELNUM=CELNUM/2
        IF(CELNUM.EQ.0) RETURN
        PARNUM=PARNUM*2
        CELDIS=CELDIS*2
        DELTAY=DELTAY/2
        GO TO 100
        END




        SUBROUTINE BC
        COMMON X(8192),IS(1025),N
        INTEGER X
        XSUM=0
C BASELINE CORRECT DATA BY MAKING ITS INTEGRAL 0
        DO 10 I=1,N
        XSUM=XSUM+X(I)
10      CONTINUE
        ISUM=XSUM/N
        DO 20 I=1,N
        X(I)=X(I)-ISUM
20      CONTINUE
        RETURN
        END


        FUNCTION IBITR(J,NU)
C BIT INVERTS THE NUMBER J AROUND NU BITS
C COPY J
        J1=J
        IBITR=0
        DO 200 I=1,NU
        J2=J1/2
C IBITR IS DOUBLED AND BIT 0 SET TO 1 IF J1 IS ODD
        IBITR=IBITR*2+ (J1-2*J2)
C HALVE J1 FOR TEST OF NEXT BIT
200     J1=J2
        RETURN
        END
```

FIGURE 4.15 *(continued)*

```
      SUBROUTINE SHUFFL(INV)
C SHUFFLES POINTS FROM ALTERNATE TO 1ST-HALF, 2ND-HALF ORDER IF INV=0
C REVERSES PROCESS IF INV=1
C ALGORITHM IS MUCH LIKE COOLEY TUKEY
C STARTS WITH LARGE CELLS AND WORKS INTO SMALLER ONES IF INV=0
C STARTS WITH SMALL CELLS AND INCREASES IF INV=1
      COMMON X(8192),IS(1025),N
      INTEGER SIZE,PARNUM,CELDIS,CELNUM,X
      IF(INV.EQ.0) GO TO 5
C USE THESE PARAMETERS IF INVERSE SHUFFL
      CELDIS=2
      CELNUM=N/4
      PARNUM=1
      GO TO 10
C HALF AS MANY CPLX POINTS
5     CELDIS=N/2
CONE CELL IN 1ST PASS
      CELNUM=1
C N/4 PAIRS PER CELL IN FIRST PASS
      PARNUM=N/4
10    I=2
      DO 100 J=1, CELNUM
      DO 200 K=1, PARNUM
      ITEMP=X(I)
C SWAP THESE TWO PAIRS
      X(I)=X(I+CELDIS-1)
      X(I+CELDIS-1)=ITEMP
      I=I+2
200   CONTINUE
C END OF CELL, ADVANCE TO NEXT ONE
      I=I+CELDIS
100   CONTINUE
C CHANGE VALUES FOR NEXT PASS
      IF(INV.EQ.1)GO TO 300
C NEW PASS STARTS HERE
C HALF AS MANY WORDS BETWEEN CELLS
C TWICE AS MANY CELLS
C HALF AS MANY PAIRS PER CELL
      CELDIS=CELDIS/2
      CELNUM=CELNUM*2
      PARNUM=PARNUM/2
      IF(CELDIS.GE.2) GO TO 10
      RETURN
C USE THESE PARAMETERS IF INVERSE SHUFFLE FROM 1ST-HALF 2ND-HALF TO
C ALTERNATE POINTS REAL-IMAG
300   CELDIS=CELDIS*2
      CELNUM=CELNUM/2
      PARNUM=PARNUM*2
      IF(CELNUM.EQ.0) RETURN
      GO TO 10
      END
```

FIGURE 4.16. Shuffle of points from imaginary alternating with real to imaginary in second half, and reverse.

periodicities in physical data, such as might come from biophysical measurements such as EEGs and EKGs. Cross-correlation has also been applied in a type of rapid-scanning NMR experiment that is described in Chapter 9.

If we have two functions x_k and h_k of N points each, their cross-correlation product is given by

$$z_k = \sum_{i=0}^{N-1} h_i x_{k+i} \tag{4.61}$$

This clearly represents another case of N^2 multiplications to arrive at a result. However, this correlation may also be carried out through the use of the forward and inverse Fourier transform by calculating

$$X_n = \sum_{k=0}^{N-1} x_k W^{nk} \tag{4.62}$$

$$H_n = \sum_{k=0}^{N-1} h_k W^{nk} \tag{4.63}$$

and then calculating their point-by-point complex product

$$Y_n = X_n H_n \tag{4.64}$$

The cross-correlogram z_k is then found by simply taking the inverse transform of Y_n:

$$z_k = \sum_{n=0}^{N-1} Y_n W^{-nk} \tag{4.65}$$

Thus, while equation (4.61) requires N^2 multiplications, calculation of a correlation function through (4.62)–(4.64) requires only $3N \log_2 N$ multiplications and is much faster for even relatively small arrays.

4.3.14. Disk-Based Fourier Transforms

Data acquired onto magnetic disk instead of into a computer's main memory can also be Fourier transformed, although at a slower rate. A 32K word Fourier transform can be performed in a minicomputer in assembly language in 30–45 sec depending on the speed of the processor and memory, but may take several minutes on a magnetic disk. This is because the data must continually be read from disk and rewritten onto disk, and most disk transfers will require that the computer wait until the disk has revolved to the beginning of a sector before the transfer can take place. The disk rotation speeds usually mean that one revolution will take 5–10 msec and that each new access of the disk will require this waiting period in addition to the transfer time of several microseconds per data word.

Disk-based Fourier transforms can be most efficiently written using subdivision and reshuffling of the points into smaller blocks by equations (4.10) and (4.11) and then recombination of the smaller subtransforms by equations (4.16) and (4.17). One method of doing such transforms has been described by Singleton.[18]

4.3.15. Hardware Fourier Processors

A procedure as simple as the discrete FFT can easily be hardwired into a unit specially designed for FFT processing. More important, since there is no vertical dependence of points within the same column of the flow graph, a *parallel* processor can be designed to carry out each cell of the transform simultaneously. This leads to very high speed processing indeed. However, since each subprocessor must have its own addition and multiplication hardware, the cost of such processors is still extremely high compared to the cost of minicomputers.

4.4. NOISE IN THE FOURIER TRANSFORM PROCESS

4.4.1. Round-Off Errors

The observable dynamic range in a time domain signal is directly related to the ADC resolution. It might appear that the longer the ADC the greater the observable dynamic range in FT spectroscopy. However, the Fourier transform itself intervenes and becomes a parameter of our measurements. The transform has associated with it certain approximations that restrict the observable dynamic range, particularly when implemented entirely in integer form. Consider the fact that there can be a one-half-bit round-off error in every multiplication and in every addition, and a loss of 1 bit of information every time that the data are divided down to prevent memory overflow. If memory is full at the outset, as might be the case when the data being transformed contain both large solvent peaks and small data peaks, there will be significant round-off due to scaling with every pass through the transform. This can lead to significant errors and much reduced dynamic range.

This problem was examined by Welch[19] for the fixed-point transform and through some theoretical considerations he arrived at equation (4.66) to describe the maximum rms error in a Fourier transform:

$$\frac{E_{\max}^{\text{rms}}}{\text{rms(result)}} = \frac{2^{(M+3)/2}2^{w}(0.4)}{\text{rms (input data)}} \tag{4.66}$$

In this equation, $M = \log_2 N$ and w is the word length as before. The dynamic range is reciprocally related to the error in the transform, and combining the rms values of the input and result into K, we get

$$D_{\max} = 2.5(2^{-(M+3)/2})\,2^{w}K \tag{4.67}$$

where $D_{\max}$ is the maximum expected dynamic range and $K = \text{rms(result)}/\text{rms(input)}$. Unfortunately equation (4.67) is not in general easily evaluated, since K is not known in advance. Recently, some experiments[8] were reported examining Welch's work and producing some practical data regarding dynamic range in the Fourier transform. These experiments synthesized some spectra having a large peak and some smaller peaks of known dynamic range as sine waves and then Fourier transformed them. The last visible frequency domain peak was recorded as the maximum dynamic range for that system.

Since Welch derived equation (4.66) on the assumption that memory was full at the outset, it was clear that this assumption is not exactly true in the NMR experiment. If the large peak were placed exactly on the Nyquist frequency and sampling occurred at its crests, then memory would be full in every channel only if no relaxation occurred. However, in real NMR data, the sine waves representing the various lines relax after the initial excitation[20] and decay exponentially with time. Thus only memory words near the beginning of the NMR free-induction decay will be nearly full and memory words will be less and less nearly full later in the spectrum. Thus, Welch's assumptions do not hold in the FT–NMR experiment, or indeed in FT–IR or FT–ICR either.

Accordingly, the data that were synthesized in these experiments were multiplied by an exponential window[21] typical of the decay in NMR experiments. The results of these experiments are shown in Table 4.4 for the 16- and 20-bit word computer word and for various transform lengths when the high-intensity peak is placed at the Nyquist frequency. Clearly, dynamic range varies both with transform size and computer word length so strongly that longer and longer ADC's are not as important as the computer word length and the length of the Fourier transform itself. In fact the popular 16-bit computer does so poorly that for some experiments it cannot be used in single precision.

Since the number of channels of memory full at the beginning of the transform affects the dynamic range, the frequency of the large peak also affects the dynamic range slightly. If the large peak is at a relatively low frequency, the number of memory words that are full is less since the sine wave has less crests in the spectrum, and more of the points lie on the sides of the wave. Therefore the dynamic range was also examined when the strong peak was placed in the center of the spectrum and at the low end. The results of these experiments are shown in Table 4.5. Clearly, the dynamic range can be improved by careful positioning of the solvent peak in the spectrum.

TABLE 4.4. Dynamic Range of Spectra with the Large Peak at the Nyquist Frequency

Size	$w = 20$		$w = 16$	
	$LB = 0.4$	$LB = 0.2$	$LB = 0.4$	$LB = 0.2$
4K	34,953	29,127	2,608	5,891
8K	43,691	43,691	3,567	3,567
16K	43,691	29,127	1,556	2,608
32K	43,691	24,966	1,725	2,608

TABLE 4.5. Effect of Word Length and Large Peak Frequency on Dynamic Range

Size	$w = 20$		$w = 16$	
	$LB = 0.4$	$LB = 0.2$	$LB = 0.4$	$LB = 0.2$
		Mid-frequency		
4,096	58,254	43,691	3,942	3,942
8,192	43,691	58,254	5,350	5,350
16,384	34,953	43,691	4,855	5,350
32,768	43,691	43,691	4,855	2,123
		Low frequency		
4,096	58,254	43,691	3,942	2,608
8,192	58,254	43,691	3,216	4,855
16,384	58,254	58,254	3,216	4,369
32,768	43,691	43,691	2,897	3,942

4.4.2. Block Averaging

One popular method for avoiding dynamic range limitations is block averaging, in which the data are transformed and further averaging takes place in a second memory block in the frequency domain. The question was raised as to whether the noise generated by the Fourier transform could be averaged out in the frequency domain by adding together successive blocks of data. It was conclusively shown, however, that the noise generated in the transform is coherent and is not reduced by block averaging, so that the dynamic ranges reported in Tables 4.4 and 4.5 are indeed the best that one can do in single-precision accumulation and transformation.

4.4.3. Double-Precision Fourier Transforms

Since the 16-bit computer fared so poorly in the dynamic range experiments discussed above, the problem of double-precision averaging and double-precision Fourier transforms must now be examined. In theory, a properly written FFT program utilizing double precision should behave just as a 32-bit word computer would.

One reason why such transforms are seldom done, however, is related to the memory limitations of the average minicomputer. Even if the scientist has an unlimited budget with which to purchase computer memory, most minicomputers have an upper limit on the amount of memory that they can hold. This is usually 32,768 words (32K) or 65,536 words (64K) in 16-bit systems. The number of data points that the computer can then hold is perhaps 4K less than this maximum, after allowing some space for the FT program itself, leaving 24K or 60K, respectively. However, since the number of data points must be a power of 2 to Fourier transform them efficiently, this reduces us to 16K and 32K, respectively. Finally, in double precision, only 8K or 16K points can be transformed since each data point requires two computer memory words. This sometimes is less resolution than workers find acceptable.

Now this problem can be overcome by using a magnetic disk to store the data in double precision and to store the intermediate data during the FT, but the programming problems of a true disk-based double-precision FT are prodigious and the project is seldom tackled except by commercial software houses.

Additional problems in implementing the double-precision FT come into play when the sine look-up table is considered. It, too, must obviously be double precision to be of value in the transform. If it is, the programmer must devise a method of multiplying together two double-precision numbers and arriving at a four-word product. This must usually be done without the full aid of multiplication hardware, since such hardware is capable only of multiplying together two single-precision numbers. This slows the program down and further complicates it. Finally, if a single precision sine table is chosen as the line of least resistance, most of the advantages of the double-precision transform are lost.

4.5. SUMMARY

In this chapter, we have examined the nature of the minicomputer, the nature of signal averaging with low and high dynamic range, and the Fourier transform process. The effects of the Fourier transform on dynamic range have been summarized and methods for dealing with high dynamic range signals have been examined. A number of fundamental papers on the subject of the Fourier transform, including references 12, 14, 17, 18, and 19 have been collected in the volume by Rabiner and Rader.[21]

ACKNOWLEDGMENTS

The Fourier transform of a real function was first explained to me lucidly by Mr. George Pawle, while we were working on correlation NMR. Figures 4.2–4.6 are reproduced with the permission of Nicolet Instrument Corporation. Some of the work leading to this review article was supported by a PRF Type G grant, administered by the ACS.

REFERENCES

1. *Introduction to Programming*, DEC PDP-8 Manual, Digital Equipment Corp., Maynard, Massachusetts, 1974.
2. J. W. Cooper, *Introduction to Programming the Nicolet 1080*, Nicolet Instrument Corp., Madison, Wisconsin, 1972.
3. J. W. Cooper, *The Minicomputer in the Laboratory: With Examples Using the PDP-11*, Wiley, New York, 1977.
4. B. P. Lathi, *Communications Systems*, p. 89, Wiley, New York, 1968.
5. E. Bartholdi and R. R. Ernst, *J. Magn. Resonance* **11**, 9 (1973).
6. J. W. Cooper, *The Computer in FT–NMR, Topics in Carbon-13 Nmr*, **2**, 392–430 (G. Levy, ed.), Wiley, New York, 1976.
7. J. W. Cooper, *Computers and Chemistry* **1**, 55–60 (1976).
8. J. W. Cooper, *J. Magn. Resonance* **22**, 345–357 (1976).
9. R. Freeman, Varian Research Report 101, Varian Associates, 1973.
10. W. A. Anderson, U.S. Patent 3,622,765.
11. J. W. Cooper, *Amer. Laboratory* **5**(9), 63 (1973).
12. J. W. Cooley and J. W. Tukey, *Math. Comput.* **19**, 297 (1965).
13. E. Oran Brigham, *The Fast Fourier Transform*, Prentice-Hall, Englewood Cliffs, New Jersey, 1974.
14. G-AE Subcommittee on Measurement Concepts, W. T. Cochran *et al.*, *IEEE Trans. Audio Electroacoust.* **AU15**, 45–55 (1967).
15. C. Hastings, *Approximations for Digital Computers*, Princeton University Press, New Brunswick, New Jersey, 1955.
16. E. O. Stejskal and J. Schaeffer, *J. Magn. Resonance* **14**, 160 (1974).
17. J. W. Cooley, P. A. Lewis, and P. D. Welch, *J. Sound. Vib.* **12**, 315–337 (1970).
18. R. C. Singleton, *IEEE Trans. Audio Electroacoust.* **AU-15**, 91–97 (1967).
19. P. D. Welch, *IEEE Trans. Audio Electroacoust.* **AU-17**, 151 (1969).
20. J. A. Pople, W. G. Schneider, and H. J. Bernstein, *High Resolution Nuclear Magnetic Resonance*, McGraw-Hill, New York, 1959.
21. L. R. Rabiner and C. M. Rader, *Digital Signal Processing*, IEEE Press, New York, 1972.

Chapter 5

Fourier Transform Infrared Spectrometry: Theory and Instrumentation

Peter R. Griffiths

5.1. INTRODUCTION

Over the past decade Fourier transform infrared spectrometry (FT–IR) has become an important tool for vibrational spectroscopists and analytical chemists. In this chapter we will discuss the theory of FT–IR and show how it relates to instrumental design. The performance of FT–IR spectrometers will be compared to that of conventional grating spectrometers, and illustrated in the next chapter through descriptions of several applications where FT–IR has been used to advantage.

5.2. THE MICHELSON INTERFEROMETER

The central component of a Fourier transform infrared spectrometer is a *two-beam interferometer*, which is a device for splitting a beam of radiation into two paths, the relative lengths of which can be varied. A phase difference is thereby introduced between the two beams and, after they are recombined, the interference effects are observed as a function of the path difference between the two beams in the interferometer. For FT–IR spectrometry, the most commonly used device is the *Michelson inter-*

Peter R. Griffiths • Department of Chemistry, Ohio University, Athens, Ohio 45701

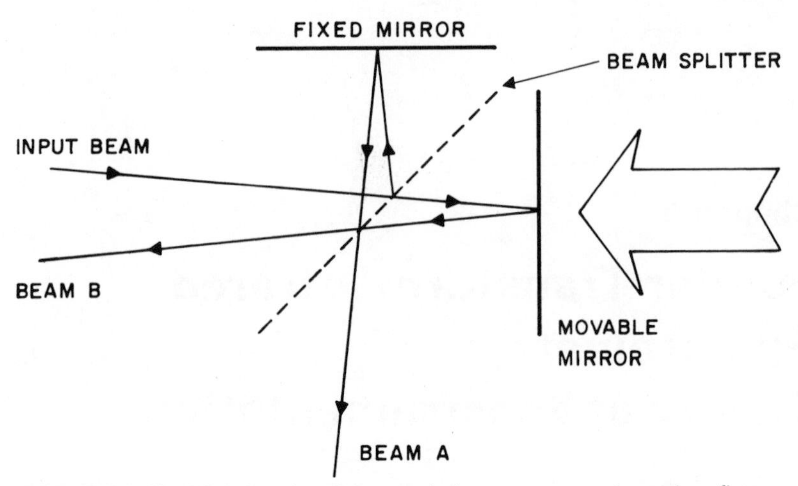

FIGURE 5.1. The Michelson interferometer: In its most commonly used configuration the input beam is perpendicular to either the fixed or the movable mirror, so that only beam *A* may be detected, while beam *B* returns to the source and is lost.

ferometer, although other types of two-beam interferometer, such as the lamellar grating interferometer, have also been used.

The Michelson interferometer, which is depicted schematically in Figure 5.1, consists of two plane mirrors, the planes of which are mutually perpendicular. One of the mirrors is stationary and the other can move along an axis perpendicular to its plane. A semireflecting film, called the *beamsplitter*, is held in a plane bisecting the planes of the two mirrors.

To see how such a device can provide spectral information, let us consider the effect of a Michelson interferometer on a collimated beam of monochromatic radiation, of frequency $\bar{\nu}$ cm^{-1} and intensity $I(\bar{\nu})$. The beamsplitter divides the beam into two paths, one of which has a fixed pathlength, while the pathlength of the other can be varied by translating the movable mirror. When the beams recombine at the beamsplitter they interfere due to their optical path difference (or *retardation*). For beam A in Fig. 5.1 constructive interference occurs if the retardation δ is an integral number of wavelengths:

$$\delta = n/\bar{\nu} \tag{5.1}$$

and destructive interference occurs if

$$\delta = (n + \tfrac{1}{2})/\bar{\nu} \tag{5.2}$$

(where n is an integer).

If the reflectance $R_{\bar{\nu}}$ of the beamsplitter and its transmittance $T_{\bar{\nu}}$ are both equal to 0.5 at $\bar{\nu}$, the beamsplitter is said to be *ideal* at this frequency, and the intensity of the beam *transmitted* through the interferometer (beam A) is given by

$$I'(\delta)_A = \tfrac{1}{2}I(\bar{\nu})(1 + \cos 2\pi\bar{\nu}\delta) \tag{5.3}$$

The intensity of the beam returning in the direction of the source (beam B) is given by

$$I'(\delta)_B = \tfrac{1}{2}I(\bar{\nu})(1 - \cos 2\pi\bar{\nu}\delta) \tag{5.4}$$

It should be noted that when there is constructive interference for beam A there is destructive interference for beam B, and that the sum of $I'(\delta)_A$ and $I'(\delta)_B$ is always equal to $I(\bar{\nu})$, the input energy.

If $R_{\bar{\nu}}$ is not equal to $T_{\bar{\nu}}$, which is the usual case, the corresponding expressions to equations (5.3) and (5.4) are

$$I'(\delta)_A = 2R_{\bar{\nu}}T_{\bar{\nu}}I(\bar{\nu})(1 + \cos 2\pi\bar{\nu}\delta) \tag{5.5}$$

$$I'(\delta)_B = 2(R_{\bar{\nu}}^2 + T_{\bar{\nu}}^2)I(\bar{\nu}) - 2R_{\bar{\nu}}T_{\bar{\nu}}I(\bar{\nu}) \cos 2\pi\bar{\nu}\delta \tag{5.6}$$

Provided that there is no absorption in the beamsplitter, i.e.,

$$R_{\bar{\nu}} + T_{\bar{\nu}} = 1 \tag{5.7}$$

it can be easily verified that, again

$$I'(\delta)_A + I'(\delta)_B = I(\bar{\nu}) \tag{5.8}$$

Usually only beam A is detected, since beam B will often return along the path of the incident beam. Occasionally, however, both $I'(\delta)_A$ and $I'(\delta)_B$ are measured by different detectors, while in an important application of FT–IR known as dual-beam Fourier transform spectroscopy, beams A and B may be passed through different cells and then measured at the same detector, in which case the difference spectrum can be measured (*vide infra*). For the present, however, we will only consider the measurement of $I'(\delta)_A$.

It may be seen from equation (5.5) that $I'(\delta)_A$ consists of two parts, a constant (DC) component equal to $2R_{\bar{\nu}}T_{\bar{\nu}}I(\bar{\nu})$, and a modulated (AC) component. The AC component is called the interferogram $I(\delta)_A$, and is obviously given by

$$I(\delta)_A = 2R_{\bar{\nu}}T_{\bar{\nu}}I(\bar{\nu}) \cos 2\pi\nu\delta \tag{5.9}$$

An infrared detector and AC-coupled amplifier converts $I(\delta)_A$ into an electrical signal:

$$V(\delta)_A = R(\bar{\nu})I(\delta)_A \qquad \text{(volts)} \tag{5.10}$$

where $R(\bar{\nu})$ is the response of the detector and amplifier to radiation of wavenumber $\bar{\nu}$, typically expressed in volts/watt.

Setting

$$B(\bar{\nu}) = 2R_{\bar{\nu}}T_{\bar{\nu}}R(\bar{\nu})I(\bar{\nu}) \tag{5.11}$$

we see that

$$V(\delta)_A = B(\bar{\nu}) \cos 2\pi\bar{\nu}\delta \tag{5.12}$$

$B(\bar{\nu})$ is the measured amplitude of the incident radiation, and it can be seen from equation (5.12) that $V(\delta)_A$ is the cosine Fourier transform of $B(\bar{\nu})$.

In practice a frequency-dependent phase lag, $\theta_{\bar{\nu}}$, is often introduced into the phase angle $2\pi\nu\delta$, either by the filters in the amplifier or by dispersion in the beamsplitter, so that equation (5.12) becomes

$$V(\delta)_A = B(\bar{\nu}) \cos (2\pi\bar{\nu}\delta - \theta_{\bar{\nu}}) \tag{5.13}$$

When the source emits polychromatic radiation, the interferogram is the integral over all frequency elements of equation (5.13), i.e.,

$$V(\delta)_A = \int_{-\infty}^{\infty} B(\bar{\nu}) \cos (2\pi\bar{\nu}\delta - \theta_{\bar{\nu}})\, d\bar{\nu} \tag{5.14}$$

This equation may equally be expressed as

$$V(\delta)_A = \int_{-\infty}^{\infty} B(\bar{\nu})\, e^{-2\pi i\bar{\nu}\delta}\, d\bar{\nu} \tag{5.15}$$

Thus in the general case the interferogram $V(\delta)_A$ is the *complex* Fourier transform of the spectrum $B(\bar{\nu})$, which can be recovered by performing the complex inverse Fourier transform of the interferogram, i.e.,

$$B(\bar{\nu}) = \int_{-\infty}^{\infty} V(\delta)_A\, e^{2\pi i\bar{\nu}\delta}\, d\delta \tag{5.16}$$

5.3. RESOLUTION AND APODIZATION

The resolving power of an interferometer depends on several factors, including

(a) the maximum retardation of the interferometer,
(b) the apodization function used in computing the spectrum,
(c) the divergence of the beam passing through the interferometer,
(d) the stability of the mirror drive system.

We will consider the effects of increasing the solid angle of the beam and of a poor mirror drive later in this chapter. In this section we will assume that a highly collimated beam of radiation is passing through the interferometer, and that the moving mirror of the interferometer does not tilt from its original plane throughout the entire scan. The effect on resolution of limiting the maximum retardation and of apodizing the interferogram will be discussed.

The limits of the integral in equation (5.16) show that in order to obtain a perfect representation of the spectrum $B(\bar{\nu})$, the interferogram must be measured with an infinitely long mirror travel. However, in practice the retardation must be restricted to some finite value Δ_{max}, and the effect of Δ_{max} being less than infinite is to limit the resolution of the measured spectrum to a value given to a good approximation by

$$(\Delta\bar{\nu}) = 1/\Delta_{max} \quad \text{cm}^{-1} \tag{5.17}$$

Resolution can be defined in several ways; the two most common criteria used in spectroscopy are the *baseline* criterion and the *Rayleigh criterion* for resolution. Consider two narrow, well-separated lines of equal intensity, each of which is broadened by the measurement characteristics of the spectrometer. As the separation of the two lines is reduced, a point is reached at which the intensity at the average frequency of the lines starts to increase; at smaller separations than this the lines are incompletely resolved. At the separation, where there is still a 100% dip between the lines but any reduction in the separation would cause the dip to decrease, the lines are said to be resolved according to the baseline criterion. As the separation is reduced below the baseline criterion the lines begin to merge. When the dip is approximately 20% of the peak intensities, the lines are said to be resolved according to the Rayleigh criterion. The value of 20% used in the definition of the Rayleigh criterion is not just an arbitrary number, and its significance will be seen later in this section.

The resolution of a spectrometer is quantitatively described by its instrument line shape function (ILS), which gives the response of the instrument to radiation of frequency $\bar{\nu}$ when the instrument measures at frequency $\bar{\nu}_i$. The ILS, $\sigma(\bar{\nu} - \bar{\nu}_i)$, would be observed experimentally by measuring the spectrum of an infinitesimally narrow emission line. For FT–IR with no apodization (boxcar truncation), the ILS is given by[1]:

$$\begin{aligned} \sigma(\bar{\nu} - \bar{\nu}_i) &= \frac{2 \sin 2\pi\Delta_{max}(\bar{\nu} - \bar{\nu}_i)}{2\pi\Delta_{max}(\bar{\nu} - \bar{\nu}_i)} \\ &= 2 \operatorname{sinc} 2\pi\Delta_{max}(\bar{\nu} - \bar{\nu}_i) \end{aligned} \tag{5.18}$$

while for triangularly apodized interferograms the ILS is given by

$$\sigma'(\bar{\nu} - \bar{\nu}_i) = \operatorname{sinc}^2 \pi\Delta_{max}(\bar{\nu} - \bar{\nu}_i) \tag{5.19}$$

The shape of these functions is shown in Figure 5.2. It may be noted that the ILS for a monochromator operating at low resolution is usually approximated by a triangle, while the ILS for a monochromator working at diffraction-limited resolution is a sinc^2 function similar to equation (5.19). The Rayleigh criterion for resolving two narrow spectral lines of equal intensity was originally developed for a sinc^2 ILS; when the maximum of one measured line coincides with the first zero value for the second measured

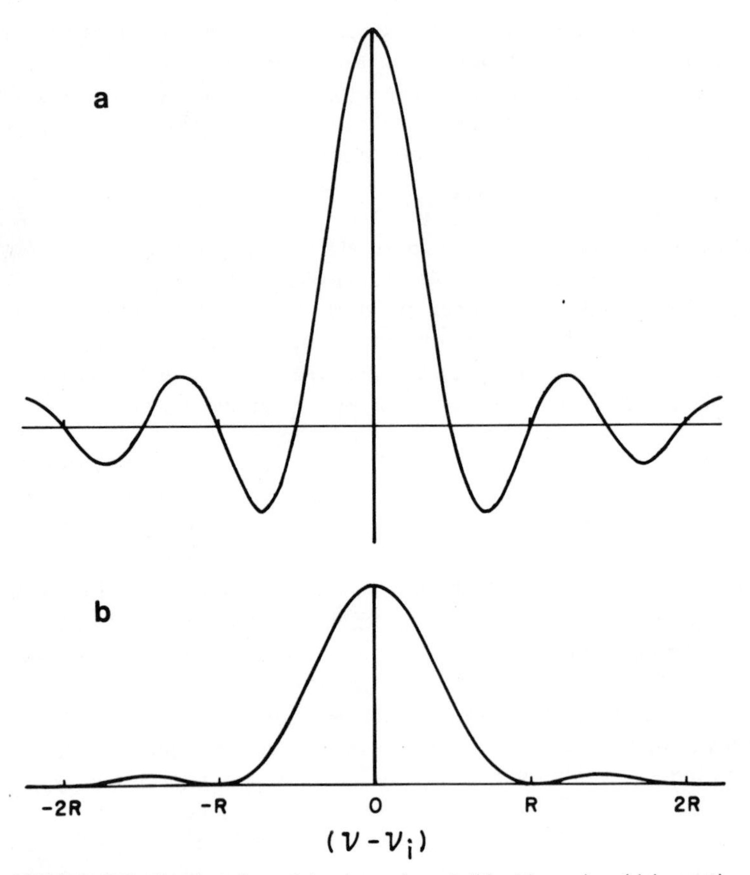

FIGURE 5.2. The functions (a) $\sigma(\nu - \nu_i)$ and (b) $\sigma'(\nu - \nu_i)$, which are the instrument line shape functions for spectra computed using no apodization and triangular apodization, respectively.

line, the resultant curve shows a dip of approximately 20% between the two maxima (see Figure 5.3).

The full width at half-height (FWHH) of the sinc ILS for a maximum retardation of Δ_{max} is $0.605/\Delta_{max}$, and the distance between the first zero values either side of the maximum is $1/\Delta_{max}$. The FWHH of the sinc^2 ILS for the same retardation is $0.88/\Delta_{max}$, and the distance between the first zero values either side of the maximum is $2/\Delta_{max}$. Thus for baseline resolution of two spectral lines, it is necessary to scan the mirror twice as far if the interferograms are triangularly apodized than if no apodization is applied. However, to achieve a 20% dip between the peaks of two partially resolved lines of equal intensity (the Rayleigh criterion for a sinc^2 ILS), the lines must be separated by approximately $0.73/\Delta_{max}$ for unapodized

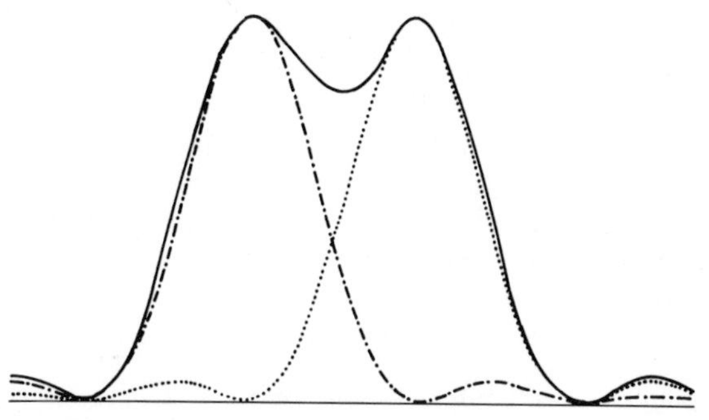

FIGURE 5.3. The Rayleigh criterion for resolution: When two very narrow spectral lines are measured with a $\text{sinc}^2 x$ ILS, the lines are said to be just resolved when the maximum of one line coincides with the first zero value of the other. It should be noted that if two lines measured with a $\text{sinc}\, x$ ILS are separated such that the maximum value of one falls at the same frequency as the first zero values of the other, the resultant curve shows no dip and the lines are completely unresolved.

interferograms and $1/\Delta_{\text{max}}$ for triangularly apodized interferograms. Thus in view of the reduction in the intensity of the sidelobes of the ILS on apodization, the small loss in practical resolution found on apodizing the interferogram will usually not be significant, and *for emission spectroscopy* it is usually beneficial to apodize the interferogram before the FFT.

For *absorption spectroscopy* apodization is again found to be beneficial for unambiguous spectral assignments when the width of the bands being measured is *less than* or equal to the FWHH of the ILS, since apodization reduces the possibility of confusing a weak real band with the sidelobes of a strong neighboring band. However Anderson and Griffiths[2] have shown that the photometric accuracy for measurements of absorption bands by FT–IR without apodization is generally higher than for the corresponding measurement carried out using triangular apodization, especially when the width of the bands is *greater than* or equal to the FWHH of the ILS. Therefore the nature of the experiment will often determine whether or not apodization should be applied.

5.4. EFFECT OF BEAM DIVERGENCE

It is fairly easy to understand why the beam passing through the interferometer must be well collimated. If there is a widely diverging beam passing through the interferometer the retardation for off-axis rays will be greater than that for on-axis rays. The maximum solid angle of the beam that can be tolerated before the resolution is degraded is that angle for which the difference in retardation for the median and extreme ray is one-half the shortest wavelength in the spectrum. It is apparent therefore that the maximum solid angle is going to depend not only on the highest frequency in the spectrum, $\bar{\nu}_{max}$, but also on the resolution desired, $\Delta\bar{\nu}$, since the difference in retardation between the median and on-axis rays will increase linearly with the retardation. The maximum solid angle Ω_{max} can be shown[3] to be

$$\Omega_{max} = 2\pi\,\Delta\bar{\nu}/\bar{\nu}_{max} \tag{5.20}$$

The greater the solid angle of the beam, the greater is the signal at the detector; it is therefore common to optimize the S/N for any measurement by placing a variable aperture at some focal point in the beam of FT–IR spectrometers to limit the solid angle to the value given by equation (5.20). (It may be noted that this aperture is equivalent to the entrance slit of a monochromator, which also serves the purpose of collimating the beam before it is dispersed by the prism or grating.)

For measurements taken at low resolution or when $\bar{\nu}_{max}$ is low (e.g., in far-infrared spectrometry) it is often found that the maximum allowed solid angle cannot be attained in practice, and the optical throughout is limited by instrumental factors such as the dimensions of the source, the detector, or the sample. In this case a throughput that is lower than the maximum allowed value has to be tolerated, and the practical sensitivity of the spectrometer is less than the maximum theoretical sensitivity.

5.5. MIRROR DRIVE TOLERANCE

If the moving mirror of the interferometer tilts during the scan, the effective path difference for the beam passing through one region of the beamsplitter may be different from that for the beam passing through a different region. Thus for a given wavelength one part of the beam can be undergoing constructive interference while another part of the beam can be undergoing destructive interference. The result is a reduction in the amplitude of the signal for that wavelength.

The difficulty in maintaining a tilt-free drive gets larger as the maximum retardation required increases. If the angle of tilt increases as the retardation increases, the effect on the interferogram of a monochromatic source is equivalent to the effect of apodization; hence the result of a poor mirror drive is usually to degrade the resolution of the spectrum.

The tolerance of the drive mechanism for the moving mirror of an interferometer depends on the same factors as the beam divergence, i.e., the resolution required and the shortest wavelength in the spectrum. As a consequence, mirror drive mechanisms for FT–IR spectrometers operating at high resolution in the near infrared must be of very high precision, while relatively low precision drives are needed for low- and medium-resolution interferometers for far-infrared spectrometry.

The drive mechanisms of three types of interferometer designed for different spectral regions nicely illustrate this point. Most far-infrared interferometers use fairly crude drive systems, with the mirror being mounted on the end of a piston, which is driven through a cylinder using grease as a lubricant. The moving mirror of Michelson interferometers used for medium-resolution ($0.1 < \Delta\bar{\nu} < 10\,\text{cm}^{-1}$) mid-infrared spectrometry are often driven using an air bearing to allow a freer, more precise motion. Even this mechanism is insufficient for most high-resolution ($\Delta\bar{\nu} < 0.05\,\text{cm}^{-1}$) measurements, and considerable effort has been expended on the development of tilt compensation devices such as the cube-corner or the cat's-eye retroreflector.[4] These devices compensate for the effect of tilt in the manner shown in Figure 5.4. For the highest-resolution measurements ($\Delta\nu \sim 10^{-3}\,\text{cm}^{-1}$), it may also be necessary to stop the movable mirror at each sampling point and adjust the plane of the mirror to ensure perfect alignment throughout the interferogram.

5.6. DYNAMIC RANGE

The shape of the interferogram $V(\delta)_A$ depends to a large extent on the nature of the spectrum $B(\bar{\nu})$. If the spectrum consists of a small number of

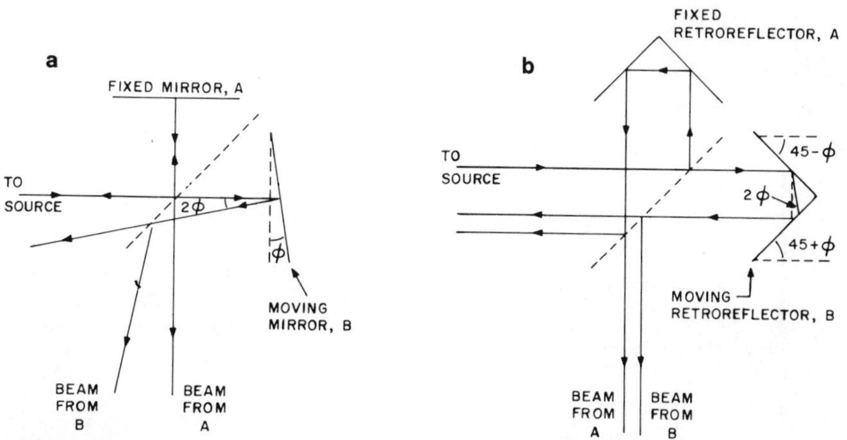

FIGURE 5.4. (a) If the moving mirror of a conventional Michelson interferometer tilts during the scan, the beam from the moving mirror (B) no longer coincides with the beam from the fixed mirror (A), and both resolution and the modulation efficiency for short-wavelength radiation will be lowered. (b) If retroreflectors are used instead of plane mirrors in the interferometer, the beams from the fixed (A) and moving (B) retroreflectors remain parallel even if the moving retroreflector tilts during the scan. Cat's-eye retroreflectors work in a similar fashion.

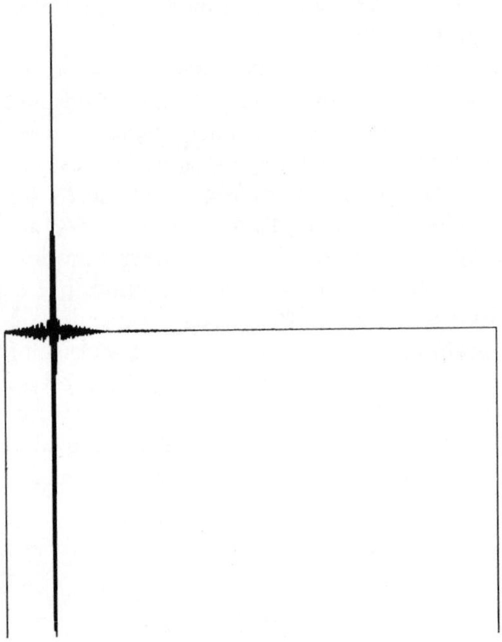

FIGURE 5.5a. A typical mid-infrared interferogram measured on a well-purged FT–IR spectrometer with a retardation of 0.25 cm. Note that the modulated signal dies away quite close to the zero retardation point.

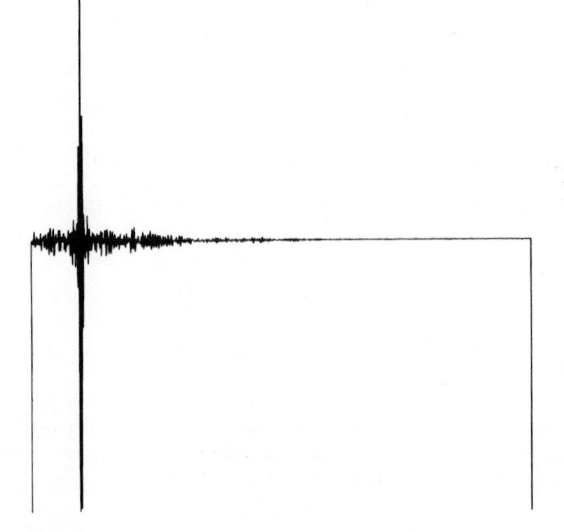

FIGURE 5.5b. An interferogram from the same FT–IR spectrometer with a 0.05-mm sheet of polystyrene in the beam. Note that appreciable modulation is seen in the interferogram to greater retardation than (a), but dies out after a retardation of about 0.15 cm; from this observation it may be deduced that the narrowest bands in the spectrum of polystyrene have widths (at half-height) of about $(0.15)^{-1}$ cm^{-1}, or 6–7 cm^{-1}.

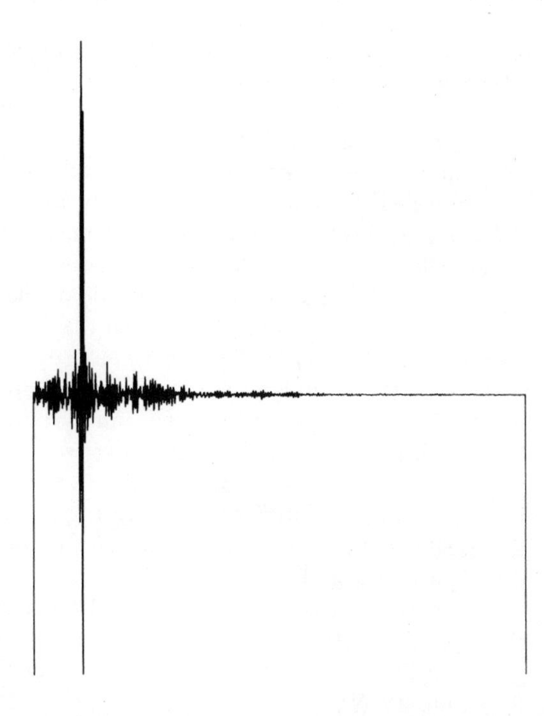

FIGURE 5.5c. The result of subtracting interferogram (a) from (b): this interferogram should only be due to radiation absorbed by the polystyrene film.

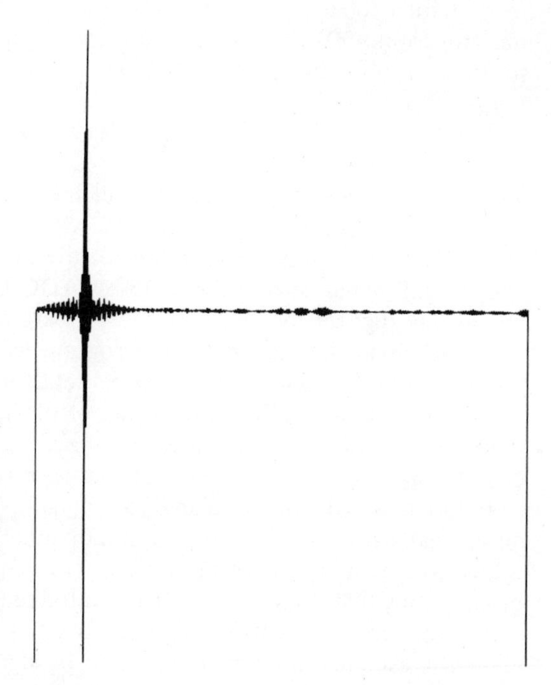

FIGURE 5.5d. The result of subtracting interferogram (a) from that of an unpurged instrument: the differences are primarily due to atmospheric water vapor. Since these lines are very narrow, appreciable modulation may be seen not only at the end of this scan (Δ_{max} = 0.25 cm), but also for much longer scans.

sharp emission lines, the interferogram will be similar in shape to that of a typical free-induction decay pattern in FT–NMR (see, for example, Figure 8.1). If the lines being measured are not very intense, the S/N of the interferogram will not be very high and few constraints will be put on the nature of the interferometer, detector, or data system.

However, the majority of spectra measured by FT–IR are absorption spectra, and most of the signal being measured at the detector is caused by emission from the intense continuous source. Assuming that $\theta_{\bar{\nu}}$ [see equation (5.13)] is small, the amplitude of the interferogram is very large near the zero retardation point, when the radiation from all frequencies is approximately in phase. Thus the interferogram has a very high S/N near $\delta = 0$, but the modulations die out quite rapidly for only small changes in retardation (see Figure 5.5a). The interferogram due to sample absorption bands will be subtracted from the interferogram due to the source, and the presence of discrete absorption bands is evidenced by small modulations in the *wings* of the interferogram (see Figure 5.5b). The sharper the bands, the greater is the retardation at which the modulations may be observed (see Figures 5.5c and d).

Absorption spectroscopy by FT–IR places far greater constraints on the data acquisition system of the spectrometer than emission spectroscopy. To illustrate the magnitude of the signal-to-noise ratio of the interferogram, $(S/N)_I$, for absorption spectroscopy, let us consider the measurement of the spectrum of a continuous source whose intensity is uniform from 4000 to 400 cm^{-1} and zero outside this range. The signal-to-noise ratio of the spectrum, $(S/N)_s$, of such a boxcar source is related to $(S/N)_I$ as

$$(S/N)_I = M^{1/2}(S/N)_s \tag{5.21}$$

where M is the number of resolution elements. Thus if we want to measure a spectrum of the above source with $(S/N)_s = 500$ at a resolution of 1 cm^{-1} ($M = 3600$) in a single scan, it can readily be seen that $(S/N)_I = 3 \times 10^4$ and the full dynamic range of a 15-bit ADC is only just large enough to adequately digitize the signal. If $(S/N)_I$ were any greater than 30,000 the "noise level" in the digitized interferogram would be set by the least significant bit of the ADC rather than by detector noise.

Since values of $(S/N)_I$ of the order of 10^4 are fairly easy to obtain from typical incandescent sources used in infrared spectrometry with scan times of just a few seconds, it is apparent that spectra with very low noise levels ($<0.1\%$) can only be measured by signal-averaging interferograms (or occasionally spectra computed from individual interferograms). Even then the word-length of the computer used to acquire the data must be sufficiently long that $(S/N)_I$ or $(S/N)_s$ is not limited in any way. For this reason minicomputers with a wordlength as large as 20 bits have to be used to allow a reasonable number of scans to be signal averaged and computed in single precision (one word per datum point) without adding noise to the spectrum. However it has recently been shown[5] that even a 20-bit word is not long enough to allow signal-averaging to be effective over a much longer period than two or three hours for some measurements, and that the use of double-precision data handling (two words per datum point) is more generally beneficial. Since the maximum number of bits per word required for reasonable measurement times (say, a weekend or less) is about 28, computers with either a 16- or a 20-bit word length can be used with equal effectiveness in the double-precision mode for FT–IR.

5.7. SCAN SPEED AND SPECTRAL MODULATION

For the measurement of spectra with only a small number of resolution elements, or where the source energy is low or discrete, long scan times can be used without encountering the problems of dynamic range discussed above. Thus whereas nearly all interferometers designed for mid-infrared absorption spectrometry are of the rapid-scanning, signal-averaging type, many interferometers designed for far-infrared spectrometry, where the source energy is low and the number of resolution elements is small, use a single slow scan.

Two types of slow-scanning interferometers have been used for far-infrared measurements—continuous-scanning and stepped-scanning instruments. The moving mirror in the continuous-scanning interferometers is translated at a constant velocity throughout the entire scan, whereas in stepped-scanning interferometers the mirror is held stationary at each sampling point for a certain time, during which the signal is integrated, and then it is rapidly translated to the next sampling point. There is little to choose between the performance of continuous-scanning and stepped-scanning interferometers, either in theory or in practice, for far-infrared spectrometry.

The effective scan speed of slow-scanning interferometers allows the retardation to be changed at a rate of between 0.5 and 500 μm s^{-1}, with 5 μm s^{-1} being a typical value. Thus it requires on the order of 30 minutes to measure an interferogram with $\Delta_{max} = 1$ cm. The moving mirror of interferometers designed for mid-infrared absorption spectrometry will scan at a few millimeters per second, and it usually requires just three or four seconds to measure an interferogram with $\Delta_{max} = 1$ cm.

The different scan-speeds for slow- and rapid-scanning interferometers have a direct effect on the need for external modulation of the beam. For a continuously scanning interferometer with a mirror velocity, V cm s^{-1}, the time, t seconds, required to scan a retardation δ cm is given by

$$t = \delta/2V \tag{5.22}$$

TABLE 5.1. Comparison of Parameters for Far-Infrared and Mid-Infrared Interferometers

Parameter	Far-infrared	Mid-infrared
$2V$	5×10^{-4} cm s^{-1}	3×10^{-1} cm s^{-1}
$\bar{\nu}_{max}$	400 cm^{-1}	4000 cm^{-1}
$\bar{\nu}_{min}$	20 cm^{-1}	400 cm^{-1}
$f_{\bar{\nu}_{max}}$	0.2 Hz	1200 Hz
$f_{\bar{\nu}_{min}}$	0.01 Hz	120 Hz

Substituting (5.22) into (5.12), we see that the signal varies as a function of time as

$$V(t) = B(\bar{\nu}) \cos 2\pi (2V\bar{\nu})\, t \tag{5.23}$$

Thus the modulation frequency $f_{\bar{\nu}}$ of the signal due to a wavenumber $\bar{\nu}$ is

$$f_{\bar{\nu}} = 2V\bar{\nu} \quad \text{Hz} \tag{5.24}$$

Table 5.1 compares the range of modulation frequencies of far-infrared radiation measured on a continuous slow-scanning interferometer and mid-infrared radiation measured on a rapid-scanning interferometer.

The modulation frequencies of mid-infrared radiation fall in the audio-frequency range and the signals can conveniently be amplified using conventional audio-frequency amplifiers. The modulation frequencies for far-infrared radiation fall well below the audio-frequency range and the beam has to be further modulated with an external chopper at frequency well above $f_{\bar{\nu}_{max}}$. For stepped-scanning instruments, where $V = 0$ at each sampling point, it is of course necessary to use an external chopper in exactly the same way as for the slow continuous-scanning instruments.

It may be noted that for rapid-scanning interferometers the amplifier usually incorporates a high-pass filter, and will detect neither the DC component of $I'(\delta)$ nor any low-frequency drifts due to the source or the electronics; the AC interferogram is measured directly. For slow-scanning interferometers with an external chopper, both the AC and DC components of the signal are measured, and low-frequency drifts will also show up in the measured signal. Before the FFT is performed on this type of interferogram, the DC signal must be calculated and subtracted from each point of the measured interferogram.

As a means of eliminating the low-frequency drifts of slow-scanning interferometers a technique known as *phase modulation* has been developed.[6-9] For phase modulation one of the mirrors of the interferometer is vibrated with an amplitude that is approximately one-half the shortest wavelength in the spectrum. Under these circumstances the measured signal is, to a good approximation, equal to the first derivative of the normal (amplitude-modulated) signal. Since phase modulation techniques are usually applied on interferometers on which the amplitude-modulated interferogram is highly symmetrical ($\theta_{\bar{\nu}} \sim 0$ for all wavenumbers), it may be seen that

$$V(\delta)_{PM} = -\int_{-\infty}^{\infty} B(\bar{\nu}) \sin 2\pi\bar{\nu}\delta \, d\bar{\nu} \tag{5.25}$$

Thus the spectrum is obtained by performing the *sine* transform of phase-modulated interferograms.

The fact that this technique obviates the need for a chopper means that twice as much radiation reaches the detector and the sensitivity is higher than the sensitivity of the corresponding amplitude-modulated experiment for all wavenumbers in the spectrum except those at the extreme low-wavenumber and high-wavenumber ends of the spectrum.[6] Since the frequency of the vibrating mirror is fairly high, the phase modulation technique discriminates against low-frequency drifts and shows real advantages over amplitude modulation for far-infrared Fourier spectrometry.

5.8. DATA ACQUISITION

For slow-scanning interferometers, where no dynamic range problem exists, $(S/N)_I$ may be increased merely by slowing down the scan speed of the moving mirror and using a longer time constant for the amplifier. However, as we discussed earlier in this chapter, when $(S/N)_I$ is large a value is reached at which it exceeds the dynamic range of the ADC. Beyond this value reducing the scan speed does not increase $(S/N)_I$, which becomes limited by *digitization noise* rather than by detector noise. If for any measurement $(S/N)_I$ exceeds the dynamic range of the ADC for a single slow scan, the mirror velocity has to be increased to reduce the scan time to the point that $(S/N)_I$ is reduced below the dynamic range of the ADC; signal-averaging techniques must then be applied to raise $(S/N)_I$ above the single-scan value.

In practice, either interferograms or spectra can be averaged; however, any time the FFT takes longer than the acquisition time of an individual interferogram, it can readily be seen that averaging spectra is an inefficient operation. On the other hand, if the FFT of a measured interferogram can be performed *during* the acquisition of the subsequent interferogram and completed before the start of the following scan, averaging spectra is no less efficient than averaging interferograms and the method has the additional advantage of allowing a spectrum to be displayed, and constantly updated, on a CRT display during the measurement.

In practice most interferometers designed for mid-infrared spectrometry signal-average *interferograms*, and special techniques have been developed to ensure that the signals are added coherently. The use of a real-time clock to provide the sampling trigger for data collection is not recommended for extensive signal-averaging, since the mirror velocity may vary by as much as $\pm 0.5\%$ during a single scan and successive interferograms start to lose phase coherence after scanning for only a few milliseconds. To ensure that phase coherence is maintained throughout the entire scan, interferograms have to be sampled at equal intervals of *retardation* rather than equal intervals of *time*. To achieve this end, a second interference signal is monitored at the same time as the main interferogram.

Slow-scanning far-infrared interferometers often monitor fringes from a Moiré grating device to provide the sampling trigger for data collection. This mechanism is sufficiently accurate to allow the signal to be digitized at approximately equal intervals of retardation at slow data rates but is not quite accurate enough to permit coherent signal-averaging with rapid-scanning mid-infrared interferometers. For this type of instrument the interferogram of a highly monochromatic source (usually a small He–Ne laser) is measured along with the main interferogram. The AC signal from this source is a pure sine wave, and by sampling the main interferogram, either at each zero value of this sinusoidal *reference interferogram* or at equal multiples of the interval between each zero crossing, it is found that interferograms can be digitized at exactly the same positions during each scan.

Coherent signal-averaging depends on the fact that the first data point is sampled at an identical retardation for every scan. To this end another interferogram, this time from a white (visible) light source is measured at the same time as the other two interferograms. The very sharp interferogram produced by this source provides an "optical fiduciary marker" at precisely the same place for each scan. Whenever the *white light interferogram* exceeds a certain threshold voltage V_t, data collection is initiated at the next zero crossing of the laser reference interferogram.

Often the white light and laser interferograms are measured through a separate *reference interferometer*, the moving mirror of which is attached to the moving mirror of the main interferometer and hence is driven at the same velocity. If the Mertz method is used for phase correction (see Section 2.6) it is necessary to sample the interferogram for a short distance on one side of zero path difference, and scan through $\delta = 0$ to the retardation required to achieve the desired resolution. The retardation at which data collection is started can be adjusted by moving the fixed mirror and beamsplitter of the reference cube relative to the moving mirror, so that the zero retardation point for the reference interferometer occurs about 100 μm before the zero retardation point for the main interferometer (see Figure 5.6).

Typical of laser-referenced rapid-scanning Michelson interferometers in use today is the Digilab Inc. (Cambridge, Massachusetts) x96 series of

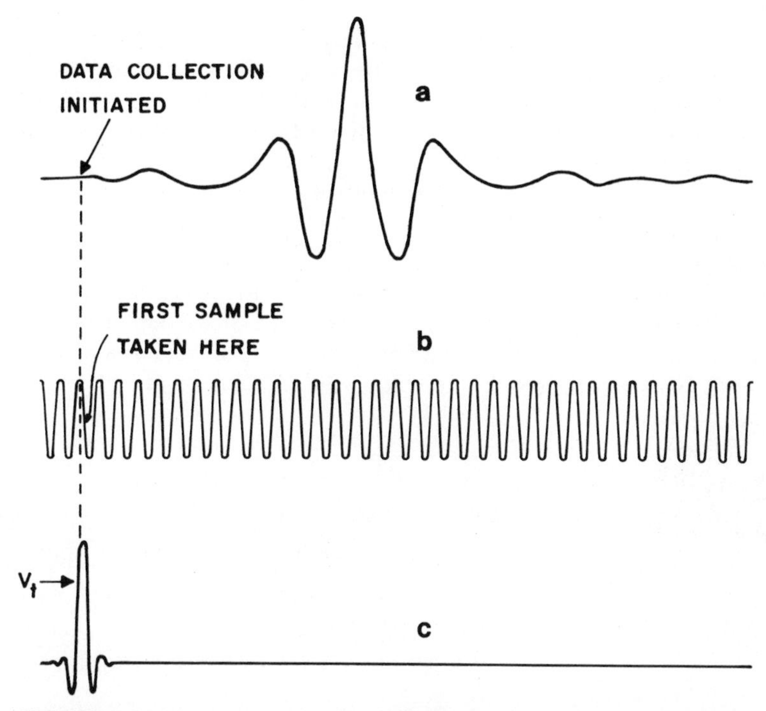

FIGURE 5.6. Signals from an interferometer with a separate reference cube for measuring the laser and white light interferograms. (a) Signal (infrared) interferogram; (b) laser interferogram; (c) white light interferogram. The zero retardation position of the reference interferometer has been displayed relative to that of the main interferometer so that data collection may be initiated a short distance before the maximum of the infrared interferogram.

instruments. The moving mirror of these interferometers is driven on an air-bearing to ensure a precise motion and the instruments should be thermostatted above ambient temperature to ensure that the mirrors and beamsplitter remain aligned. A small reference interferometer is used to generate the sampling trigger. The maximum retardation available on any of the Digilab interferometers is 10 cm, but if slightly higher resolution is required the Nicolet Model 7001 interferometer may be used. This instrument permits a maximum retardation of approximately 17 cm, and its drive system is stabilized through a dual air-bearing configuration.

5.9. BEAMSPLITTERS

It can be seen from equation (5.5) that the amplitude of the modulated beam is greatest when the product $R_{\bar{\nu}} T_{\bar{\nu}}$ is a maximum, that is, when

$$R_{\bar{\nu}} = T_{\bar{\nu}} = 0.5 \tag{5.26}$$

Of course, in practice the reflectance of any material is rarely exactly equal to 0.5, so that beamsplitters generally produce less than the maximum theoretical modulation of the interferogram. The efficiency of the beamsplitter $\eta_{\bar{\nu}}$ at any frequency $\bar{\nu}$ has been defined as $4R_{\bar{\nu}}T_{\bar{\nu}}$, so that

$$\eta_{\bar{\nu}} \leq 1.00 \tag{5.27}$$

However, if absorption is low and $R_{\bar{\nu}}$ and $T_{\bar{\nu}}$ are between 0.35 and 0.65, $\eta_{\bar{\nu}}$ exceeds 0.90; any material fulfilling these criteria can be used efficiently as a beamsplitter provided that it can be made into a thin flat film.

The reflectance R_0 of a thin film depends on the refractive index of the material n, the thickness of the film d, the angle of incidence of the beam θ, and the single-surface reflectance R, in the following way:[10]

$$R_0 = \frac{2R(1 - \cos \varepsilon)}{1 + R^2 - 2R \cos \varepsilon} \tag{5.28}$$

R is given by the Fresnel equations[10] for radiation whose electric vector is polarized parallel (p) and perpendicular (s) to the plane of incidence:

$$R_p = \frac{\sin^2(\theta - \theta_t)}{\sin^2(\theta + \theta_t)} \tag{5.29}$$

$$R_s = \frac{\tan^2(\theta - \theta_t)}{\tan^2(\theta + \theta_t)} \tag{5.30}$$

while the angle ε is given by

$$\varepsilon = 4\pi\bar{\nu}nd \cos \theta_t \tag{5.31}$$

where θ_t is the angle between the ray and the normal *inside* the material. Thus the beamsplitter efficiency for a nonabsorbing material is

$$\eta = \frac{8R(1 - R)^2(1 - \cos \varepsilon)}{1 + R^2 - 2R \cos \varepsilon} \tag{5.32}$$

For unpolarized light the beamsplitter efficiency is merely the average of η_s and η_p.

It can be seen from equations (5.28) and (5.31) that η takes a zero value when $\cos \varepsilon = 1$, that is, when

$$\varepsilon = m\pi \tag{5.33}$$

where m is an integer. Substituting equation (5.33) into (5.31) it can be seen that this occurs when

$$\bar{\nu} = m(4nd \cos \theta_t)^{-1} \tag{5.34}$$

The efficiency curve shows a series of "hoops" that have zero values when $m = 0, 1, 2, \ldots$. For each hoop the curve is symmetrical about the frequency given by

$$\bar{\nu}_0 = \frac{2n + 1}{8nd \cos \theta_t} \tag{5.35}$$

(i.e., the frequency midway between the zero values bounding that particular hoop). In practice only the first hoop is usually used for FT–IR spectroscopy.

For far-infrared spectrometry, beamsplitters are usually thin films of polyethylene terephthalate, for which $n \sim 1.75$; this material usually goes under the trade names of Mylar in the U.S. and Melinex in the U.K. The variation of the efficiency of Mylar beamsplitters with angle of incidence is shown in Figure 5.7. It can be seen that when $\theta = 45°$ (the usual value for a Michelson interferometer), the efficiency never exceeds 0.65, and Mylar can only be used over a rather limited range with an efficiency greater than 0.5.

The highest efficiency over a wide spectral range is found when $\theta > 75°$, but the difficulties in constructing a Michelson interferometer with an incidence angle much greater than 45° have meant that far-infrared Fourier spectrometers using Mylar beamsplitters are usually operated at less than their optimum efficiency. An interferometer with $\theta = 30°$ has been constructed by Polytec GmbH (Karlsruhe, Germany); the efficiency of this

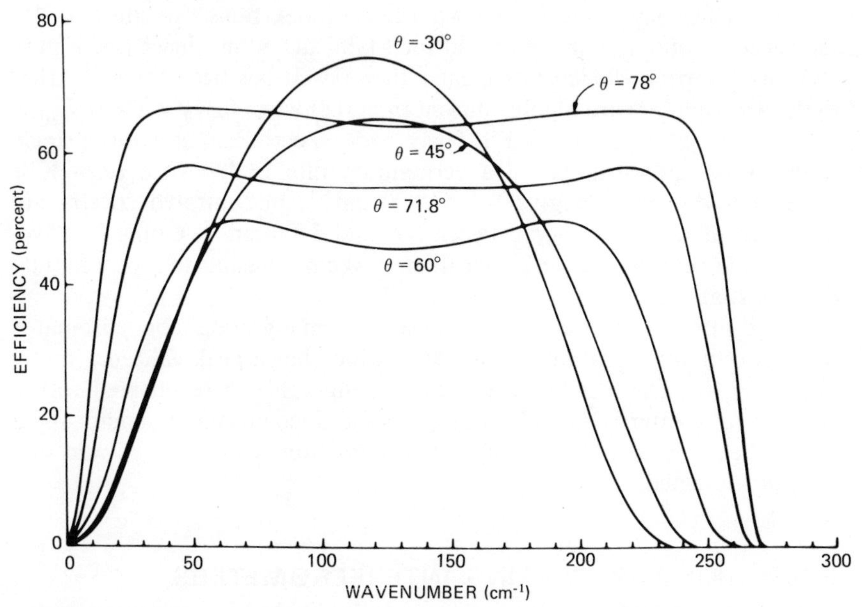

FIGURE 5.7. Beamsplitter efficiency calculated for a 12.5-μm-thick Mylar beamsplitter used at different incidence angles. [Reproduced from *Applied Spectroscopy* **30**, 303 (1976) by permission of the Society for Applied Spectroscopy; copyright 1976].

instrument is greater than that of interferometers with $\theta = 45°$, but they can only be used over a very limited spectral range. For this reason the Polytec FIR-30 spectrometer is equipped with a *beamsplitter wheel*, which is a device on which several beamsplitters of different thickness are mounted. These beamsplitters can be rapidly interchanged without breaking the vacuum in the spectrometer, so that the whole far-infrared region can be covered at high efficiency.

Metal mesh has also been suggested[11] as a beamsplitter material for far-infrared spectrometry. Even though metal mesh beamsplitters appear to give higher efficiency over wider spectral ranges for 45° incidence than Mylar, they have not been used in practice to any great extent.

Mylar cannot be used efficiently as a beamsplitter material for mid-infrared spectrometry for two principal reasons. First, polyethylene terephthalate shows strong absorption bands across its mid-infrared spectrum; second the mechanical strength of Mylar films of the thickness required for mid-infrared measurements ($d < 1\ \mu m$) is very low. The most commonly used beamsplitter materials for mid-infrared Fourier spectrometry are thin layers of germanium ($n = 4.0$) or silicon ($n = 3.6$) deposited on flat alkali halide substrates. A compensator plate of the same material at the same thickness is usually placed over the layer of germanium or silicon. For near-infrared measurements a layer of Fe_2O_3 ($n = 3.0$) on CaF_2 or quartz has been successfully used.

The efficiency curve for high-refractive-index films operating at 45° incidence is similar in shape to that for Mylar at 78° incidence (see Figure 5.7), but the peak efficiency is greater than 0.9. It has been shown[12] that these beamsplitters are highly efficient ($\eta > 0.8$) when $0.3\bar{\nu}_0 < \bar{\nu} < 1.7\bar{\nu}_0$. It is therefore possible to cover a fairly wide spectral range with a single beamsplitter. For example, if a germanium film is deposited on a KBr plate at a thickness to give $\bar{\nu}_0$ at 2000 cm^{-1}, mid-infrared spectra are measured at high efficiency between 600 and 3400 cm^{-1}. Going to wavenumbers lower than 600 or higher than 3400 cm^{-1} results in a very rapidly deteriorating efficiency.

Mid-infrared spectra would be measured on a grating monochromator using two or three gratings, each one of which has a peak efficiency rarely exceeding 0.8. The fact that the spectral range able to be covered with a single beamsplitter is so wide is quite lucky, since automatic interchanges for the heavy mid- and near-infrared beamsplitters have not yet been successfully designed.

5.10. LAMELLAR GRATING INTERFEROMETERS

Several types of interferometers other than the Michelson interferometer have been suggested for spectroscopic measurements, but most of

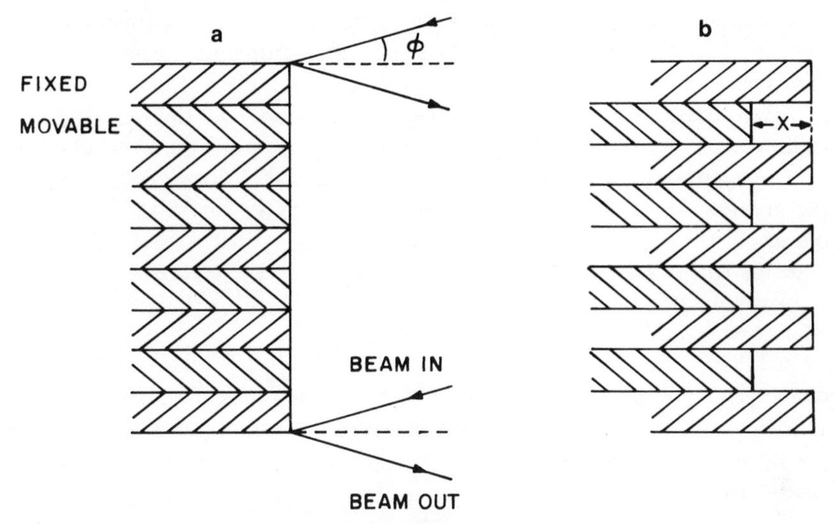

FIGURE 5.8. Schematic diagram of a simple lamellar grating interferometer. One set of facets remains in a fixed position while the other set is moved either continuously or in a stepwise fashion. (a) Zero retardation; (b) retardation is $2X \sec \Phi$.

these have been designed for quite specialized applications and are rarely used for chemical spectroscopy. For *mid-infrared* measurements only the Michelson interferometer is in common use. For *far-infrared* spectroscopy, however, the *lamellar grating interferometer* has certain advantages over the Michelson interferometer, primarily because of the low efficiency and the polarization properties of Mylar beamsplitters.

In lamellar grating interferometers, a path difference between two beams of light is created by adjusting the position of two interleaved facets of a mirror in the fashion shown in Figure 5.8. Interference is caused by wavefront division due to reflection from the moving and stationary facets, as opposed to amplitude division, which is the interference mechanism in a Michelson interferometer. Provided that the angle of incidence on the facets of the grating is fairly high so that "shadowing"[13] does not take place to any great extent, the efficiency of a lamellar grating interferometer can be very high. However, upper and lower frequency limits can be calculated,[14] outside of which the efficiency of lamellar grating interferometers drops off rapidly so that they have never been successfully used for mid-infrared spectroscopy.

5.11. DETECTORS FOR FT-IR

There is, of course, nothing intrinsically different between a detector used on a Fourier transform spectrometer and a detector used on a grating spectrometer. However the difference in the modulation frequency and in the signal-to-noise ratio for the two types of instruments means that some care has to be taken in the choice of detectors for FT–IR spectrometry.

Two parameters have been used to characterize the sensitivity of radiation detectors. The first is the *specific detectivity* D^*, the value of which to a first approximation is independent of the dimensions of the detector; the greater the value of D^*, the more sensitive is the detector. The second parameter is the *noise equivalent power* (NEP), which is the ratio of the detector rms noise voltage V_n in volts hertz$^{-1/2}$, to the voltage responsivity, $R(\bar{v})$ in volts per watt. It can be shown that

$$D^* = (A_D)^{1/2}/\text{NEP} \quad (5.36)$$

where A_D is the area of the detector in cm^2. The *lower* the NEP, the more sensitive is the detector. To achieve maximum sensitivity one should select a detector whose dimensions are exactly equal to the dimensions of the smallest possible focus of the beam.

5.11.1. Far-Infrared Detectors

The most commonly used detector for far-infrared spectroscopy is the Golay pneumatic detector.[15,16] This detector operates at room temperature, and recent developments[17] have reduced the NEP to an average of about 7×10^{-11} W Hz$^{-1/2}$. The Golay cell is an excellent detector for use with slow-scanning far-infrared interferometers, but its sensitivity falls when the modulation frequency is raised much above 20 Hz. Therefore, Golay detectors cannot be used effectively with rapid-scanning interferometers.

When rapid-scanning interferometers are used for far-infrared spectrometry, a detector with a better response to radiation modulated at high frequency (see Table 5.1) must be used. The type of detector that has been most commonly used with rapid-scanning interferometer is the pyroelectric bolometer, and the most commonly used detector in this category has been the triglycine sulfate (TGS) detector. The TGS detector is not as sensitive as the Golay cell—the NEP of a 3 × 3 mm TGS detector is approximately 1×10^{-9} W Hz$^{-1/2}$—although it is somewhat more reliable. In addition, since rapid-scanning interferometers with TGS detectors are used almost exclusively for mid-infrared absorption spectroscopy, the conversion from a far-infrared spectrometer to a mid-infrared spectrometer can be achieved very rapidly by changing beamsplitters and detectors. TGS is rather hygroscopic, and it is advisable that separate detectors (one with a KBr window and one with a polyethylene window) are used for mid- and far-infrared spectrometry, respectively.

It is possible that strontium barium niobate (SBN), another pyroelectric material but one that is not hygroscopic and therefore does not need

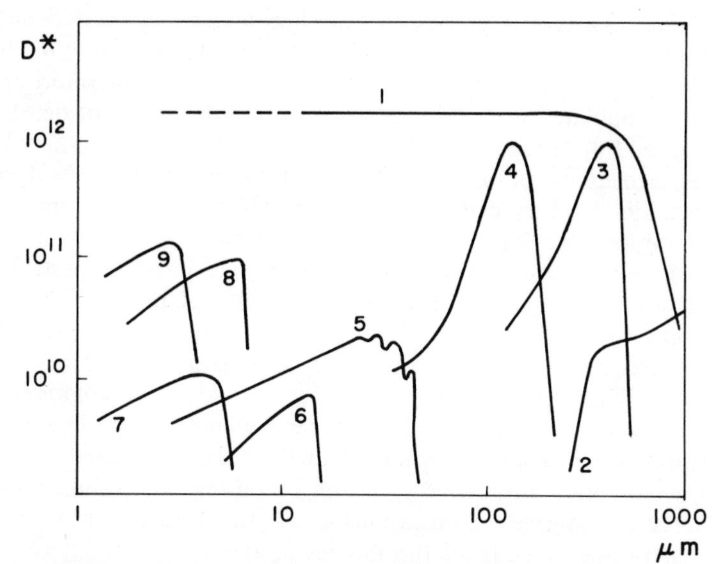

FIGURE 5.9. Specific detectivity D^* for several far- and mid-infrared detectors useful for FT–IR spectrometry: (1) silicon bolometer, 1.4 K; (2) indium antimonide, 2 K; (3) gallium arsenide, 4.2 K; (4) gallium-doped germanium, 4.2 K; (5) copper-doped germanium, 4.2 K; (6) mercury cadmium telluride, 77 K; (7) lead selenide, 295 K; (8) indium antimonide, 77 K; (9) lead sulfide, 295 K.

to be sealed behind an infrared transmitting window, may replace TGS for spectroscopic measurements. However at the present time, the NEP of SBN is higher than that of TGS, and SBN is not commonly used.

For measurements in which very high sensitivity is required, gallium-doped germanium or gallium arsenide detectors (both operating at 4.2 K), indium antimonide (operating at 2 K), or best of all, a silicon bolometer (operating at 1.4 K) can be used for far-infrared spectrometry (see Figure 5.9). The silicon bolometer is more than three orders of magnitude more sensitive than the Golay detector.

5.11.2. Mid- and Near-Infrared Detectors

As mentioned above, the most commonly used detector for mid-infrared absorption spectroscopy is the TGS pyroelectric bolometer with a KBr window. At the modulation frequencies necessary to prevent the interferogram from becoming digitization noise limited when incandescent continuous sources are measured, it is found that the D^* of conventional thermal detectors, such as thermocouples and bolometers, falls off drastically compared to their D^* at, say, 15 Hz. Therefore even though a thermocouple used in a typical grating spectrophotometer may be over an order of magnitude more sensitive than a TGS detector for radiation modulated at 15 Hz, the TGS detector is more sensitive than the thermocouple for radiation modulated at 1 kHz, and is therefore the standard detector for mid-infrared Fourier spectrometers.

The sensitivity of photoconductive detectors, such as lead sulfide, lead selenide, mercury cadmium telluride (MCT), and lead tin telluride, is much better than that of TGS, but these detectors have a limited wavelength range and cannot be used to cover the mid-infrared spectrum from 4000 to 400 cm^{-1}. Unlike TGS, the sensitivity of photoconductive detectors *increases* as the modulation frequency increases, and this property makes these detectors very suitable for use with rapid-scanning interferometers. The properties of these detectors, and of the commonly used photovoltaic detector indium antimonide, are summarized in Figure 5.9.

These detectors are so sensitive that they cannot be used for absorption spectroscopy without the spectrum becoming digitization noise limited, unless the sample only transmits a very small fraction of the incident radiation or the velocity of the moving mirror of the interferometer is speeded up to reduce the S/N of a single scan. On the other hand, photodetectors can be used very profitably for absorption spectroscopy of highly absorbing or scattering samples or for emission spectroscopy of low temperature or discrete sources.

Mercury-doped germanium bolometers at 4.2 K or silicon bolometers at 1.4 K can be used for very high sensitivity mid-infrared measurements. The sensitivity of these detectors is limited by statistical fluctuations in the background radiation emitted from surroundings at room temperature. Therefore, to achieve the maximum sensitivity possible with these detectors, the entire interferometer has to be cooled with liquid nitrogen. A good illustration of the use of a liquid-helium-cooled bolometer with a liquid-nitrogen-cooled interferometer has been given by Moehlmann *et al.*[18]

5.11.3. Ultraviolet–Visible Spectroscopy

Interferometers have not been used to any great extent for ultraviolet or visible spectroscopy, largely because of the nature of the noise of the detectors used in this spectral region. Photomultiplier (PM) tubes are very sensitive detectors, but the noise from a PM detector increases with the square root of the incident signal. (This is different from the case for thermal, pneumatic, pyroelectric, and photodetectors, the noise of which is independent of the signal.)

The signal dependence of the noise of PM detectors precisely offsets Fellgett's advantage for FT spectroscopy with continuous sources, but does not necessarily offset Fellgett's advantage for discrete sources. For example if a single discrete line were measured, the photon noise would only be seen close to the frequency of the line on spectra measured on a grating spectrometer but would be distributed across the spectrum (and would therefore be lower in the region of the line) if the spectrum were measured with a Michelson interferometer. Hirschfeld[19] has called this property the *distributive advantage* of Fourier spectroscopy.

5.12. AUXILIARY OPTICS

5.12.1. Source Optics

For all measurements made by FT–IR spectrometry, the radiation from the source must be collimated to the extent given by equation (5.20), and then passed through the interferometer and focused on a detector. For absorption and reflection spectroscopy, an additional focus, at which the sample is held, must be located somewhere between the source and the detector.

For far-infrared spectrometry at medium resolution, Ω_{max} is usually so great that the solid angle of the beam is limited in practice to a value less than Ω_{max} by the size of either the source, the sample or the detector. However, for mid-infrared spectrometry $\bar{\nu}_{max}$ is usually so large that the components of the spectrometer do not restrict the solid angle to a value less than Ω_{max}, and an aperture has to be placed at a focal plane in order that the desired resolution is attained across the entire spectral range. A typical optical arrangement whereby the solid angle of the beam passing through the interferometer can be changed is shown in Figure 5.10. The diameter of the aperture is either selected by the operator or by the computer software using the values of $\bar{\nu}_{max}$ and $\Delta\bar{\nu}$ for that measurement. It should be noted that this aperture stop serves exactly the same purpose as the entrance slit of a monochromator.

For emission spectroscopy a similar stop is needed, and a provision for changing the solid angle of the beam should always be included somewhere in the path from the source to the detector. When *remote* emission spectra are measured using a telescope to collect the radiation, this field stop is best located in the telescope itself.

5.12.2. Absorption Spectroscopy

For absorption spectroscopy the sample is held at a focus that is commonly formed using a long-focal-length off-axis paraboloidal mirror. The diameter of the focus at the sample position is usually determined by the field stop in the source optics, and it may be noted that the image is round, as opposed to the rectangular image found with monochromators. By a judicious choice of the focal length of the paraboloidal mirror, sampling accessories (beam condensers, ATR units, long path gas cells, etc.) that have

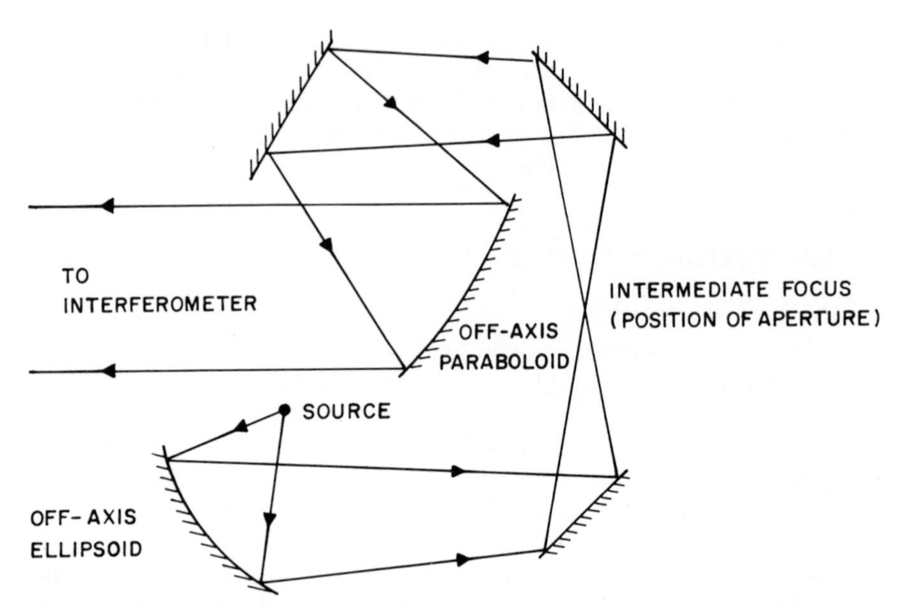

FIGURE 5.10. Source optics for the Digilab FTS® spectrometers; by placing apertures of the appropriate diameter at the intermediate focus, the throughput can be optimized for any combination of $\bar{\nu}_{max}$ and $\Delta\bar{\nu}$.

been designed for grating spectrometers can be used with FT–IR spectrometers.

The sample focus is usually formed in the path between the interferometer and the detector (rather than between the source and the interferometer) for two principal reasons. First, the beamsplitter can sometimes act as a filter for undesirable wavelengths. For example, Mylar beamsplitters will absorb much of the ultraviolet radiation emitted from the high-pressure mercury lamps used for far-infrared spectroscopy. Similarly, germanium or silicon beamsplitters will not transmit near-infrared radiation so that the sample is kept much cooler than if it were held directly in front of the source. Second (and perhaps more important), by placing the sample after the interferometer, any radiation emitted from the sample is not modulated by the interferometer and is therefore not measured as part of the AC interferogram. This is of particular importance when hot samples or cells are being measured, for example in GC–IR (see Chapter 6).

5.12.3. Reflection Spectroscopy

Reflection spectra can be measured on most instruments designed for absorption spectroscopy merely by placing an accessory for specular reflectance or attenuated total reflectance (ATR) in the beam at the sample focus. Some far-infrared Fourier spectrometers have been designed to allow reflection spectra to be measured *without* the need for a sampling accessory.[20]

The measurement of *diffuse* reflection spectra, i.e., reflection spectra from samples that scatter the incident radiation over a wide angle, can perhaps best be performed on a special-purpose instrument using an integrating sphere. The sample is held at the surface of a sphere that is coated with a diffusely reflecting layer of a material of very high reflectance. Radiation reflected from the sample is thereby "integrated" over the surface of the sphere. A detector is positioned somewhere in the sphere and even though only a small proportion of the radiation reflected from the sample is measured, it has been shown that very accurate measurements of reflectance spectra can be made. A Fourier transform spectrometer for diffuse reflectance spectroscopy using an integrating sphere is commercially available from the Willey Corporation (Melbourne, Florida).

5.13. DATA SYSTEMS

Data system requirements for FT–IR spectrometers vary tremendously with the nature of the experiment being performed. For example, interferograms for far-infrared spectroscopy at low resolution (say 4 cm^{-1}) will usually require less than 1000 data points, interferograms for mid-infrared spectroscopy at medium resolution (say 0.5 cm^{-1}) can require 32,000 data points to be collected, while interferograms for ultrahigh resolution ($\sim 10^{-3}$ cm^{-1}) can require more than 10^6 data points. In each case different data systems are required.

5.13.1. Far-Infrared Spectroscopy

With the relatively small transforms required for far-infrared spectroscopy, there is little need for a large computer memory in the data system. Typically data systems for far-infrared FT spectrometers have 4–8K of core, and a small magnetic tape recorder (often a cassette) may be interfaced if spectra are to be stored.

For slow-scanning interferometers, a *real-time Fourier transform*[20,21] can be performed between each input point, so that the spectrum may be displayed and constantly updated during data acquisition. One of the advantages of this method of computation is that a complete long interferogram may be stored while a small frequency region can be observed. A valid decision can therefore be made as to when the scan should be stopped in terms of the trade-off between resolution and S/N. When the scan is terminated, the interferogram is zero-filled[22] to produce $2N$ data points (N is an integer), and a fast Fourier transform is performed so that the complete spectrum from zero to $\bar{\nu}_{max}$ is computed. This type of data system is found on the Polytec FIR-30 far-infrared spectrometer.

5.13.2. Mid-Infrared Spectroscopy

In view of the length of most mid-infrared interferograms, the memory requirements for mid-infrared FT spectrometers are more severe than for far-infrared FT spectrometers. Two approaches have been taken with respect to the hardware of these systems, each requiring the use of a minicomputer and a disk. The first[(23)] involves the use of a minicomputer with a large core (40K words of 20-bit length), so that transforms of up to 16K points can be carried out in core; the disk in this case is used primarily for the purpose of spectral storage. The second[(24)] approach is to use less core memory (often only 4K words of 16-bit length) and to use the disk much more extensively. Such a system is more versatile than the first system, since any length of interferogram may be transformed in single or double precision, but it is also considerably slower than the first, core-based system. When interferograms with more than 16K data points are to be transformed on systems of the first type, data do have to be transferred to and from the disk during the FFT, but far less frequently than for small-core systems, so that for a given length of interferogram large core data systems invariably require the shortest computation times.

Recently a system has been introduced[(25)] in which a hardwired FFT is incorporated. This system can transform a 256K word $\times$ 32-bit array in 3 minutes, and finish all processing including phase correction, interpolation and unpacking in less than 10 minutes. In addition to the hardwired FFT processor, this system has 32K of core and 5 million words of disk storage.

5.13.3. Ultra-High-Resolution Spectroscopy

The highest resolution FT–IR spectrometers can require more than a million input points to be transformed. While such computations can be performed on a dedicated data system, perhaps using several disks, most of the work to date has been performed off-line on large computers. DeLouis[(26)] has described Fourier transform programs of large (2048K) interferograms using an IBM 360/75 computer with 256K $\times$ 32-bit words. A very fast program for performing transforms of up to 128K real samples is available for this computer, so that the complete data array is split up into subsequences each of which has 128K words and can be transformed in core using this program. The complete spectrum can be obtained very rapidly by applying the *decimation in time* technique of Connes[(27)] and using a direct-access disk storage unit to carry the rest of the data.

5.14. DUAL-BEAM FOURIER TRANSFORM SPECTROSCOPY

One of the main problems for chemical spectroscopists using FT–IR is the detection of extremely weak ($<0.1\%$) mid-infrared absorption bands. If slow-scanning interferometers are used with almost any type of detector or if rapid-scanning interferometers are used with photodetectors, the interferogram will almost certainly be digitization noise limited unless a very narrow band pass filter is placed in the beam to eliminate the contribution to the signal at zero retardation of these wavelengths away from the band of interest. The use of a filter is rarely of any practical application since it limits the region in which one can observe the band and also reduces Fellgett's advantage of an interferometer. In practice this results in the fact that weak bands are usually measured using a rapid-scanning interferometer with the rather insensitive TGS detector and signal-averaging over a long period.

There exists a little-used method by which weak absorption bands can be measured using an intense continuous source, a rapid-scanning interferometer, and a sensitive photodetector, without encountering the problem of digitization noise. This method is either known as *optical subtraction* or *dual-beam Fourier transform infrared spectroscopy*. As the name implies, both beams emerging from the interferometer (A and B in Figure 5.1) are passed onto the same detector and the signal from one beam at least partially cancels the signal from the other beam.

If a sample of tansmittance $T(\bar{\nu})$ is placed in beam B, the two individual interferograms, $V(\delta)_A$ and $V(\delta)_B$, are given by

$$V(\delta)_A = \int_{-\infty}^{+\infty} B(\bar{\nu}) \cos(2\pi\bar{\nu}\delta - \theta_{\bar{\nu}})\, d\bar{\nu} \tag{5.37}$$

$$V(\delta)_B = -\int_{-\infty}^{+\infty} T(\bar{\nu}) B(\bar{\nu}) \cos(2\pi\bar{\nu}\delta - \theta_{\bar{\nu}})\, d\bar{\nu} \tag{5.38}$$

Thus the resultant interferogram is given by

$$V(\delta)_A + V(\delta)_B = \int_{-\infty}^{+\infty} [1 - T(\bar{\nu})]\, B(\bar{\nu}) \cos(2\bar{\nu}\delta - \theta_{\bar{\nu}})\, d\bar{\nu} \tag{5.39}$$

The greater the transmittance of the sample, the more closely the resultant dual-beam interferogram approaches zero. Thus the amplitude of a dual-beam interferogram at zero retardation is primarily a measure of the radiation *absorbed* by the sample (as opposed to the radiation transmitted in the single-beam case). Thus the greater the transmittance of the sample, the smaller is $(S/N)_1$, so that for weakly absorbing samples sensitive detectors can be used without limiting the detection limit by digitization noise.

In practice, a perfect optical subtraction is rarely, if ever, achieved when no sample is present in either beam, primarily because of absorption of the input radiation by the beamsplitter. However $(S/N)_1$ can still be reduced to such an extent that the spectrum is not digitization noise limited with reasonably sensitive photodetectors, such as mercury cadmium telluride. The dual-beam technique is of particular importance in *rapid* trace analysis, and should be applied to an increasing extent in the on-line measurement of the infrared spectra of components of mixtures separated by gas chromatography and high-performance liquid chromatography.

REFERENCES

1. R. J. Bell, *Introductory Fourier Transform Spectroscopy,* Chapter 5, Academic Press, New York, 1972.
2. R. J. Anderson and P. R. Griffiths, *Anal. Chem.* **47**, 2339 (1975).
3. R. J. Bell, *Introductory Fourier Transform Spectroscopy,* Chapter 11, Academic Press, New York, 1972.
4. W. H. Steel, Interferometers for Fourier spectroscopy, *Aspen Int. Conf. Fourier Spectrosc., 1970,* Air Force Cambridge Research Laboratories Special Report No. 114 (April, 1971), p. 43.
5. T. Hirschfeld, Signal–noise ratios in Fourier transform spectrometry, *1976 Pittsburgh Conf. Anal. Chem. Appl. Spectrosc. (Cleveland, Ohio),* paper no. 385.
6. J. Chamberlain, *Infrared Phys.* **11**, 25 (1971).
7. J. Chamberlain and H. A. Gebbie, *Infrared Phys.* **11**, 57 (1971).
8. J. Connes, P. Connes, and J. P. Maillard, *J. Phys.* **28**, C2: 120 (1967).
9. G. Guelachvili and J. P. Maillard, Fourier spectroscopy from 10^6 samples, *Aspen Int. Conf. Fourier Spectrosc., 1970,* Air Force Cambridge Research Laboratories Special Report No. 114 (April, 1971), p. 151.
10. M. Born and E. Wolf, *Principles of Optics,* Macmillan, New York, 1964.
11. P. Vogel and L. Genzel, *Infrared Phys.* **4**, 257 (1964).
12. H. Sakai, Consideration of the signal-to-noise ratio in Fourier spectroscopy, *Aspen Int. Conf. Fourier Spectrosc., 1970,* Air Force Cambridge Research Laboratories Special Report No. 114 (April, 1971), p. 19.
13. P. R. Griffiths, *Chemical Infrared Fourier Transform Spectroscopy,* pp. 120–125, Wiley-Interscience, New York, 1975.
14. P. L. Richards, *J. Opt. Soc. Am.* **54**, 1474 (1964).
15. M. J. E. Golay, *Rev. Sci. Instrum.* **18**, 347, 357 (1947).
16. J. E. Stewart, *Infrared Spectroscopy: Experimental Methods and Techniques,* Chapter 11, Marcel Dekker, New York, 1970.
17. Cathodeon Ltd., Nuffield Road, Cambridge, CB4 1TF, England.
18. J. G. Moehlmann, J. T. Gleaves, J. W. Hudgens, and J. D. MacDonald, *J. Chem. Phys.* **60**, 4790 (1974).
19. T. Hirschfeld, *Appl. Spectrosc.* **30**, 68 (1976).
20. F. Levy, R. C. Milward, S. Bras, and R. leToullec, "Real-time" far infrared Fourier spectroscopy using a small digital computer, *Aspen Int. Conf. Fourier Spectrosc., 1970,* Air Force Cambridge Research Laboratories Special Report No. 114 (April, 1971), p. 331.
21. H. Yoshinaga, S. Fujita, S. Minami, Y. Suemoto, M. Inoue, K. Chiba, K. Nakano, S. Yoshida, and H. Sugimori, *Appl. Opt.* **5**, 1159 (1966).
22. P. R. Griffiths, *Appl. Spectrosc.* **29**, 11 (1975).
23. Nicolet Instrument Corp., 5225 Verona Road, Madison, Wisconsin, 53711.

24. Digilab Inc., 237 Putnam Avenue, Cambridge, Massachusetts, 02139.
25. IDAC Division of Carson Systems, Inc., 4630 Campus Drive, Newport Beach California, 92660.
26. H. DeLouis, Fourier transformation of a 10^6 samples interferogram, *Aspen Int. Conf. Fourier Spectrosc., 1970,* Air Force Cambridge Research Laboratories Special Report No. 114 (April, 1971), p. 145.
27. J. Connes, Computing problems in Fourier spectroscopy, *Aspen Int. Conf. Fourier Spectrosc., 1970,* Air Force Cambridge Research Laboratories Special Report No. 114 (April, 1971), p. 83.

Chapter 6

Infrared Fourier Transform Spectrometry: Applications to Analytical Chemistry

Peter R. Griffiths

6.1. FT–IR VERSUS GRATING SPECTROPHOTOMETERS

Before the applications can be discussed for which the use of FT–IR spectrometers give the greatest advantage over grating spectrometers, we will first discuss the various factors leading to differences in the performance of the two types of spectrometer. Obviously the performance of *any* type of spectrometer is dependent on the nature of its components, but in this section we shall only compare typical commercially available mid-infrared spectrometers. Thus for the purpose of illustration, the FT–IR spectrometer will contain a rapid-scanning Michelson interferometer with 2-in.-diameter mirrors, a Ge : KBr beamsplitter and a TGS detector. The grating spectrometer will be an optical null instrument with 5.2 × 5.2 cm interchangeable gratings and a thermocouple detector.

6.1.1. Fellgett's Advantage

Fellgett's advantage is, of course, one of the principal advantages of FT–IR over grating spectrometers. As discussed in Chapter 3, for measurements taken at *equal resolution* an in an *equal measurement time* with the *same detector* and on an instrument with the *same optical throughput and efficiency*, the S/N of spectra taken on the Fourier spectrometer will be

Peter R. Griffiths • Department of Chemistry, Ohio University, Athens, Ohio 45701

$M^{1/2}$ times greater than the S/N of spectra measured on the grating spectrometer (neglecting computing and plotting time), where M is the number of resolution elements. Of course, FT–IR and grating spectrometers do not use the same detector as grating spectrometers, nor do they have equal optical throughput or equal efficiency. The effect of these differences has been discussed in some detail by Griffiths *et al.*[1] and will be briefly discussed in this section. Before proceeding to this discussion, however, the implications of Fellgett's advantage to chemical infrared spectrometry should be mentioned.

While measurements made in the same time will theoretically yield an advantage of $M^{1/2}$ in S/N for FT–IR spectrometers, measurements at the same S/N can be made M times faster on a Fourier spectrometer. Therefore if there is a limited time for the measurement, or if the measurement takes an extremely long time on a grating spectrometer, there is a very definite advantage in using a Fourier spectrometer. Thus the best results for on-line GC–IR have been found using interferometers, and it is probable that when the infrared spectra of components of mixtures separated by HPLC are measured on-line, Fourier spectrometers will again be used for this application. Kinetic studies can also be performed using FT–IR spectrometers. Applications where the source is weak and the measurement time on grating spectrometers can be inordinately long include astronomical spectroscopy, remote sensing, and far-infrared spectroscopy, and FT–IR has been used beneficially for these and similar applications.

6.1.2. Jacquinot's Advantage

There is another important theoretical advantage of an interferometer over a grating spectrometer; this one relates to the increased *optical throughput* of an interferometer that can be tolerated before the spectral resolution is degraded. It is known as Jacquinot's advantage, in honor of the French scientist who first recognized its spectroscopic implications. The throughput of an interferometer Θ^{I} is given by the product of $\Omega_{\max}$ [given in equation (5.20)] and the area of the mirrors in the interferometer A^{I}:

$$\Theta^{\mathrm{I}} = \frac{2\pi \Delta\bar{\nu} A^{\mathrm{I}}}{\bar{\nu}_{\max}} \quad (\mathrm{cm}^2\ \mathrm{sr}) \tag{6.1}$$

The throughput of a grating spectrometer Θ^{G} is given by

$$\Theta^{\mathrm{G}} = \frac{h\,\Delta\bar{\nu} A^{\mathrm{G}}}{f\bar{\nu}^2 a} \quad (\mathrm{cm}^2\mathrm{sr}) \tag{6.2}$$

where h is the height of the entrance and exit slits of the monochromator, f the focal length of the collimator, a the grating constant, and A^{G} the area of the grating.

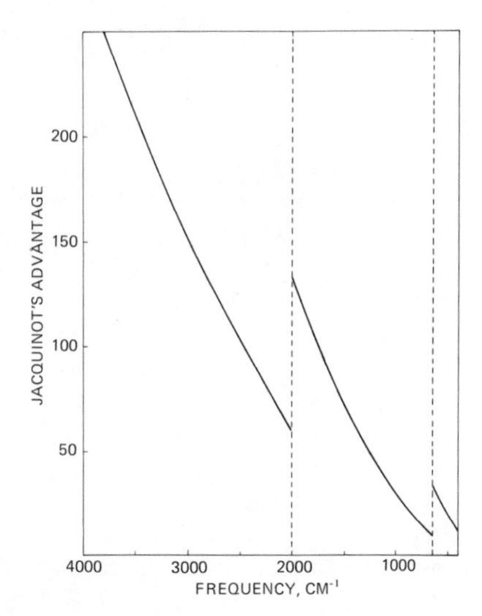

FIGURE 6.1. Ratio of the throughput of a commercial Fourier spectrometer (Digilab FTS® series) to that of a commercial dispersive spectrometer (Beckman 4200 series). (Reproduced from ref. 1 by permission of the Society for Applied Spectroscopy; copyright 1977.)

Thus the value of Jacquinot's advantage at any frequency $\bar{\nu}$ is given by

$$\frac{\Theta^I}{\Theta^G} = \frac{2\pi A^{\mathrm{I}} f a \bar{\nu}^2}{h A^{\mathrm{G}} \bar{\nu}_{\mathrm{max}}} \tag{6.3}$$

All the terms in the above expression remain constant except for a (since most grating spectrometers use interchangeable gratings) and $\bar{\nu}$, and it should be observed that the numerical value of Jacquinot's advantage is independent of resolution provided that the solid angle of the beam passing through the interferometer is given by equation (5.20) and the throughput is not limited by any other components of the spectrometer. The magnitude of Jacquinot's advantage for two commercial spectrometers is given in Figure 6.1.

6.1.3. Effect of Detector Performance

It is apparent that the combination of Fellgett's and Jacquinot's advantages should lead to an enormous difference between the performance of FT–IR and grating spectrometers, especially at high frequency. However, in practice this large advantage may not be realized for mid-infrared spectrometers due to the difference in the performance of TGS and thermocouple (TC) detectors.

For both detectors the D^* falls off as the modulation frequency increases, even though for a given wavelength the D^* of both detectors is approximately independent of wavelength. The frequency at which the beam of a grating spectrometer is modulated is constant for all wavelengths and typically has a value of approximately 15 Hz. The modulation frequencies of the components of an interferogram from a rapid-scanning interferom-

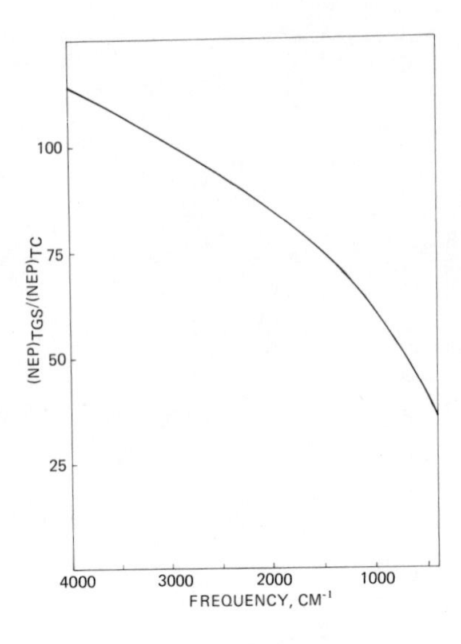

FIGURE 6.2. Ratio of the NEP of the triglycine sulfate detector normally used in FT–IR spectrometers to that of the thermocouple detector normally used in grating spectrometers. The NEP of each detector is approximately constant for all mid-infrared wavelengths if the modulation frequency is constant, but falls as the modulation frequency increases. The range of modulation frequencies of radiation passing through the interferometer is given in Table 5.1, while all radiation is modulated at the same frequency in the grating spectrometer. (Reproduced from ref. 1 by permission of the Society for Applied Spectroscopy; copyright 1977.)

eter vary from about 100 Hz to greater than 1 kHz (see Table 5.1). At *low* modulation frequencies the TC detector is about an order of magnitude more sensitive than TGS, and this advantage further increases for high-frequency radiation (see Figure 6.2). The difference in the performance of the two types of detectors goes a long way toward offsetting the large value of Jacquinot's advantage shown in Figure 6.1.

6.1.4. Other Differences

There are several other factors that cause smaller additional differences in the performance of FT–IR and grating spectrometers. The most important difference is between the efficiency of beamsplitters and gratings. As discussed in the previous chapter, beamsplitters have a high efficiency over a wide range (typically between 700 and 3800 cm^{-1} for mid-infrared interferometers), but outside this range their efficiency drops off markedly. Gratings have smaller ranges over which they operate at high efficiency but can be automatically interchanged in order to keep the efficiency at a fairly high level across the complete spectrum. Automatically interchangeable beamsplitters have not yet been designed for mid-infrared FT–IR spectrometers, so that the efficiency of gratings will usually be greater than that of beamsplitters at the high and low ends of the spectrum, but is somewhat lower than the efficiency of beamsplitters across the central portion of the spectrum.

Other factors contributing to performance differences between the two instruments include the fact that whereas transmittance spectra are usually measured directly on grating spectrometers, in FT–IR spectrometry the sample and reference interferograms must be measured at separate times, effectively doubling the measurement time. The overall measurement time is also increased by having to compute and plot the spectrum after the interferograms have been measured, although several systems now let these steps take place during the measurement of the interferogram of a subsequent sample. The measurement time for interferograms is also increased by the "dead time" in which the mirror is retraced during signal-averaging.

An additional disadvantage for the *monochromator* is seen in the fact that optical filters are required to minimize stray light, whereas in mid-infrared FT–IR spectrometers no optical filter is generally used, since efficient electronic filtering of the signal avoids folding effects, and no other source of stray light is present. The sum of all the advantages and disadvantages of a commercial Fourier spectrometer over a commercial grating spectrometer have been calculated,[1] and the result is shown for three values of resolution in Figure 6.3.

6.1.5. Implications

From the above discussion it can be seen that there is little advantage to using an FT–IR spectrometer over a grating spectrometer for low- and medium-resolution measurements of "conventional" samples such as KBr disks, mulls, capillary films, and solutions in infrared transmitting solvents such as CCl_4 and CS_2, since these spectra can be measured quite rapidly on a grating spectrometer. The shorter measurement time on the interferometer is often offset by the time taken to compute and plot the spectrum. For most other situations, however, FT–IR spectrometers can be used to

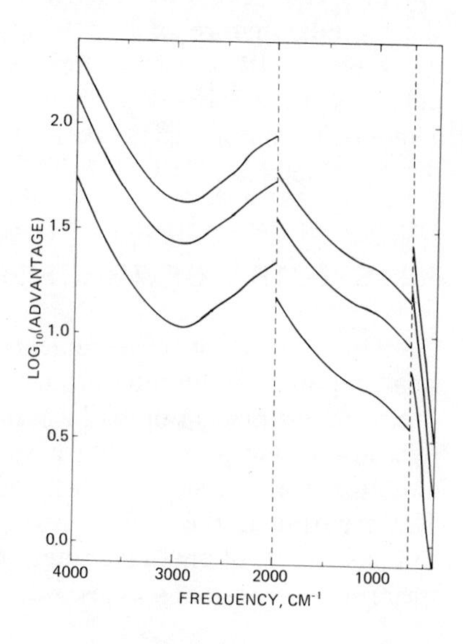

FIGURE 6.3. The total advantage of an interferometer over a grating spectrometer at 8-cm^{-1} (lower trace), 4-cm^{-1} (center trace), and 2-cm^{-1} resolution (upper trace) calculated using the components and dimensional parameters of the Digilab FTS-14 and Beckman 4240 as models. (Reproduced from ref. 1 by permission of the Society for Applied Spectroscopy; copyright 1977.)

advantage. The applications where Fourier spectrometers can best be used are summarized below:

(a) measurements of transient samples or sources, where the measurement time is limited and the spectrum can be computed and plotted *after* the sample has disappeared;

(b) measurements where the energy is very low, either because of the nature of the source (as in far-infrared spectrometry, low-temperature emission spectrometry, remote measurements, or astronomical spectroscopy) or because of samples that either absorb or scatter the incident radiation; in this case the *data acquisition time* required to attain a certain S/N on a grating spectrometer can be substantially reduced by FT–IR, and the acquisition time is usually still much longer than the computation and plotting time;

(c) measurements at high resolution where Fellgett's advantage is very large;

(d) measurements where the excellent frequency reproducibility of a laser-referenced Fourier spectrometer is important.

It has been frequently suggested that the presence of a dedicated data system gives FT–IR spectrometers another advantage, since array operations (subtraction, division, etc.) can be performed on spectra that have just been measured. In truth, however, dedicated data systems can also be interfaced to grating spectrometers, and precisely the same operations can be performed on spectra measured on grating or FT–IR spectrometers.

Nevertheless there are other, perhaps less important advantages derived through the use of FT–IR spectrometers. These include the lack of stray light in the spectrum, the ease by which the instrument line shape function can be controlled (by changing the apodization function), and the modular nature of FT–IR spectrometers. This last property enables users of FT–IR spectrometers to construct optical layouts for special purposes, such as GC–IR and remote measurements, much more easily than if a monochromator had to be removed from a double-beam grating spectrophotometer.

6.2. SPECTRA OF TRANSIENT SPECIES

One of the first measurements of a transient *source* ever taken with a small rapid-scanning interferometer was made in 1961 when the emission spectrum of a nuclear fireball was measured using an airborne instrument.[2] It is interesting to note that a very large monochromator that was also mounted in the same aircraft failed to record the spectrum of the fireball. This experiment, the results of which were classified at the time, was a fine demonstration of the advantages of an interferometer for measuring the spectrum of transient species.

6.2.1. GC–IR

Analytical chemists have long recognized that rapid-scanning interferometers can be used to measure the infrared spectra of peaks eluting from a gas chromatograph without trapping the sample (GC–IR). For some years the *feasibility* of GC–IR was demonstrated using both the early low-resolution interferometers[3,4] and later laser-referenced higher resolution interferometers.[5,6] The early experiments with both types of instrument demonstrated detection limits between 1 and 10 μg for most organic compounds. Only very recently[7] has the feasibility of routinely measuring spectra from 100 ng of most compounds (without trapping the sample) been demonstrated.

The central component of most GC–IR systems is a long (25–100 cm), narrow (1.5–3.0 mm diameter) gold-coated flow-through gas cell. This cell is usually known as a *light-pipe* in view of the fact that the infrared beam undergoes several reflections as it travels down the cell. The volume of the cell must be less than the volume between the half-height points of the narrowest chromatographic peak so that GC peaks that are only just resolved by the chromatograph do not merge in the light-pjpe. Since the sample concentration is always very low in GC–IR, the pathlength of the light-pipe must be long to yield an observable absorption spectrum.

The smaller the diameter of the light-pipe, the more the beam is attenuated by reflection losses. With long, narrow light-pipes, the S/N of GC–IR spectra measured with a TGS detector may be too low to allow the detection of weak absorption bands. However if the light-pipe dimensions are chosen correctly, the use of an MCT detector will increase $(S/N)_I$ to a value just less than the full dynamic range of a 15-bit ADC, and therefore allow much weaker bands to be detected (see Figure 6.4). Work that is currently in progress[8] suggests that the application of dual-beam FT–IR techniques to GC–IR will result in a reduction in the detection limits to below 50 ng.

To a certain extent GC–IR can be considered to be a competitive technique to GC–MS, even though infrared spectra and mass spectra provide complementary information. If sensitivity were the only consideration, GC–MS has to be favored over GC–IR since on-line mass spectra can be measured at subnanogram levels. However, mass spectra are often difficult to interpret, and the structural information given by an infrared spectrum is often very helpful in determining the identity of a compound. The *combination* of GC–IR and GC–MS will provide the analytical chemist with a powerful tool for the qualitative analysis of complex mixtures.

6.2.2. LC–IR

Another chromatographic technique for which on-line spectroscopic information is often required is high-performance liquid chromatography.

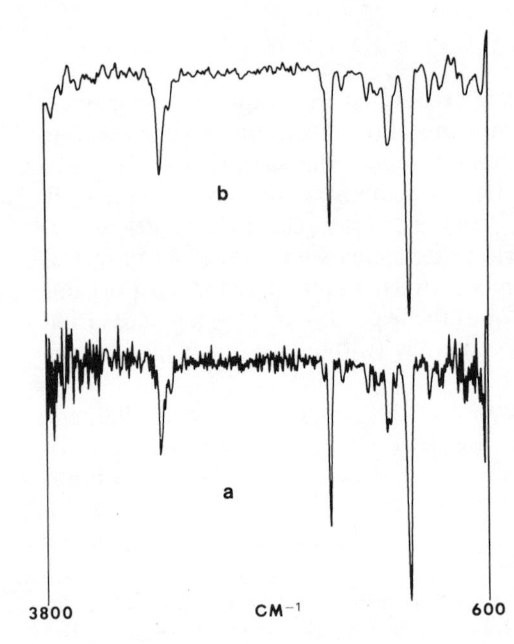

FIGURE 6.4. GC–IR spectrum of isobutyl methacrylate measured without trapping the sample, (a) with no spectral smoothing and (b) after smoothing. The spectrum was measured after 0.2 μl of a solution containing 1 μg/μl of the sample was injected into the chromatograph, and represents about 100 ng of sample in the light-pipe (200 ng of total sample). Note the smoothing allows several of the low-frequency bands to be unequivocally assigned but masks the isopropyl doublet near 1310 cm^{-1}. (Figure courtesy of D. Wall and A. W. Mantz, Digilab, Inc., from work to be published.)

The use of a Fourier spectrometer for the identification of HPLC peaks has been termed LC–IR. The principal difference between LC–IR and GC–IR lies in the transmission characteristics of the mobile phase. Whereas the mobile phase for GC is completely transmitting, the mobile phase for liquid chromatography never transmits 100%, and in the region of absorption bands the transmission may be very low. Consequently if solute spectra of HPLC peaks are to be obtained over the complete infrared spectrum, the pathlength of the flow-through cell has to be kept rather short. In practice a compromise has to be made between sensitivity and useful spectral range, since high solute sensitivity is only able to be achieved in "windows" in the solvent spectrum. At a pathlength of 100 μm, only a few bands in the spectrum of nonpolar solvents will absorb more than 90% of the incident radiation, so that solute bands will be able to be observed over most of the spectrum with this pathlength provided the solute is present at a high enough concentration.

Very little on LC–IR has yet been published although it is being studied in several laboratories. Kizer *et al.*[9] have shown some on-line spectra, and it appears from this and other unpublished work that the detection limits of LC–IR spectra measured with a Fourier spectrometer and a TGS detector are rarely less than about 100 μg for solutes eluted isocratically. When gradient elution techniques are used for HPLC, solvent compensation becomes more difficult and detection limits increase still further.

Two developments may increase the sensitivity of LC–IR measurements; both are necessary because of the strong absorption of the mobile phase. With a 100-μm cell, the transmittance in the window regions of the solvent spectrum is so great that the spectrum becomes digitization-noise limited if the interferograms are measured with an MCT detector. If the pathlength of the cell is increased, the high transmittance of most organic solvents between 1800 and 2700 cm^{-1} (where very few solute bands absorb) means that the $(S/N)_I$ will remain greater than the dynamic range of the ADC. Thus the small increase in sensitivity in the window regions is more than offset by the reduction in useful spectral range.

However the use of *dual-beam* FT–IR techniques should enable thin cells to be used (allowing a wide spectral range) while weak bands can be detected without encountering dynamic range problems. A disadvantage of the dual-beam system is that absorption due to the solvent must be precisely cancelled by placing a cell of the same thickness filled with pure solvent in the reference beam. With this condition it will be almost impossible to use such a system when gradient elution techniques are being used to elute the solutes.

The obvious answer to this problem is to remove the solvent, and once again work is in progress to this end. One approach that has been proposed is a pseudocontinuous on-line method in which the effluent from the chromatograph is sprayed into one of a series of light-pipes held in a carousel. The solute is deposited on the walls of the light-pipe while the solvent is evaporated and carried through by the nitrogen spray gas. The carousel is rotated after each peak elutes and the reflection–absorption spectrum of each deposited sample is measured after the carousel is rotated and while the next peak is being eluted. No systems of this type are currently commercially available, but it is very likely that some type of LC–IR system (not necessarily based on this principle) will be developed in the next couple of years.

6.2.3. Reaction Kinetics

In the past, kinetic information has been acquired using infrared prism and grating spectrophotometers by measuring the variation of absorbance of a single band as a function of time. This technique allows only one component to be monitored during a reaction, and the effect on the product and reactant concentrations of experimental parameters varying from run to run is difficult to follow. However with the use of a rapid-scanning FT–IR spectrometer, interferograms can be measured at approximately one-second intervals so that the variation of absorbance of bands due to *all* components in the reaction mixture can be monitored as a function of time. For rapid reactions, each successive interferogram is stored but for slower reactions several interferograms are signal-averaged before they are stored. Of course, when FT–IR is applied to kinetic measurements a considerable amount of

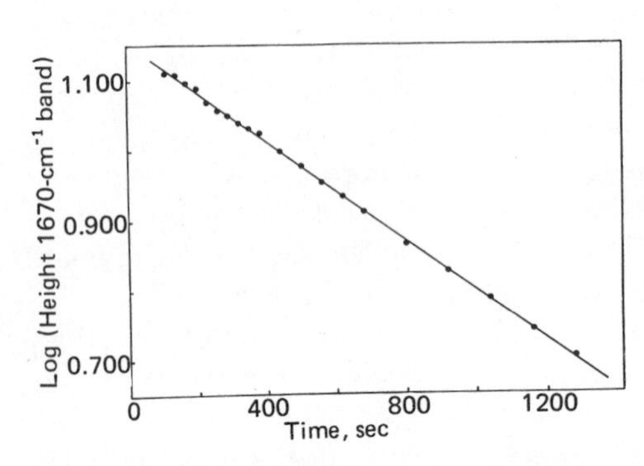

FIGURE 6.5. Plot of log (peak absorbance of 1670-cm^{-1} band) vs. time for the oxygen exchange reaction between ^{18}O–acetone and D_2O; the standard deviation of the least-squares slope is less than 1%. The value of the pseudo-first-order rate constant of the reaction calculated from the slope agreed well both with the value determined by monitoring the increase of the 1697-cm^{-1} ^{16}O–carbonyl band and the value found by noninfrared methods. (Reproduced from ref. 11, by permission of the Society for Applied Spectroscopy; copyright 1975.)

time is involved at the end of the experiments to compute the spectra from the interferograms that were measured during the reaction.

The first study in which a rapid-scanning interferometer was used to monitor a chemical reaction was described by Low *et al.*,[10] who showed that the spectrum of *cis*-1,2-dimethyldiborane changed shortly after its preparation, and that a previously reported spectrum of this compound was incorrect. More recently, Oertel *et al.*[11] have accurately determined the pseudo-first-order rate constant for the relatively slow exchange reaction between ^{18}O enriched acetone and $D_2{}^{16}O$ by monitoring the doublet in the carbonyl stretching region due to $C{=}^{18}O$ (at 1670 cm^{-1}) and $C{=}^{16}O$ (at 1697 cm^{-1}). A semilog plot of the variation of the peak absorbance of the 1670-cm^{-1} band with time was linear, with a standard deviation of less than 1%, and gave a value for the rate constant which was in good agreement with other methods (see Figure 6.5).

Some elegant work on faster reactions occurring in the gas phase has been reported in two papers by Lephardt and Vilcins. In their first paper,[12] the reaction of nitrogen dioxide with butadiene was monitored by measuring the interferograms at intervals of 6.5 s. Not only could the formation of the reaction product be followed but the disappearance of NO_2, N_2O_4, and butadiene could be monitored simultaneously. In a later report,[13] these authors demonstrated that methyl nitrite, which had been reported as a gas phase component of cigarette smoke, was in fact not present in fresh smoke but was formed as a result of aging. This work verified that the reaction probably involves the conversion of nitric oxide to nitrogen dioxide by air oxidation, followed by the reaction of nitrogen dioxide with methanol to form methyl nitrite. The variation of the absorbance of NO, NO_2, CH_3OH, and CH_3ONO as a function of time is shown in Figure 6.6.

Liebman *et al.*,[14] have studied the formation of pyrolysis and combustion products from polymer samples as a function of time. Polymer samples could either be very rapidly degraded, in which case the measurement time would be limited to a few seconds, or slowly pyrolyzed by gradually raising

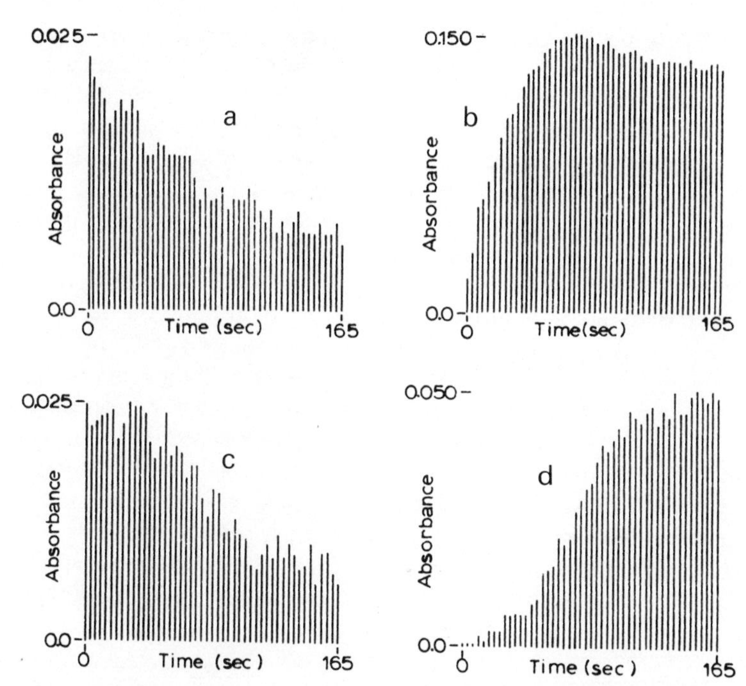

FIGURE 6.6. Aging process of cigarette smoke from nitrate-enriched burley tobacco: (a) Nitrogen oxide (1910 cm^{-1}), (b) Nitrogen dioxide (1629 cm^{-1}), (c) Methanol (1055 cm^{-1}), and (d) methyl nitrite (822 cm^{-1}). Corresponding lines for each component were obtained from the same spectrum. The spectra demonstrate that methyl nitrite is not present in fresh smoke and that methanol appears to be an intermediate in the formation of CH_3ONO. (Reproduced from ref. 13, by permission of the Society for Chemical Industry and the author; copyright 1975.)

the temperature from ambient to several hundred degrees in a fashion similar to temperature programming in gas chromatography. In each case a gas (helium or nitrogen for pyrolysis or air for combustion experiments) was passed over the sample, and the gaseous reaction products were flushed through a GC–IR light-pipe. The variation in absorbance of up to seven or eight products could be monitored using this technique.

Reactions with a half-life of approximately 10 s can be studied with commercial rapid-scanning interferometers. With faster reactions than this, the concentration of components can change appreciably during each scan. The effect will be to distort the absorption bands in the same way as if the frequency components in the interferogram where absorption due to the rapidly changing component is significant were multiplied by a function that varies with time in the same fashion as the absorption of the component. Lephardt and Vilcins[12] have termed this effect *frequency-dependent apodization*, although it may be noted that only if the component is being removed during the scan would the result be equivalent to apodization (i.e., reduction in the amplitude of the side lobes of the instrument line shape function). If the component is being rapidly generated, the effect is to increase the amplitude of any side lobes that are evident in the spectrum.

Recently, very fast reactions have been studied by FT–IR using a technique known as *time-resolved spectrometry.* The method was first studied by Murphy and Sakai,[15] who used a stepped-scanning interferometer to study the relaxation of vibrationally excited CO_2. At each position of the mirror, CO_2 molecules in the ground state were excited, and the emitted energy was monitored over a period of about 0.1 s; the signal was sampled four times during this period. Thus at the end of the measurement, four complete interferograms had been collected and the reaction had been initiated and allowed to proceed to completion at each stationary position of the mirror. Rapid-scanning interferometers are now being modified in order that reactions with half-lives of considerably less than 1 ms can be studied. Using this technique it is possible to measure the spectrum of components that are present in a flow-through reaction vessel as shortly as 50 μs after the initial excitation.[16] An example of the type of spectra that can be measured by this technique is shown in Figure 6.7. Although time-

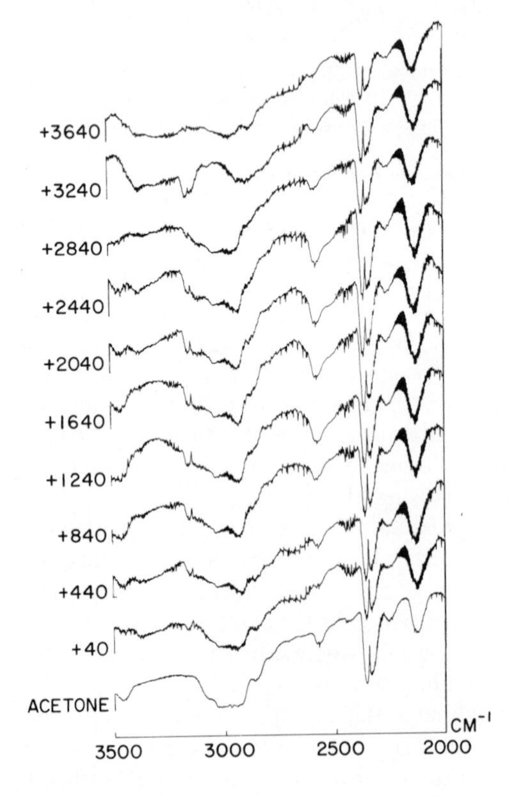

FIGURE 6.7. Portions of time-resolved spectra from photolyzed acetone at 85 Torr flowing through a gas cell at 45 cm^3/min and subjected to ultraviolet flash photolysis at 250 flashes per second. Interferogram points were collected, over a 2-μs window, 40, 840, 1240 μs, etc., after each flash. The rapid formation of CO (fine structure centered at 2149 cm^{-1}) is evident, and the feature around 3138 cm^{-1} has been assigned to ketene. The features between 2500 and 2800 cm^{-1}, which were not assignable to any species, show the interesting property of first appearing in emission; after 1.24 ms the species causing this structure is absorbing but at 3.64 ms the system is emitting again. [Reproduced from A. W. Mantz, *Appl. Spectrosc.* **30**, 459 (1976), by permission of the author and the Society for Applied Spectroscopy; copyright 1976.]

resolved spectrometry is still in its infancy, it is probable that this technique will become an important tool for studying very rapid reactions.

6.3. LOW-ENERGY ABSORPTION SPECTROMETRY

Many measurements have been made using FT–IR spectrometers for which the energy reaching the detector is so low that, were a grating spectrometer used, either a low *S/N* or an unnecessarily long measurement time would have resulted. It is beyond the scope of this chapter to report on all experiments of this type for which FT–IR spectrometers have been used to advantage. Instead a relatively brief discussion will be given with selected examples to provide the reader with a "feel" for the type of measurements that can be made.

6.3.1. Far-Infrared Spectrometry

The far-infrared region was the first spectral region in which FT–IR was used by a large number of chemists. Sources of far-infrared radiation are notoriously weak, and the severe optical filtering that must be used with far-infrared monochromators to prevent radiation diffracted from higher grating orders from reaching the detector further served to reduce the performance of far-infrared grating spectrometers. On the other hand, the very high optical throughput allowed for far-infrared interferometers coupled with the rather low mechanical tolerances make it relatively easy to measure far-infrared spectra interferometrically with an adequate signal-to-noise ratio. Since fewer than 1000 data points are usually required to achieve a reasonable resolution for much chemical spectrometry in the far infrared, the computing time was short even in the mid-1960s before the Cooley–Tukey fast Fourier transform algorithm had been applied to FT–IR. Thus data on topics as diverse as pure rotation spectroscopy, [17,18] collision-induced spectra,[19] internal rotation and potential barriers,[20,21] hydrogen bonding,[22] and charge transfer complexes[23] were obtained before 1970 using relatively inexpensive interferometers with off-line computers.

Perhaps the most important area to chemists that has been studied using far-infrared spectrometers is inorganic chemistry, and the study of metal–ligand vibrational modes has been particularly important. These modes often absorb below 200 cm^{-1}, which is the low-frequency limit on many modern grating spectrometers. Because low-frequency modes were sometimes neglected, spectra of several complexes have been improperly assigned in the past; Goldstein and Unsworth[24] have discussed the importance of correctly assigning low-frequency modes and the desirability of measuring the spectra of inorganic complexes interferometrically. Now that automated

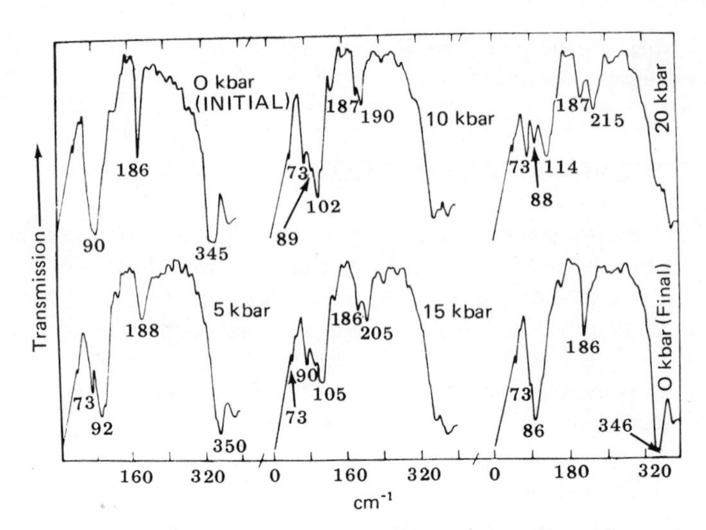

FIGURE 6.8. Far-infrared spectra of K_2PtCl_6 at pressures of up to an average applied pressure of 25 kbar. The spectra are unratioed outputs from a Beckman-RIIC FS-720 Fourier spectrometer; the band at 73 cm^{-1} is due to polyethylene, ν_3 (ν Pt–Cl) is at 345 cm^{-1}, ν_4 (δ Pt–Cl) is at 185 cm^{-1}, and the band at 90 cm^{-1} at zero applied pressure is a lattice mode. (Reproduced from reference 25 by permission of the Society for Applied Spectroscopy; copyright 1973.)

far-infrared FT–IR spectrometers are commercially available, many inorganic chemists are using these instruments on a routine basis. Among the recent work of interest in this area is a study of the effect of pressure on the absorption bands of complexes using a high-pressure diamond cell;[25] some spectra measured in this study are shown in Figure 6.8. Guillory and Smith[26] have studied the photolysis of matrix-isolated GeH_2Br_2 using a far-infrared Fourier spectrometer. Many more results of importance to inorganic chemistry are appearing in the literature each year.

Far-infrared spectra of gaseous species in the stratosphere have been measured using an airborne phase-modulated interferometer equipped with an indium antimonide detector by the group at the National Physical Laboratory in England.[27,28] Published spectra, which were measured at a resolution of approximately 0.06 cm^{-1}, showed more than 100 resolved emission lines from species present in the stratosphere including O_2, O_3, H_2O, HNO_3, NO_2, N_2O, and (possibly) SO_2.

6.3.2. Mid-Infrared Absorption Spectrometry

As we discussed earlier, FT–IR spectrometers can be used about as effectively as grating spectrometers for the measurement of the spectra of KBr disks, mulls, and other types of samples that are commonly measured by infrared spectrometers, since the reduced data acquisition time of an

FT–IR spectrometer will usually be offset by its computing and plotting time. In addition the number of these samples that can be run per hour is usually limited by the sample preparation time. Only in the analytical services laboratory of large companies employing several technicians to prepare samples of this type will the use of an FT–IR spectrometer prove beneficial for routine spectroscopy. However, when samples only transmit a small proportion of the incident energy, the use of an FT–IR spectrometer becomes more advantageous, since data acquisition time will usually far exceed the time taken to prepare the sample or to compute and plot the spectrum.

An important type of sample that falls into this category is one whose diameter is considerably less than the dimensions of the beam at the focus of a beam condenser. For example, if the spectrum of submicrogram quantities of organic compounds is required, the sample is often prepared as a 0.5-mm-diameter KBr microdisk, whose area is usually at least ten times smaller than the area of the focus. For nanogram quantities of sample, the spectrum will usually have to be displayed using large ordinate scale expansion so that a very low baseline noise level is needed. Several good examples of the type of spectra that can be measured from microsamples using a Fourier spectrometer have been given by King.[(29)] Anderson and Wilson[(30)] have described a technique for preparing samples on KBr microdisks of even smaller diameter by using aluminum oxide watch jewels of 0.35–0.15 mm diameter. A useful technique for measuring the spectra of small specks of materials or of inhomogeneities in polymer films is to use a mask with a small diameter behind which the sample is aligned. King[(29)] has shown the spectrum of a polyethylene film measured using a 50-μm-diameter aperture and a $6\times$ beam condenser on a Fourier spectrometer (see Figure 6.9). Most published spectra to date have been measured using a TGS detector. The use of an MCT detector should allow samples as small as 10 μm to be measured, although diffraction effects may start to become important.

When spectra of small amounts of volatile samples are to be measured, the use of KBr microdisks is not recommended since a large proportion of the sample can evaporate before the disk is pressed.[(31)] Samples of this type should be dissolved in a small amount of CCl_4 or CS_2 and measured in a solution cell with a small cross-sectional area but a fairly long (say 1 mm) pathlength.

Another type of cell for which a large proportion of the incident energy is not transmitted is the diamond cell. Diamond cells are not only useful for measuring the spectra of samples under very high pressure, but can also be used to reduce scattering from translucent samples and for flattening samples whose thickness is not uniform. This latter property may be of great application in forensic analyses. For example, the hair oil on a single strand of hair may be identified[(32)] by cutting the hair in two and washing the oil

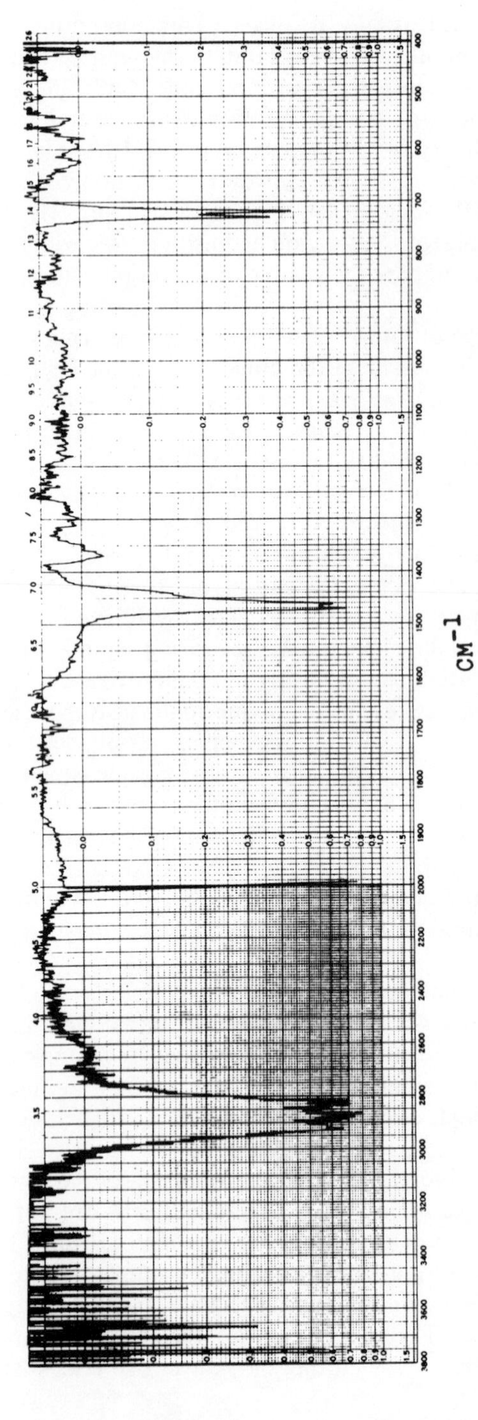

FIGURE 6.9. The spectrum of a polyethylene film masked with a 50-μm-diameter aperture held at the focus of a 6 × beam condenser measured using a Digilab FTS-14 spectrometer (with a TGS detector) after averaging 300 scans. (Reproduced from reference 29 by permission of the American Chemical Society; copyright 1973.)

off one piece. The spectra of each half are measured after compressing them in the diamond cell, and the spectrum of the hair oil may be found by using scaled absorbance subtraction routines (*vide infra*, pages 161–163) to remove bands due to the hair protein.

Another type of sample that can remove a large proportion of the incident radiation is illustrated by carbon-filled polymers where the filler strongly absorbs mid-infrared radiation. Koenig and Tabb[33] have published spectra of a carbon-filled elastomer showing that certain bands in the spectrum of the elastomer are shifted slightly on addition of the carbon black. Spectra of the same sample measured using a grating spectrometer have to be measured with the slits so wide (to allow sufficient energy to reach the detector) that the bands cannot be plotted with enough definition to allow this effect to be observed.

Other samples will scatter rather than absorb the incident radiation, and many catalyst studies have been made (using both grating and FT–IR spectrometers) where the sample scatters most of the radiation, especially at short wavelength. For example, Low *et al.* have measured the spectra of SO_2 adsorbed on CaO[34] and MgO[35] using a Fourier spectrometer, and like Koenig and Tabb,[33] these authors also state that they were unable to obtain results of the same quality with a grating spectrometer. The transmittance spectra of samples on TLC plates have been measured directly[36] by depositing the adsorbent on an infrared-transmitting substrate. The early spectra measured in this work were not of very high quality in the short-wavelength region, largely because of the poor performance of the TGS detector at high modulation frequencies. In later work[37] the scattering was reduced by treating the adsorbent with a mulling oil (Nujol or Fluorolube) of similar refractive index. Spectra of some chlorinated pesticides measured in this way show differences in the C–H stretching region, which have been correlated with the mechanism of the adsorption (see Figure 6.10).

The spectra of solvents do not usually have a low *average* transmittance; however, FT–IR spectrometers can give certain advantages to the measurement of solvent-compensated spectra of solutes, since solvent spectra will often have a very low transmittance in the region of absorption bands. Good spectra of solutes have been measured even when the peak transmittance of the solvent band is less than 0.1 %. The spectra of species in aqueous solution are perhaps more difficult to measure than spectra in any other solvent, especially in the important region around 1640 cm^{-1}, where the H–O–H bending mode of water absorbs. Although special cells have been developed for keeping the pathlength less than 10 μm in order to keep the transmittance of the 1640-cm^{-1} band greater than 5 %, they are not commercially available and in practice it is difficult to purchase sealed cells with a pathlength less than 20 μm.

We were forced to use just such a cell for some quantitative studies of proteins in aqueous solution.[38] In spite of the fact that the peak absorbance

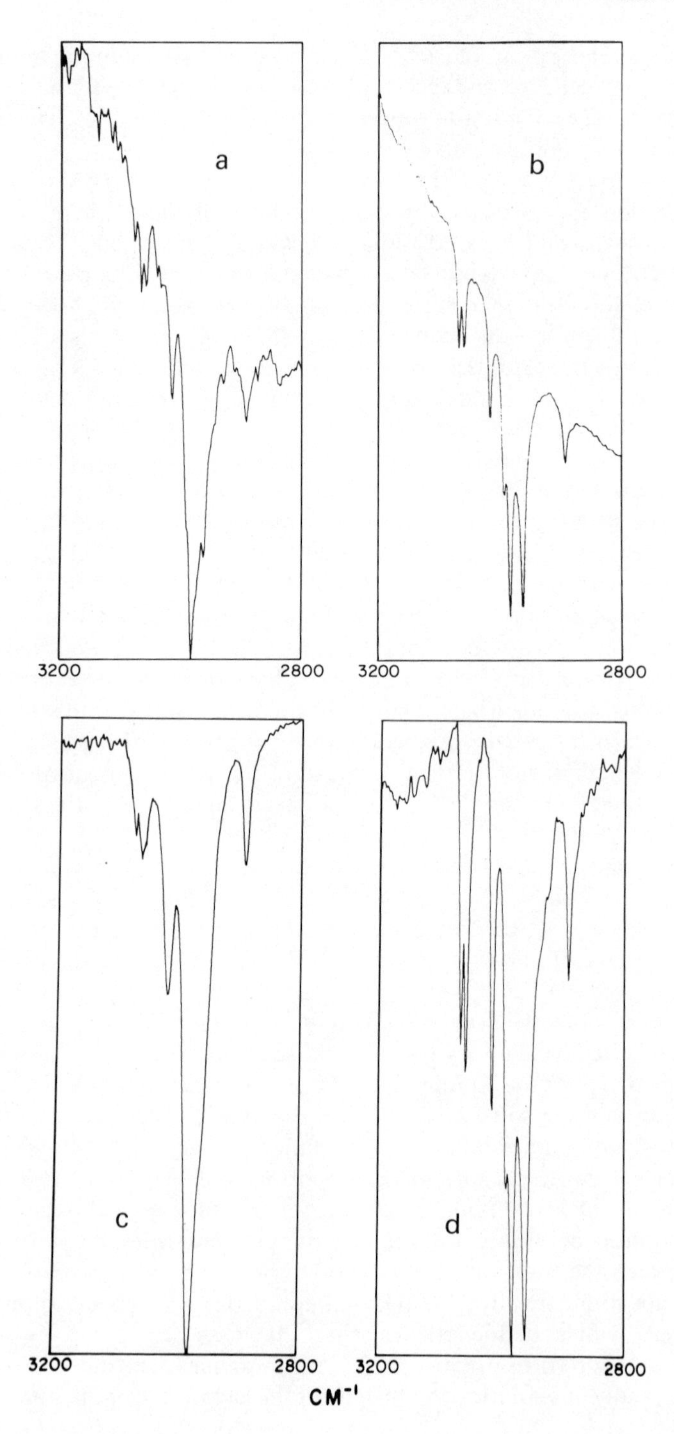
a
b
3200
2800
3200
2800
c
d
3200
2800
3200
2800
CM⁻¹

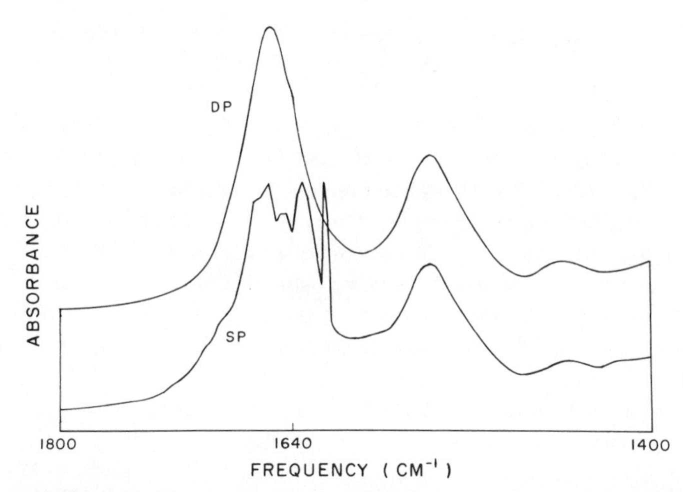

FIGURE 6.11. The spectrum of a 1% aqueous solution of myoglobin with the water absorption removed using the scaled absorbance subtraction routine with single-precision (SP) and double-precision (DP) arithmetic on a computer with a 16-bit word. Both spectra were measured under identical conditions (4-cm^{-1} resolution, 400 scans) in all other respects. The improvement in S/N where the background energy has been decreased by the 1640-cm^{-1} water band is most evident on changing to DP arithmetic.

of water at 1640 cm^{-1} was 2.3 using our cell, we were able to obtain accurate absorbance data on the amide I band, which absorbs near 1650 cm^{-1}. In this work the strong absorption due to the solvent was compensated by measuring the spectrum of water in the same cell and applying scaled absorbance subtraction techniques to compensate for the water bands. Initially, when all data were collected and computed in single precision on a data system with a 16-bit word, a rather poor S/N was found in the resulting protein spectra near 1640 cm^{-1}, even though our calculations showed that neither the interferograms nor the spectra should have been digitization-noise limited. When all the data were collected and computed in double precision, however, the S/N near 1640 cm^{-1} improved dramatically (see Figure 6.11). This improvement is probably caused by the improved preci-

FIGURE 6.10. The spectrum of the pesticide Endrin: (a) sorbed on silica gel, (b) sorbed on alumina, (c) dissolved in carbon tetrachloride, (d) dispersed in a KBr disk, measured on a Digilab FTS-14 spectrometer. The silica gel and alumina were deposited on AgCl sheets, and the sample was treated with Fluorolube to reduce scattering. The similarity of spectra (a) and (c) to (b) and (d), respectively, give evidence on the nature of the interaction of the chlorinated pesticides with silica gel and alumina. (Reproduced from reference 37 by permission of the Society for Applied Spectroscopy; copyright 1976.)

sion in calculating the absorbance of the solvent and solution spectrum when the transmittance is low gained through the use of double-precision techniques.

These data suggest that many problems in biochemistry for which infrared spectrometry has not been used in the past may now be studied through the use of FT–IR spectrometers. Not only can good spectra of relatively strong absorption bands (such as the amide I band) be measured in the region of intense water absorption, but also weak bands can be measured in regions where the water absorption is not quite so strong. For example, Alben *et al.* have carried out quantitative studies involving S–H bands of proteins[39,40] and also bands due to carbon monoxide complexed with hemoglobin.[40]

The use of Fourier spectrometers has also been shown to be beneficial for determining trace atmospheric components using long-path gas cells. For this application the property of FT–IR spectrometers of measuring spectra at a high S/N for measurements at a fairly high resolution is made use of, so that lines in the vibration–rotation spectra of different components in the same frequency region may be distinguished. Hanst *et al.*[41] have developed multiple-pass gas cells that allow an effective pathlength of nearly 1 km. With more than 50 reflections, reflection losses are quite severe, and when the full pathlength of these cells is used the energy is attenuated to such an extent that the use of a mercury cadmium telluride detector (for the region from 700 to 2000 cm^{-1}) and an indium antimonide detector (for the region from 2000 to 4000 cm^{-1}) is recommended.

These authors have estimated the detection limits of most common pollutants using 50 passes through a 10-m cell, assuming 98% reflectivity for the mirrors. The lowest detection limit is at a level of 0.02 ppb acetylene using the band at 735 cm^{-1}. Only for hydrogen sulfide (50 ppb) and nitric oxide (2 ppb) is the predicted detection limit greater than 1 ppb. For several molecules the detection limits are far less than the usual concentration of the compound in "unpolluted" air. For example, the detection limit for CH_4 is calculated as 0.6 ppb even though the atmospheric concentration never falls below 1 ppm. Similarly the detection limit for CO is 80 times less than its minimum atmospheric concentration.

With the combination of an interferometer, a long-path gas cell, and a sensitive detector, many different studies are made possible. By surrounding the cell with banks of lights, photochemical reactions can be studied for which the initial concentration of the reactant is in the ppm range. The formation of photochemical smogs in Pasadena, California, has been studied by installing a system on the roof of a laboratory at California Institute of Technology,[42] and a mobile unit with a slightly shorter cell has been used in other locations.

6.4. DIFFERENCE SPECTROSCOPY

In trace analysis it is generally necessary to remove the spectral information due to the major component(s) before the bands of the minor component(s) can be observed. In the past the spectrum of the major component has usually been compensated by placing an exactly equal optical thickness of the pure component in the reference beam of a double-beam dispersive spectrometer, so that only the absorption bands of the minor component are measured. This experiment is usually carried out for the identification of solutes by placing a variable-pathlength cell containing the pure solvent in the reference beam of the spectrometer and adjusting the pathlength of the cell until the solvent bands disappear from the spectrum. When the absorption due to trace amounts of solutes is very low ($<1\%$), large ordinate scale expansion is required to permit observation of the solute bands; thus the thickness of the variable-pathlength cell has to be controlled to much better than 0.1 μm. Most commercial variable-pathlength cells cannot be controlled this finely, in which case the sensitivity of the experiment may not be determined by the S/N of the spectrometer but rather by the mechanical precision of the sampling accessories.

Major components in the gas phase (in particular, atmospheric water vapor) have been compensated by placing a gas cell of the same pathlength in the reference beam and adjusting the partial pressure of the interfering component until all bands are compensated. On the other hand, if the sample is a solid (e.g., a polymer film), it is often very difficult to prepare a reference sample of the same thickness. Therefore, the application of difference spectroscopy using dispersive spectrometers has been primarily limited to the identification of components in the liquid or gas phase.

Since most FT–IR spectrometers have a computerized data system as an integral component of the instrument, it is a relatively simple matter for programs to be written that operate on spectra as soon as they have been computed from the interferogram. One of the most important of these programs is the *scaled absorbance subtraction* routine, which has been mentioned twice earlier in this chapter. This program can be used to achieve the same result as placing a sample in the reference beam of the spectrometer, but the thickness of the reference cell or sample does not need to be controlled as accurately. The typical experimental procedure is as follows:

(1) The transmittance spectrum of the mixture is measured, usually against an air reference, converted to the linear absorbance format, and stored.

(2) The absorbance spectra of each component of the sample that is to be compensated are then measured in the same way and stored.

(3) Starting with the spectrum of the component that is present in the

mixture at the highest concentration, the absorbance spectrum of each pure component is multiplied by a scaling factor and subtracted from the spectrum of the mixture; the scaling factor is chosen so that when the scaled spectrum is subtracted from the spectrum of the mixture, no bands due to that component are seen in the resulting spectrum. It may be shown that the best results are obtained when the thickness of each component is approximately equal to the effective thickness of that component in the sample cell. When Beer's law is obeyed by all the components in the mixture, the spectra of three or four components may be subtracted from the spectrum of a mixture in order to identify a trace component.

Koenig[43] has discussed the application of scaled absorbance subtraction routines to various problems in polymer chemistry. For example, the separation of the spectra of reacted and unreacted species has been illustrated by studies on the irradiation damage of polyethylene in air and nitrogen atmospheres, the oxidation of polybutadiene, and the vulcanization of rubber. Intermolecular interactions have been investigated by studying the effect of varying the concentration of plasticizers in poly(vinyl chloride). Spectra of the crystalline phase in heterophase polymers have been obtained by first measuring the absorbance spectrum of a sample in which the polymer is almost entirely in the amorphous phase (spectrum I). The absorbance spectrum of a second sample that is partially crystalline is then measured (spectrum II). Spectrum I is then scaled by a suitable factor and subtracted from spectrum II, leaving the spectrum of the crystalline phase

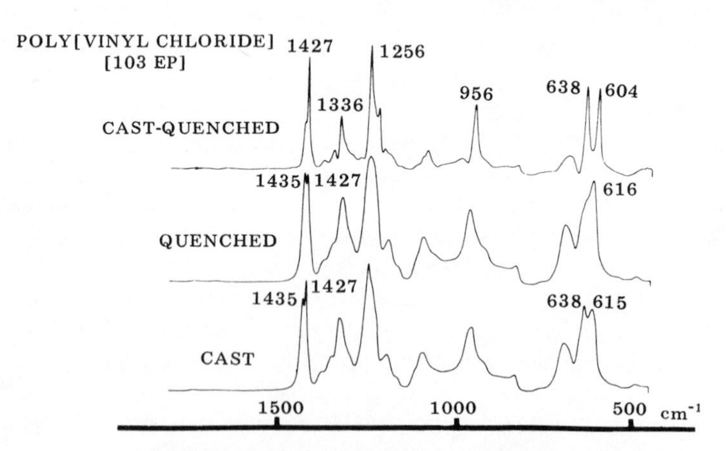

FIGURE 6.12. Absorbance spectra of poly (vinyl chloride). Bottom: cast film containing both the amorphous and crystalline phases. Middle: quenched film containing a greater proportion of the amorphous phase. Top: spectrum of crystalline PVC computed from these two spectra using scaled absorbance subtraction routines. (Reproduced from reference 43 by permission of the Society for Applied Spectroscopy; copyright 1975.)

(see Figure 6.12). In this way a spectrum of the crystalline phase may be obtained even though a completely crystalline sample of the polymer may never be prepared. In the same paper, Koenig has also given examples of the subtraction of surface effects from bulk effects and the study of spectral changes arising from mechanical effects.

It should be stressed that the use of scaled absorbance subtraction techniques is not limited to FT–IR spectrometers. The method can be used with any spectrometer with good frequency reproducibility and photometric accuracy that is interfaced to a data system. Similar programs have been written for grating spectrometers by Mattson.[44] However, most of the published work where this technique has been used was done using FT–IR spectrometers.

Hirschfeld and Kizer[45] have shown that when intermolecular interactions distort the bands of one or more componets, complete subtraction cannot be achieved. If the bands of any component shift, change their half-width, or change their absorptivity due to interactions with a second component, the result will be seen in the difference spectrum. For example if a band shifts slightly, "derivative-like" bands will be seen in the difference spectrum. Hirschfeld[46] has demonstrated how this effect can be used to determine the wavenumber reproducibility of any spectrometer. A spectrum of a sample with a sharp isolated absorption band is measured and stored. Spectra measured subsequently are subtracted from this spectrum, and the distance between the maximum and minimum values of the difference band allows the wavenumber shift to be calculated.

Gendreau and Griffiths[47] have further applied this effect to the determination of the proportions of closely related isomers whose spectra show shifts of less than 1 cm^{-1} for all bands. The absorbance spectrum of each pure isomer is measured at a known concentration and stored in the data system of the spectrometer; the absorbance spectrum of the unknown mixture of isomers at the same total concentration as the samples of the pure isomers is then measured. The two difference spectra found on subtracting the spectrum of each pure isomer from that of the mixture (with no scaling factor applied) are then calculated and plotted, and the ratio of the two isomer concentrations may be determined from the relative amplitudes of the maxima and minima at any frequency in the difference spectra.

6.5. REFLECTION SPECTROMETRY

There are several different types of reflection measurements that can be made by infrared spectroscopy, each of which has been studied by FT–IR. These may be defined as follows:

Specular reflection: reflection from the surface of smooth samples, such that the angle of incidence is equal to the angle of reflection.

Diffuse reflection: reflection from the surface of rough samples, where the reflected beam is scattered over a larger solid angle than the incident beam.

Attenuated total reflection: absorption of the evanescent wave of a beam that is totally reflected as it passes down the interior of crystal of high refractive index by species on the exterior surface.

Reflection–absorption: absorption of radiation by species on the surface of a highly reflecting specular reflector, especially metal surfaces.

It may be argued that the last two categories are not truly reflection measurements, but it is most convenient to discuss them along with specular and diffuse reflection.

Specular reflection spectra may only be obtained on fairly large crystals, and unless the reflection bands are very weak or spectra are required at high resolution there is little advantage to using a Fourier spectrometer for mid-infrared measurements. Solid-state chemists and physicists (especially the latter) have used far-infrared spectrometers fairly extensively for measuring specular reflection spectra, but this technique has few, if any, applications in analytical chemistry.

The type of samples of interest to analytical chemists for which reflection spectra might be useful tends not to be large and flat but rather are materials that are either difficult to characterize by absorption spectroscopy due to their physical nature or must be measured "as is," without sample preparation. These may include powders of different materials, crushed rocks, paper, or woven fibers. Each of these samples will scatter incident radiation and may be studied by diffuse reflection spectrometry.

For example, Low[48] has measured the reflection spectra of crushed minerals using an early rapid-scanning interferometer. The use of a diffuse reflection FT–IR spectrometer with an integrating sphere[49] has been demonstrated for a variety of samples including pharmaceutical products, rocks and minerals (see Figure 6.13), paper, and cloths. This type of instrument is the only commercial spectrometer on which *total* reflectance data

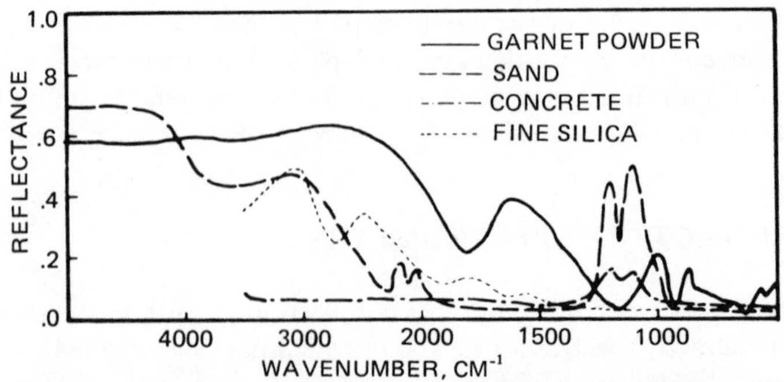

FIGURE 6.13. Total reflectance spectra of several minerals measured using the Willey Model 318 Reflectance Spectrometer. (Reproduced by permission of the Willey Corporation.)

can be accurately obtained in the mid-infrared. It is photometrically very accurate and appears to have considerable potential for engineering studies involving calculations of heat transfer. The quantitative analysis of mixtures in powder form by infrared diffuse reflection spectrometry has not yet been reported, but it is believed by this author that the method may well be applied for quality control and quality assurance in the pharmaceutical and chemical industries.

Many applications of attenuated total reflection (ATR) measurements are well known to analytical spectroscopists. In contrast to specular and diffuse reflection spectra, for which band shapes or intensities may be quite different from those in the corresponding absorption spectrum, ATR spectra appear similar to absorption spectra and so are more easily interpreted than specular and diffuse reflection spectra. ATR spectra of most samples are fairly easily measured on a grating spectrometer, and the use of FT–IR spectrometers is not particularly beneficial, not only because of the computing and plotting time required but also because the shape of the entrance face of most multiple-reflection ATR crystals is the same shape as the image of the slit of a monochromator. When the round focus of a Fourier spectrometer is focused onto the rectangular face of such crystals, the image is usually larger than the crystal face and some light is lost.

Nevertheless, if the energy passing through the crystal is attenuated, either by the nature of the sample or by having a polarizer in the beam, or if very weak bands are to be studied, the use of FT–IR spectrometers for the measurement of ATR spectra can prove beneficial. A good illustration of the use of a Fourier spectrometer for a difficult ATR measurement has been reported by Jakobsen,[50] who has measured polarized spectral bands due to monolayer coverage of stearic acid on the surface of a germanium ATR crystal.

Reflection–absorption measurements are best used for studying materials on metal surfaces. The simplest type of measurement of this type is for the identification of plastic films on the interior surface of beverage cans.[51] Studies of very thin layers are more difficult, and a very high angle of incidence may be required to increase the effective pathlength of the beam through the surface layer. Multiple reflectance, preferably still at a high incidence angle, also increases the intensity of the bands due to the adsorbed species. An example of this kind of measurement has been given by Harkness,[52] who measured the spectrum from a partial monolayer (5% coverage) of carbon monoxide chemisorbed on palladium.

Of the three principal sampling techniques used in infrared spectroscopy (absorption, reflection, and emission), it is probably true to say that reflection measurements have been the least studied using Fourier spectrometers. There are several problems in analytical chemistry that are difficult to study by any other technique, and further developments and applications in this field may be expected in the future.

6.6. EMISSION SPECTROMETRY

The field of infrared emission spectrometry has been revolutionized by the application of Fourier transform spectrometers. Excellent emission spectra of gases at temperatures as low as 60°C can be measured at 1-cm^{-1} resolution using a rapid-scanning interferometer and a TGS detector in one or two minutes.[53] Low and Coleman [54] have even published low-resolution emission spectra of minerals held 4°C *below* ambient temperature with a detector at room temperature. Spectra of weak remote sources such as the hot gases emerging from a smokestack have been measured.[55,56] Even more remote sources have been studied using Michelson interferometers. Instruments have been mounted in spacecraft and satellites for monitoring both terrestrial and planetary atmospheres, and infrared astronomy has been made much more sensitive by the application of FT–IR techniques. It is interesting to note that several of the pioneers of FT–IR spectrometry (in particular, Mertz in America and Connes in France) were astronomers who became interested in FT–IR because of the sensitivity advantage over other spectrometric techniques.

Infrared emission spectrometry has also been used in laboratory environments for the identification of heated samples. While absorption spectrometry is usually preferred over emission spectrometry for laboratory measurements because of its increased sensitivity, some samples cannot be mounted in a suitable configuration for absorption spectrometry. In such cases it is sometimes possible to configure the experiment for emission spectrometry. The sample should be similar in dimensions to samples measured by absorption spectrometry, and should be either self-supporting, held between windows of high transmittance or on the surface of a substrate with high reflectance. The thickness of condensed-phase samples has to be approximately the same as for the measurement of the corresponding absorption or reflection–absorption spectrum.

It should be recognized that no sample can emit more radiation at any frequency than a blackbody at the same temperature. The ratio of the intensity of radiation emitted by a sample at a given wavenumber to the intensity emitted by a blackbody of the same geometry at that wavenumber is called emissivity ε. For samples of low reflectance, it can be shown that

$$\varepsilon = 1 - \tau \tag{6.4}$$

where τ is the transmittance of the sample. Thus the thickness of the sample should be very similar to samples prepared for absorption spectroscopy, or else the emission bands will either be too weak for detection or too intense, in which case the spectrum appears similar to the spectrum of a blackbody.

One difficulty associated with the measurement of emission spectra of condensed-phase samples is that the temperature of the sample has to be uniform or else radiation emitted from molecules situated well below the surface will be absorbed by cooler molecules near the surface.[53] The effect of this "self-absorption" is to flatten out the bands at their peaks and even to show self-reversal for sharp lines or for samples with very large temperature gradients.

One of the most elegant applications of chemical infrared emission spectrometry has been reported by Lauer and Peterkin,[57] who measured the emission spectrum of a lubricant that was heated by the frictional energy generated when a steel bearing was rotated in contact with a diamond window. Spectra of aromatic films as thin as 0.1 μm were obtained between 630 and 930 cm^{-1} using a modified Beckman FS-720 interferometer and a Golay detector.

Another important type of analytical measurement that can be performed by infrared emission spectrometry is the remote sensing of the gases emerging from smokestacks. As early as 1967, Low and Clancy[55] reported measuring the emission spectrum of the stack gases from a power plant using a low-resolution interferometer. Spectral bands due to SO_2 were very apparent in their spectra; however, the low resolution of their instrument probably prevented weaker bands from several other components being distinguished from the background. More recently, Prengle *et al.*[56] used a higher-resolution interferometer to monitor pollutants in plumes from a gas-fired power plant using low-sulfur fuel, and reported determining CO, NO, NO_2, unburned hydrocarbons, and combustion product olefins to an accuracy of $\pm 28\%$. There has been some doubt expressed as to whether the method of determining the temperature of the stack used in this work is generally able to be applied to all stack gas measurements, and other methods are being developed in which the rotational fine structure of measured emission bands is used as a "spectroscopic thermometer."

A completely different type of experiment for which infrared emission spectrometry is applied to a chemical problem involves measuring the spectrum of radiation emitted during a chemical reaction, or *infrared chemiluminescence.* For chemiluminescence studies the emitted radiation is so weak that the interferometer has to be cooled with liquid nitrogen and the signal is measured using a liquid-helium-cooled bolometer, as described in Section 5.11.2.

The quality of the spectra that can be obtained may be illustrated by the work of McDonald *et al.*,[58-60] who have studied abstraction and substitution reactions between molecular beams of atomic fluorine and various olefins. The spectra shown in Figure 6.14 were obtained with a photon flux of only about 10^7 photons/second,[58] which is a fairly large number if photomultipliers are used for uv–visible spectrometry, but a very small number even for the most sensitive mid-infrared detectors. The use of cooled interferometers is now enabling infrared chemiluminescence to be measured

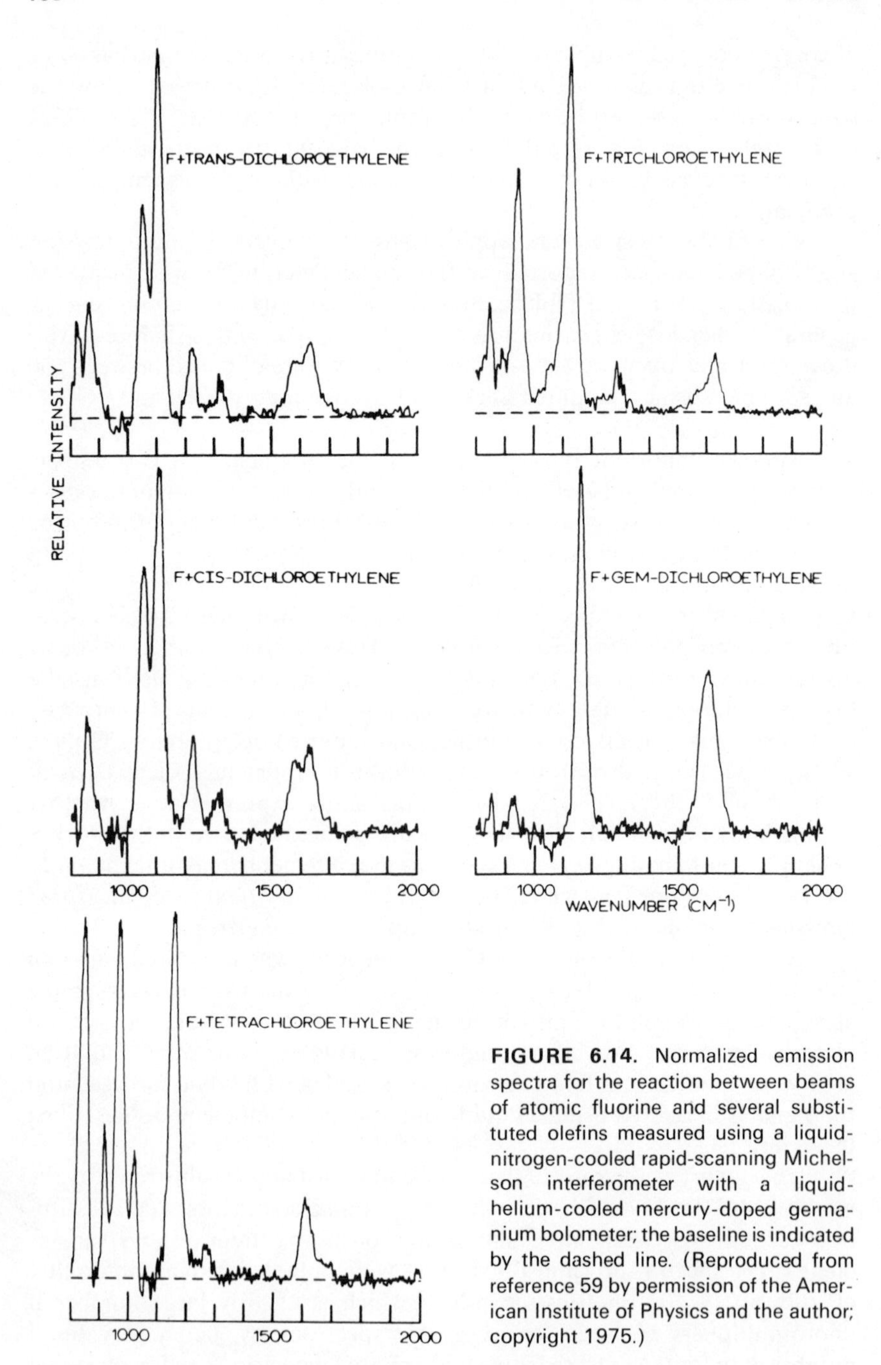

FIGURE 6.14. Normalized emission spectra for the reaction between beams of atomic fluorine and several substituted olefins measured using a liquid-nitrogen-cooled rapid-scanning Michelson interferometer with a liquid-helium-cooled mercury-doped germanium bolometer; the baseline is indicated by the dashed line. (Reproduced from reference 59 by permission of the American Institute of Physics and the author; copyright 1975.)

at higher resolution than in the past, when extremely low-resolution spectral information had to be acquired using narrow-band-pass filters.

6.7. ATOMIC SPECTROMETRY

Multiplex methods have found little application to electronic spectroscopy when photomultiplier detectors are used for measuring the signal, since detector noise is seldom the limiting factor and shot noise and/or fluctuation noise are dominant. There is certainly little or no *sensitivity* advantage to using Fourier spectrometers for absorption spectroscopy with continuous sources and there is no *multiplex* advantage when a hollow cathode lamp is used since only one spectral element is usually of interest.

Horlick and Yuen[61] have discussed some of the possible advantages of measuring *atomic emission* spectra using Fourier transform spectrometers. Among the advantages cited are the ease by which high-resolution spectra may be obtained, the wavenumber accuracy of laser-referenced Michelson interferometers, and the simplicity of the instrumentation. Among the disadvantages is the fact that although the distribution of noise across the spectrum (see Section 5.11.3) may result in strong emission lines being measured at a higher S/N than is possible on a grating spectrometer, *weak* lines may become buried in this distributed noise. Thus the use of Fourier spectrometers for atomic emission spectrometry may be beneficial for trace analysis when no element is present at high concentration, but multiplex spectrometers should not be used for the detection of a trace element in the presence of other elements at high concentration. (In this respect, it is quite possible that interferometers for atomic spectroscopy will find their greatest application for *atomic fluorescence* measurements.)

Horlick and Yuen showed results illustrating the fact that good emission spectra can be measured from solutions of the alkali metals using flame excitation, but they did not compare the practical sensitivity of grating and Fourier spectrometers. They reduced the number of data points required for their interferograms by using a rather long sampling interval and folding lines from several spectral regions into the region over which the output spectrum was computed. This technique is very useful for emission spectrometry of discrete sources with a relatively few spectral lines, but it is not clear to this author how interferograms acquired in this fashion can be properly phase corrected to obtain photometrically accurate spectra, since the phase angle θ_ν does not vary slowly with frequency, unless double-sided transforms are used.

Winefordner *et al.*[62] have carried out a detailed theoretical comparison of the use of four different types of spectrometer for atomic spectroscopy:

Sequential linear scanning spectrometers: instruments that use a monochromator with a single entrance slit and a single exit slit, and that scan through the spectrum at a constant rate.

Sequential slewed scanning spectrometers: conventional monochromators, but on which only the frequency regions of interest are scanned for any appreciable period of time.

Multiplex spectrometers: Fourier and Hadamard transform spectrometers.

Multichannel-detector systems: spectrometers where more than one spectral element is acquired simultaneously and independently, such as direct reading spectrographs, photodiode arrays and television spectrometers.

They show that, when the same detector is used in all four types of spectrometer, the multichannel approach is best and the sequential slewed scanning system is nearly as good for relatively simple spectra. However, it should be noted that multichannel systems using photodiode arrays or vidicons are less sensitive than spectrometers with photomultiplier detection. The very detailed analysis of Winefordner *et al.* arrives at conclusions that are qualitatively similar to the more intuitively-based discussion of Horlick and Yuen.

One type of atomic emission measurement where Fourier spectrometers are useful is when relatively large amounts of material are available (so that the comparative sensitivity of Fourier and other spectrometers is not important because all types of instruments can measure the spectra at an adequate S/N), but so many lines are present in the spectrum that the measurement must be taken at high resolution. Fourier spectrometers should also prove useful for measuring atomic emission spectra in the near infrared, where photomultipliers cannot be used. Conway *et al.*[63] have described a measurement that falls into both of these categories; they measured the emission spectrum of ^{244}Cm using the very high-resolution interferometer of Connes *et al.*[64]

An electrodeless lamp containing 50 μg of CmI_3 was run for 12 hours, and an 800,000-point interferogram was collected. Spectra were recorded

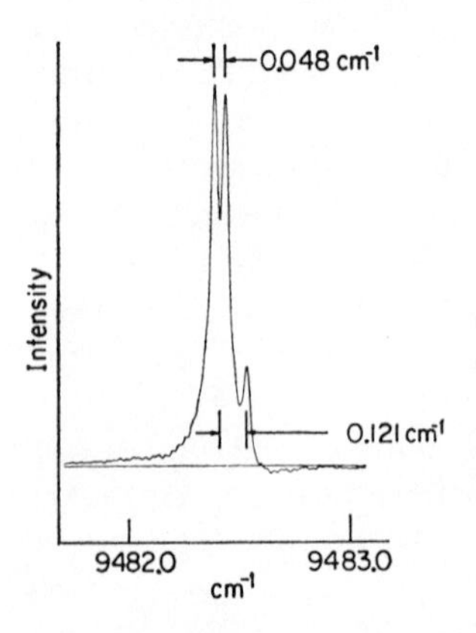

FIGURE 6.15. The curium emission spectrum near 9482.4 cm^{-1} illustrating self-reversal and the isotope shift of ^{246}Cm. This is a very small region from a spectrum containing 1743 lines due to curium measured using a very high-resolution interferometer and computed from an 800,000 point interferogram. (Reproduced from reference 63 by permission of Pergamon Press; copyright 1976.)

at a resolution of 0.02 cm^{-1} between 3700 and 11800 cm^{-1}, and a total of 1743 lines were ascribed to curium, the great majority of which were assigned to transitions between known energy levels. The measurement was made about seven months after the curium was separated from plutonium and other elements, and 116 lines of ^{240}Pu (the daughter element of radioactive ^{244}Cm, which has a half-life of 17.6 years) were found in the spectrum. Figure 6.15 shows a very small region of the measured spectrum in which self-reversal of one of the ^{244}Cm lines is evident and the isotope shift for ^{246}Cm is seen.

For work at such high resolution (the maximum resolving power is 5×10^5) performed without the use of a photomultiplier, it is apparent the Michelson interferometer gives quite a large advantage over a grating spectrometer. Future work will determine whether multiplex spectrometers of any sort will be applied to more conventional measurements of atomic absorption, atomic emission, or atomic fluorescence spectra.

REFERENCES

1. P. R. Griffiths, H. J. Sloane, and R. W. Hannah, *Appl. Spectrosc.* **31**, 485 (1977).
2. A. T. Stair, Jr., Fourier spectroscopy at the Air Force Cambridge Research Laboratories, *Aspen Int. Conf. Fourier Spectrosc., 1970*, Air Force Cambridge Research Laboratories Special Report, No. 114 (April, 1971), p. 127.
3. M. J. D. Low and S. K. Freeman, *Anal. Chem.* **39**, 194 (1967).
4. M. J. D. Low, *Anal. Letters* **1**, 819 (1968).
5. M. J. D. Low, H. Mark, and A. J. Goodsel, *J. Paint Technol.* **43**(562), 49 (1971).
6. K. L. Kizer, *Am. Lab.* **5**(6), 40 (1973).
7. L. V. Azarraga, Improved sensitivity of on-the-fly GC–IR spectroscopy, Paper No. 334, *Pittsburgh Conf. Anal. Chem. Appl. Spectrosc.*, Cleveland, Ohio (1976).
8. P. R. Griffiths, unpublished work (1976).
9. K. L. Kizer, A. W. Mantz, and L. C. Bonar, *Am. Lab.* **7**(5), 85 (1975).
10. M. J. D. Low, R. Epstein, and A. C. Bond, *J. Chem. Phys.* **48**, 2386 (1968).
11. R. P. Oertel, H. C. Smitherman, and A. J. Fehl, *Appl. Spectrosc.* **29**, 195 (1975).
12. J. O. Lephardt and G. Vilcins, *Appl. Spectrosc.* **29**, 221 (1975).
13. G. Vilcins and J. O. Lephardt, *Chem. Ind.*, p. 974 (November, 1975).
14. S. A. Liebman, D. H. Ahlstrom, and P. R. Griffiths, *Appl. Spectrosc.* **30**, 355 (1976).
15. R. E. Murphy and H. Sakai, Application of the Fourier spectroscopy technique to the study of relaxation phenomena, *Aspen Int. Conf. Fourier Spectrosc., 1970*, Air Force Cambridge Research Laboratories Special Report, No. 114 (April, 1971), p. 301.
16. A. Mantz, Time resolved spectroscopy, *in: Fourier Transform IR: Applications to Chemical Systems* (J. R. Ferraro and L. J. Basile, eds.), Academic Press, New York, 1978.
17. R. T. Hall and J. M. Dowling, *J. Chem. Phys* **45**, 1899 (1966).
18. H. A. Gebbie, N. W. B. Stone, G. Topping, E. K. Gora, S. A. Clough, and F. X. Kneizys, *J. Mol. Spectrosc.* **19**, 7 (1966).
19. D. R. Bosomworth and H. P. Gush, *Can. J. Phys.* **43**, 751 (1965).
20. T. R. Borgers and H. L. Strauss, *J. Chem. Phys.* **45**, 947 (1966).
21. J. A. Greenhouse and H. L. Strauss, *J. Chem. Phys.* **50**, 124 (1969).
22. R. F. Lake and H. W. Thompson, *Proc. Roy. Loc. (London)* **A291**, 469 (1966).
23. R. F. Lake and H. W. Thompson, *Spectrochim. Acta* **24A**, 1321 (1968).
24. M. Goldstein and W. D. Unsworth, *Spectrochim. Acta* **28A**, 1297 (1972).

25. D. M. Adams, S. J. Payne, and K. Martin, *Appl. Spectrosc.* **27**, 377 (1973).
26. W. A. Guillory and G. R. Smith, *Appl. Spectrosc.* **27**, 137 (1973).
27. J. E. Harries, N. P. L. Report No. DES 16 (1972).
28. J. E. Beckman and J. E. Harries, *Appl. Opt.* **14**, 470 (1975).
29. S. S. T. King, *J. Ag. Food Chem.* **21**, 526 (1973).
30. D. H. Anderson and T. E. Wilson, *Anal. Chem.* **47**, 2482 (1975).
31. P. R. Griffiths and F. Block, *Appl. Spectrosc.* **27**, 432 (1972).
32. K. L. Kizer, Digilab Inc., unpublished work (1976).
33. J. L. Koenig and D. L. Tabb, *Can. Res. Develop.* **7**, 25 (1975).
34. M. J. D. Low, A. J. Goodsel, and N. Takezawa, *Env. Sci. Technol.* **5**, 1191 (1971).
35. A. J. Goodsel, M. J. D. Low, and N. Takezawa, *Env. Sci. Technol.* **6**, 268 (1972).
36. C. J. Percival and P. R. Griffiths, *Anal. Chem.* **47**, 154 (1975).
37. M. M. Gomez-Taylor, D. Kuehl, and P. R. Griffiths, *Appl. Spectrosc.* **30**, 447 (1976).
38. J. T. Stoklosa, J. Rydzak, and P. R. Griffiths, unpublished work (1975).
39. J. O. Alben, G. H. Bare, and P. A. Bromberg, *Nature (London)* **252**, 736 (1974).
40. G. H. Bare, J. O. Alben, and P. A. Bromberg, *Biochemistry* **14**, 1578 (1975).
41. P. L. Hanst, A. S. Lefohn, and B. W. Gay, *Appl. Spectrosc.* **27**, 188 (1973).
42. P. L. Hanst, W. E. Wilson, R. K. Patterson, B. W. Gay, L. W. Chaney, and C. S. Burton, A spectroscopic study of Pasadena smog, EPA Report No. 650/4-75-006 (February, 1975).
43. J. L. Koenig, *Appl. Spectrosc.* **29**, 293 (1975).
44. J. S. Mattson and C. A. Smith, An on-line minicomputer system for infrared spectrophotometry, *in: Computers in Chemistry and Instrumentation,* Vol. 7 (J. S. Mattson, H. B. Mark, Jr., and H. C. MacDonald, Jr., eds.), Marcel Dekker, New York, 1975.
45. T. Hirschfeld and K. Kizer, *Appl. Spectrosc.* **29**, 345 (1975).
46. T. Hirschfeld, *Appl. Spectrosc.* **29**, 524 (1975).
47. R. M. Gendreau and P. R. Griffiths, *Anal. Chem.* **48**, 1910 (1976).
48. M. J. D. Low, *Appl. Opt.* **6**, 1503 (1967).
49. R. R. Willey, *Appl. Spectrosc.* **30**, 593 (1976).
50. R. J. Jakobsen and J. P. Crowley, The use of FT–IR and ATR for the study of organic thin films. II. Adsorption of blood plasma proteins on polymer surfaces, Paper No. 386, *Pittsburgh Conf. Anal. Chem. Appl. Spectrosc.,* Cleveland, Ohio (1976).
51. M. J. D. Low, A. J. Goodsel, and H. Mark, The modification and some uses of a commercial infrared Fourier transform spectrometer, *in: Molecular Spectroscopy 1971* (P. Hepple, ed.), Institute of Petroleum, London, 1972.
52. J. B. L. Harkness, Ph.D. Dissertation, MIT, 1970.
53. P. R. Griffiths, *Appl. Spectrosc.* **26**, 73 (1972).
54. M. J. D. Low and I. Coleman, *Appl. Opt.* **5**, 1453 (1966).
55. M. J. D. Low and F. K. Clancy, *Env. Sci. Technol.* **1**, 73 (1967).
56. H. W. Prengle, C. A. Morgan, C.-S. Fang, L.-K. Huang, P. Campani, and W. W. Wu, *Env. Sci. Technol.* **7**, 417 (1973).
57. J. L. Lauer and M. E. Peterkin, Infrared emission spectra from lubricant films in operating bearings by Fourier methods, Paper No. 333, *Pittsburgh Conf. Anal. Chem. Appl. Spectrosc.,* Cleveland, Ohio (1976).
58. J. G. Moehlmann, J. T. Gleaves, J. W. Hudgens, and J. D. MacDonald, *J. Chem. Phys.* **60**, 4790 (1974).
59. J. G. Moehlmann and J. D. McDonald, *J. Chem. Phys.* **62**, 3052 (1975).
60. J. G. Moehlmann and J. D. McDonald, *J. Chem. Phys.* **62**, 3061 (1975).
61. G. Horlick and W. K. Yuen, *Anal. Chem.* **47**, 775A (1975).
62. J. D. Winefordner, R. Avni, T. L. Chester, J. J. Fitzgerald, L. P. Hart, D. J. Johnson, and F. W. Plankey, *Spectrochim. Acta* **31B**, 1 (1976).
63. J. G. Conway, J. Blaise, and J. Vergés, *Spectrochim. Acta* **31B**, 31 (1976).
64. J. Connes, H. Delouis, P. Connes, G. Guelachvili, J. P. Maillard, and G. Michel, *Nouv. Rev. Opt. Appl.* **1**, 3 (1970).

Chapter 7

Hadamard Transform Analytical Systems

Martin Harwit

7.1. INTRODUCTION

A spectrometer sorts electromagnetic radiation into its component colors. Each color corresponds to a particular value of a parameter—wavelength, energy, or frequency—and the resulting spectrum displays the intensity of radiation for each value of this parameter.

There are many technically distinct ways of obtaining a spectrum. In some instances, however, the variety of spectrometric methods is limited by a lack of image-forming detectors—photographic plates, vidicons, and so on. We then have two alternatives. Either we sequentially analyze different component colors in the beam, or else we label each color component by means of some modulating technique and analyze the beam simultaneously for all color components. Such spectromodulation techniques have been discussed in an earlier review paper by Harwit and Decker.[1]

Their primary purpose is the maximization of the amount of radiation that can be gathered onto the optical detector. If the detector's performance is limited by noise intrinsic to the detector, maximization of the radiant power incident on the detector can lead to improved signal-to-noise ratios (S/N).

Two quite distinct modulation techniques have come to be used in the past. In both these methods numerous different spectral (or spatial) elements can be modulated in distinct ways. This simultaneous independent modulation of large numbers of distinct optical elements is called multi-

---

Martin Harwit • Center for Radiophysics and Space Research, Cornell University, Ithaca, New York 14853

plexing. The first method depends on the wave character of radiation and makes use of interferometry. The other technique makes use of dispersing spectrometers in which entrance and exit slits are replaced by opaque or transmitting masks. We will mainly discuss this second method in this chapter.

The basic multiplexing instrument consists of four essential components: an optical separator, an encoding mask, a detector, and a processor (Figure 7.1). More complex instruments may make use of additional components. In some devices, for example, two masks are used.

The separator might be nothing more than a lens that produces a focused image at the mask. The lens acts to separate light arriving from different spatial elements of a scene. The separator might equally well be a dispersing system that separates different spectral components of a beam and focuses them onto different locations on the mask.

A particular location on the mask transmits light to the detector, absorbs the light, or reflects it toward a reference detector. In this way, the intensity of an element of the separated beam is modulated. If the readings that are recorded are the differences in the intensity of light reaching the main detector and reference detector, the intensity of the element is respectively multiplied by $+1, 0,$ or -1.

If there is just one detector in the system, the modulation consists solely of $+1$'s and 0's, and the recorded intensity is just the intensity of radiation transmitted by the mask.

When m intensity values are to be determined, at least m different detector readings corresponding to m linearly independent mask arrays are required.

Three important questions in designing such an instrument are: (a)

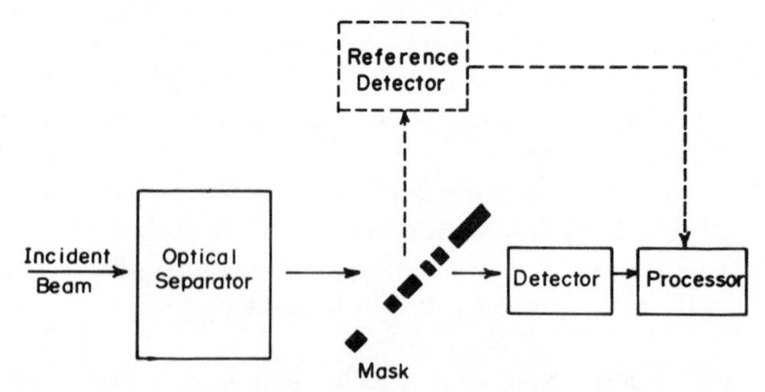

FIGURE 7.1. Basic modulation instrument. Radiation from a source is passed through an optical separator and a mask before impinging on a detector that records the transmitted intensity. An (optional) reference detector records the amount of reflected radiation for each intensity measurement. The processor then reconstructs an image and/or spectrum from the intensity and mask position data. More complex instruments make use of several masks.

How should the mask be chosen? (b) How much does the instrument improve the quality of the measurements? (c) How close to optimum is the chosen mask design?

Such questions have been studied for many years in statistics under the name of *weighing designs*. The application to optics was first pointed out by Sloane and Harwit.[2]

7.2. WEIGHING DESIGNS AND OPTICAL MULTIPLEXING

Yates[3] seems to have been the first to point out that by weighing several objects in groups, rather than separately, it may be possible to determine the weights of the individual objects more accurately.

Suppose, for example, that four objects are to be weighed, with a balance that makes an error of approximate magnitude e each time it is used. Assume that this error is a random variable with mean value zero and variance σ^2.

First, let us weigh the objects separately. Let the unknown weights be x_1, x_2, x_3, x_4. The actually measured values, obtained with the balance, are y_1, y_2, y_3, y_4; and the errors made by the balance are e_1, e_2, e_3, e_4. Then the four weighings give four equations:

$$y_1 = x_1 + e_1, \qquad y_2 = x_2 + e_2$$

$$y_3 = x_3 + e_3, \qquad y_4 = x_4 + e_4$$

The best estimates $\hat{x}_i$ of the unknown weights x_i are the measurements themselves:

$$\hat{x}_1 = y_1 = x_1 + e_1, \qquad \hat{x}_2 = y_2 = x_2 + e_2, \quad \ldots$$

These are unbiased estimates:

$$E\hat{x}_1 = x_1, \qquad E\hat{x}_2 = x_2, \quad \ldots$$

where E denotes expected values. The variance, or mean square error, as already mentioned is

$$E(\hat{x}_1 - x_1)^2 = Ee_1^2 = \sigma^2, \quad \ldots$$

Now suppose that the balance is a chemical, two-pan balance and that the four weighings are obtained in the following way:

$$\begin{aligned} y_1 &= x_1 + x_2 + x_3 + x_4 + e_1 \\ y_2 &= x_1 - x_2 + x_3 - x_4 + e_2 \\ y_3 &= x_1 + x_2 - x_3 - x_4 + e_3 \\ y_4 &= x_1 - x_2 - x_3 + x_4 + e_4 \end{aligned} \qquad (7.1)$$

We note that in the first weighing all four objects are placed, say, in the left-hand pan, and in the other weighings two objects are in the left pan and two in the right. We can invert the coefficient matrix on the right to solve for x_1, x_2, x_3, x_4. We then see that the best estimate for x_1 is

$$\hat{x}_1 = \tfrac{1}{4}(y_1 + y_2 + y_3 + y_4) = x_1 + \tfrac{1}{4}(e_1 + e_2 + e_3 + e_4)$$

Since the variance of a sum of independent random variables is the sum of the individual variances, we see that the variance of $\hat{x}_1$ (and also of $\hat{x}_2, \hat{x}_3, \hat{x}_4$) is now $4\sigma^2/16 = \sigma^2/4$. By weighing the objects together we have reduced the mean square error by a factor of 4.

Next, suppose we make use of a spring balance with only one pan. Here only coefficients 0 and 1 can be used. A good method of weighing the four objects is

$$\begin{aligned} y_1 &= \phantom{x_1 +{}} x_2 + x_3 + x_4 + e_1 \\ y_2 &= x_1 + x_2 \phantom{{}+ x_3 + x_4} + e_2 \\ y_3 &= x_1 \phantom{{}+ x_2} + x_3 \phantom{{}+ x_4} + e_3 \\ y_4 &= x_1 \phantom{{}+ x_2 + x_3} + x_4 + e_4 \end{aligned} \tag{7.2}$$

In this case the variances of x_1, x_2, x_3, x_4 can be shown to be $4\sigma^2/9$, $7\sigma^2/9$, $7\sigma^2/9$, $7\sigma^2/9$, respectively, a smaller improvement than in the previous case.

In general, for p unknowns $x_1, \ldots, x_p$, and N measurements $y_1, \ldots, y_N$ respectively involving errors $e_1, \ldots, e_N$, we have a set of equations

$$y_i = w_{i1}x_1 + \cdots + w_{ip}x_p + e_i, \qquad i = 1, \ldots, N$$

which can be written in matrix form,

$$y = Wx + e \tag{7.3}$$

The individual matrices y, x, and e of equation (7.3) are column matrices,

$$y = (y_1, \ldots, y_N)^{\mathrm{T}}, \qquad x = (x_1, \ldots, x_p)^{\mathrm{T}}, \qquad e = (e_1, \ldots, e_N)^{\mathrm{T}}$$

where T denotes transpose. A particular choice for the $N \times p$ coefficient matrix $W = (w_{ij})$ is called a *weighing design*.

The connection with multiplex optics is straightforward. In the optical analog the x_i represent intensities of individual spatial and/or spectral elements in a beam of radiation. In contrast to scanning instruments, which would measure the intensities one at a time, the multiplex optical system measures (i.e., weighs) several intensities of x_i simultaneously.

The two types of weighing design—the chemical balance design (with coefficients w_{i1}, which may be -1, 0, or $+1$), and the spring balance design (in which the coefficients must be 0 or 1)—are realized in optical systems by masks W, which use reflected, absorbed, or transmitted light in the first case, or simply open or closed slots in the second case.

7.3. HISTORICAL BACKGROUND OF MULTIPLEXING BY MEANS OF MASKS

The development of optical multiplexing long preceded the formal realization of the full mathematical analogy to weighing designs. In the late 1940s Golay[4] wrote a series of remarkable papers in which he showed that infrared radiation passing through a dispersing spectrometer with wide open entrance or exit apertures could be analyzed by passage through modulating mask patterns placed in the entrance and/or exit planes (Figure 7.2). Because the widened apertures passed increased amounts of radiation, the modulated beam could be used to obtain spectral information at a far higher rate or with a larger signal-to-noise ratio than possible in a conventional dispersing instrument.

These papers were far ahead of their time. Golay realized the importance of binary orthogonal digital codes, more than a decade before digital com-

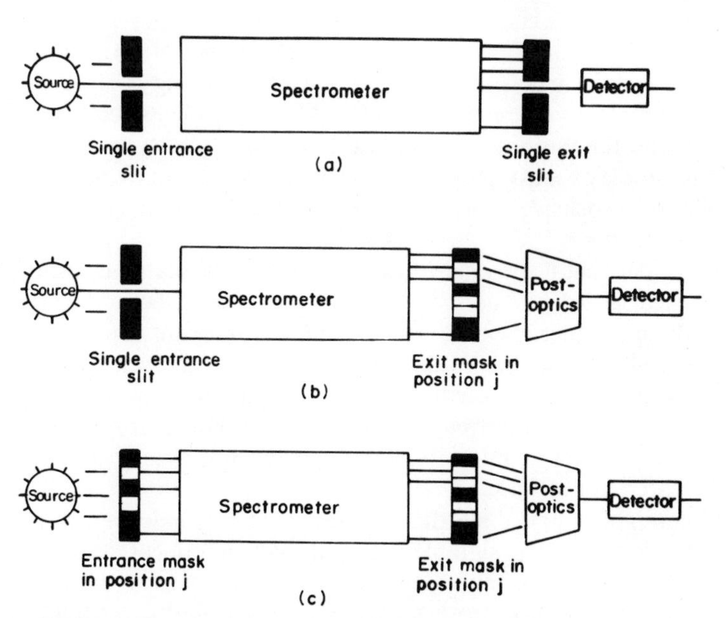

FIGURE 7.2. Essential features of modulated spectrometers. Light passing through a conventional grating instrument (a) is constrained to pass through two narrow slits, one in the entrance focal plane, the other at the exit. A mask, or set of masks, may be placed either at the entrance or exit focal plane (b) and the beam intensity incident on the detector is then measured for each mask configuration. This constitutes a modulation of the beam by the mask patterns. In an alternative version (c) such masks can be placed both at the entrance and exit focal planes. The transmitted radiation is thus increased even more.

puters had become practical realities and more than two decades before minicomputers were to become standard laboratory equipment.

Some two decades after Golay's pioneering work, many of the essential features of his instrumentation were rediscovered by Ibbett *et al.*[5] and by Decker and Harwit.[6] Almost simultaneously, Gottlieb[7] was able to show that digital modulation—or multiplexing—by means of opaque masks could improve the reconstruction of images. In his method an optical instrument was used to isolate individual spatial rather than spectral elements in a beam of radiation, but the mathematical treatment of the multiplexing scheme remained essentially unchanged.

Gottlieb noted two important points: (1) There exist cyclic codes intimately related to orthogonal binary digital codes. (2) These codes can be folded into a two-dimensional array, much as the one-dimensional lines of printed words are folded into a two-dimensional array of printing on this page.

Folding is currently an advantage in that one-dimensional codes are mathematically far better understood than their two-dimensional analogs. The advantage of cyclic codes is mainly economic. They permit the construction of large single masks, overlapping portions of which form essentially orthogonal arrays. The problem of constructing a large variety of independent masks is therefore avoided. Mask production costs are greatly lowered, and the multiplexing method immediately becomes economically attractive. Golay's mask patterns had not been cyclic and would have been exorbitantly expensive in any apparatus dealing with a large number of discrete elements.

It is clear that if the intensities of N spectral or spatial elements are to be obtained, at least N different measurements must be performed, independently of whether or not multiplexing techniques are applied. If a multiplex method is used, the N different intensities can then be derived by solving N linear algebraic equations. These equations actually are linear only if the detector employed is a linear device. However, many detectors behave linearly at low intensity levels, where multiplexing techniques are of the greatest importance.

One might at first think that any set of linearly independent measurements would be equally suitable in a multiplexing optical device. After all, N linearly independent equations suffice for solving for the required N intensity values. In fact, however, there are multiplexing schemes that are far poorer than others, in that they do not adequately minimize the effects of random errors, or noise, that accompany any realistic measurement. Proper minimization of such noise effects can be achieved through multiplexing by means of orthogonal codes. The concept of orthogonality will be explained below.

Sloane *et al.*[8] examined the question of optimum mask codes, basing their work on prior communication theoretical studies that dealt, for example, with the optimum coding for pulsed radar beams (cf. Golomb[9]).

7.4. MATHEMATICAL DEVELOPMENT

Let us choose a simple example. Suppose seven different spectral elements are resolved in the exit plane of an ordinary grating spectrometer. We place seven slits in this exit plane and open or close these slits in some manner to be prescribed. The radiation passing through the open slits is gathered at the detector. Let the intensity of the jth spectral element be ψ_j, where j can assume values from 1 to 7. Now consider a sequence of slits described by the array

$$1\ 0\ 0\ 1\ 0\ 1\ 1 \tag{7.4}$$

where the number 0 signifies a closed slit and 1 signifies an open slit. With this sequence the intensity reaching the detector is

$$\psi_1 + 0 + 0 + \psi_4 + 0 + \psi_6 + \psi_7 = \psi_1 + \psi_4 + \psi_6 + \psi_7 \tag{7.5}$$

We can think of the sequence (7.4) as a seven-dimensional vector that can be used to obtain a scalar product with a seven-dimensional vector $(\psi_1, \psi_2, \ldots, \psi_7)$. That product is the intensity (7.5) incident on the detector.

In order to solve for the individual intensities ψ_j, we must take seven linearly independent combinations of open and closed slits, and obtain an intensity measurement for each of these. One sequence of such slit combinations is given by each row of the following matrix W.

$$W = \begin{bmatrix} 1 & 0 & 0 & 1 & 0 & 1 & 1 \\ 1 & 1 & 0 & 0 & 1 & 0 & 1 \\ 1 & 1 & 1 & 0 & 0 & 1 & 0 \\ 0 & 1 & 1 & 1 & 0 & 0 & 1 \\ 1 & 0 & 1 & 1 & 1 & 0 & 0 \\ 0 & 1 & 0 & 1 & 1 & 1 & 0 \\ 0 & 0 & 1 & 0 & 1 & 1 & 1 \end{bmatrix} \tag{7.6}$$

Note that this matrix is cyclic. The second row is obtained by taking the first row, moving each element one place to the right, and moving the last element of the first row into the leftmost position in the second.

The next few paragraphs will closely follow the treatment of Sloane *et al.*[8] They postulate that in a real spectrometer the measurement process consists of observing the spectrum through M filters, or masks, the energy through the ith mask being $\sum_{j=1}^{N} \omega_{ij}\psi_j$, where $W_i = (\omega_{i1}, \ldots, \omega_{iN})$ is the ith vector of weights. The mask weights are assumed such that $\omega_{ij} = 0$ or 1 and correspond to attenuation or transmission. The photodetector adds a random noise n_i to the signal $\sum_{j=1}^{N} \omega_{ij}\psi_j$ and yields a measurement

$$\eta_i = n_i + \sum_{j=1}^{N} \omega_{ij}\psi_j, \qquad i = 1, 2, \ldots, M \tag{7.7}$$

With the notation $\langle\ \rangle$ for ensemble averages, the noise n_i was assumed to have the following properties: $\langle n_i \rangle = 0$; n_i is independent of the signal; $\langle n_i^2 \rangle = \sigma^2$; successive measurement noises are assumed to be uncorrelated ($\langle n_i n_j \rangle = 0$ if $i \neq j$).

In order to estimate the set of true energies $\{\psi_j\}$ by estimates (functions of the observations) $\{\hat{\psi}_j\}$, one needs at least as many measurements (M) as there are unknowns (N). Furthermore, at least N distinct masks $\{W_i\}$ are needed if one hopes to estimate the spectral shape. Hence, assume a vector of observations $x = (x_1, \ldots, x_M)$, spectral energies $\psi = (\psi_1, \ldots, \psi_N)$, measurement noises $n = (n_1, \ldots, n_M)$, and a matrix of masks $W = (W_1^{\mathrm{T}}, \ldots, W_M^{\mathrm{T}}) = (\omega_{ij})$. (The T stands for transpose.) With this notation

$$\eta = \psi W + n \tag{7.8}$$

An estimate $\hat{\psi}$ of ψ is a function of the observations, $\hat{\psi}(\eta)$, hopefully lying close to ψ. As a measure of the accuracy of the estimate, we adopt the mean square error criterion: minimize $\varepsilon = \langle (\hat{\psi} - \psi)(\hat{\psi} - \psi)^{\mathrm{T}} \rangle$.

Sloane *et al.* go on to show that in the case $M = N$, when there are as many measurements as unknowns, the matrix $W = (\omega_{ij})$ of mask weights should be chosen so that $\mathrm{Tr}[W^{-1}(W^{-1})^{\mathrm{T}}]$ is as small as possible. The notation Tr stands for "trace" of a matrix. Three possible choices for the matrix W are given here.

A Hadamard matrix H of order N is an $N \times N$ matrix of $+1$'s and -1's such that[10]

$$HH^{\mathrm{T}} = NI \tag{7.9}$$

H may always be normalized so that the first row and column consist entirely of $+1$'s. If G denotes the remaining $(N-1) \times (N-1)$ matrix, then

$$H = \begin{bmatrix} 1 & 1 & \cdots & 1 \\ 1 & & & \\ \vdots & & G & \\ 1 & & & \end{bmatrix} \tag{7.10}$$

Let row i and row j be any two distinct rows of H other than the first row. Then one can show that, assuming $N \geq 4$,

$$\left.\begin{array}{l} \text{row } i \text{ has } +1 \text{ and row } j \text{ has } +1 \\ \text{row } i \text{ has } +1 \text{ and row } j \text{ has } -1 \\ \text{row } i \text{ has } -1 \text{ and row } j \text{ has } +1 \\ \text{row } i \text{ has } -1 \text{ and row } j \text{ has } -1 \end{array}\right\} \text{ in } N/4 \text{ places} \tag{7.11}$$

Thus the scalar product of any two rows of this matrix is zero. This is what is meant by "orthogonality" for a matrix. N is a multiple of four, and if at least one of the following conditions is also satisfied:

(i) $N = p + 1$, p prime
(ii) $N = p(p + 2) + 1$, p and $p + 2$ prime
(iii) $N = 2^m$

then G can be chosen to be a cyclic matrix, that is, a matrix in which the $(i + 1)$th row is obtained by shifting the ith row cyclically one place to the right. For example, when $N = 8$, G may be taken to be

$$G = \begin{bmatrix} - & - & - & + & - & + & + \\ + & - & - & - & + & - & + \\ + & + & - & - & - & + & - \\ - & + & + & - & - & - & + \\ + & - & + & + & - & - & - \\ - & + & - & + & + & - & - \\ - & - & + & - & + & + & - \end{bmatrix} \tag{7.12}$$

where $+$ stands for $+1$ and $-$ for -1.

From equation (7.11) we calculate the dot products of *any* two rows of H or of G:

$$\begin{aligned} H:&\quad \text{row } i \cdot \text{row } j = \begin{cases} 0, & i \neq j \\ N, & i = j \end{cases} \\ G:&\quad \text{row } i \cdot \text{row } j = \begin{cases} -1, & i \neq j \\ N-1, & i = j \end{cases} \end{aligned} \tag{7.13}$$

Also each row of G contains $[(N/2) - 1]$ $+1$'s and $(N/2)$ -1's.

The first choice for W is the matrix H^{T}. From equation (7.9) one obtains $H^{-1} = N^{-1}H^{\mathrm{T}}$, and $\mathrm{Tr}[(H^{-1})^{\mathrm{T}} H^{-1}] = \mathrm{Tr}(N^{-2}HH^{\mathrm{T}}) = \mathrm{Tr}(N^{-1}I) = 1$.

The second choice for W is the matrix G^{T}:

$$\mathrm{Tr}[(G^{-1})^{\mathrm{T}}(G^{-1})] = 2 - \frac{2}{N} \tag{7.14}$$

The last choice for W is S^{T}, where S is the matrix obtained from G by replacing $+1$'s by 0's and -1's by $+1$'s. Clearly each row of S contains $[(N/2) - 1]$ 0's and $N/2$ $+1$'s, and from equation (7.11)

$$S:\quad \text{row } i \cdot \text{row } j = \begin{cases} N/4, & i \neq j \\ N/2, & i = j \end{cases} \tag{7.15}$$

$$\mathrm{Tr}[(S^{-1})^{\mathrm{T}}(S^{-1})] = [2 - (2/N)]^2 \tag{7.16}$$

The three possible choices for the matrix W of mask weights, together with the corresponding values of $\mathrm{Tr}[W^{-1}(W^{-1})^{\mathrm{T}}]$, are summarized as follows:

Matrix W	H^{T}	G^{T}	S^{T}
Trace	1	$2-\dfrac{2}{N}$	$4-\dfrac{8}{N}+\dfrac{4}{N^2}$
Elements	$+1, -1$	$+1, -1$	$+1, 0$

(7.17)

A mutliplex spectrometer that uses the S code works in the following way. Radiation enters the instrument through a single slit, is rendered parallel, and then directed toward the dispersive element (grating or prism). After dispersion, the radiation is collimated and focussed upon the multislit mask at the exit focal plane of the spectrometer. The radiation passed by the mask is collected onto a detector. By sequentially stepping through M mask positions at the exit plane and recording the intensity measured by the detector for each mask position, we obtain the spectrum essentially by means of an inversion of equation (7.8):

$$\hat{\psi} = \eta(S)^{-1} \tag{7.18}$$

Here we have made use of the postulated mean zero expectation value of the noise.

Following the work of Sloane *et al.*, Nelson and Fredman[11] were able to show, in 1971, that the Hadamard code provides an absolute optimum that cannot be excelled by masks with elements $|\omega_{ij}| \leq 1$. For the utilization of this code, one would have to make use of reflecting masks, which cast their light onto a secondary detector that subtracts from the intensity recorded by a primary detector gathering transmitted radiation. Such optical devices have not yet been built. Sloane and Harwit[2] were able to show that the result of Nelson and Fredman is identical to a theorem due to Hotelling,[12] which had been widely used in communication theory ever since its proof in 1944. In the same paper, Sloane and Harwit also showed that for large numbers of slits, the S matrix approaches the optimum that can be achieved with any mask having solely transmitting and blocking—as distinct from reflecting—slits. As shown in equation (7.17), that is still a factor of 4, in mean square deviation, or a factor of 2 in S/N poorer than the results that could be optimally obtained in the ideal multiplexing that can be theoretically achieved with a Hadamard code.

We still note, before going on, that the S codes are known in communication theory as the Simplex codes.

7.5. VARIETIES OF ENCODED SPECTROMETERS

There are basically three modes in which these digitally encoded instruments can operate. For convenience we will call them Hadamard transform spectrometers—a phrase coined by J. A. Decker, Jr.—even though most of the codes in actual use are only derived from the Hadamard code. The simplest model has a single entrance slit and n exit slits. It produces a spectrum having n spectral elements. A more complex instrument[13,14] is one having m entrance slits and n exit slits, $m < n$. It can be operated in two different modes. If the radiation incident on the spectrometer is diffuse and homogeneous, the prime advantage of having several entrance slits is an increase in the instrument's energy throughput; but the final product still is a single spectrum having $\sim n$ spectral elements. Alternatively, if the radiation incident on the instrument is not homogeneous, one may wish to obtain a separate spectrum for light incident on each entrance slit position.[15,16] These instruments have largely been used in the infrared, where additional detector noise is introduced by added detector size. The spectrometers constructed have therefore largely made use of dedispersion methods to reduce the size of the final image and of the required detector. A typical instrumental layout is shown in Figure 7.3.

The appearance of an encoding mask is shown in Figure 7.4. This is a mask corresponding to the S code having 255 modulation elements. Only about one-half the total number of elements shown is actually used at any given time. In any given mask position the remaining portion is blocked by blocking masks as shown in Figure 7.5. The total number of coding elements shown in Figure 7.4 is 509, consisting of 255 elements in use in any given mask position, plus 254 added elements that enable the mask to be cycled through a total of 255 different positions.

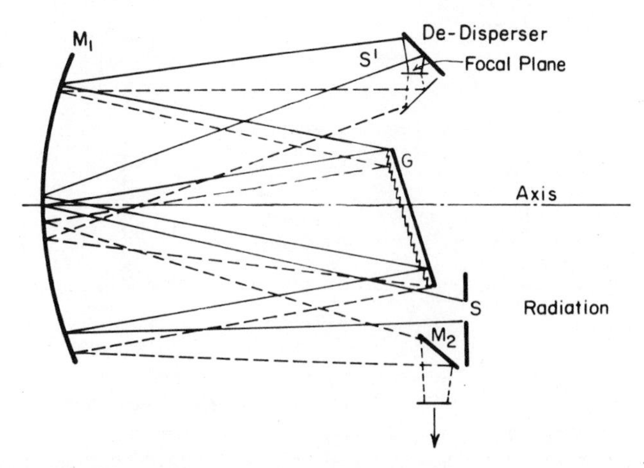

FIGURE 7.3. Light path through a Hadamard transform spectrometer. The dedispersion stage is used to collect the radiation back onto a small detector.

FIGURE 7.4. Encoding mask for a Hadamard transform spectrometer.

Decker[17] has shown experimentally that the theoretically expected S/N improvements are actually realized in practice. He constructed a 255-element, single-entrance slit spectrometer, and obtained excellent agreement. Similar results have also substantiated theoretical estimates for the more complex modes of operation that use masks, both at the entrance and exit of the spectrometer. The performance of a commercial (Spectral Imaging, Inc.) single entrance, 255-exit slit instrument constructed by Decker[18] is shown in Figure 7.6. It displays the 7.5 to 15 μm transmission spectrum of a polystyrene film. Figure 7.7 shows that spectral resolution is not lost when a spectrometer has 15 entrance slits, instead of just one, and has 255 exit slits, in addition.

Most of the equipment described is portable. The 16–25 μm atmospheric transmission spectrum shown in Figure 7.8 was obtained by Phillips and Briotta[19] atop Mt. Lemmon in Arizona. These authors also obtained spectra of Jupiter clearly showing the ammonia absorption bands detected through the atmospheric 8–13 μm transmission window. The equipment constructed for astronomical purposes by Tai *et al.*[16] permits the spectrom-

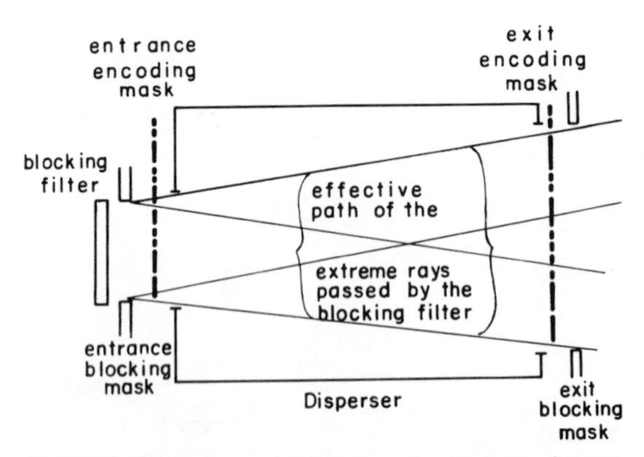

FIGURE 7.5. Schematic diagram of a Hadamard transform spectrometer. In the most general system light enters at the left, is encoded by a modulating mask, and exits at the right after a second stage of encoding. In some instruments there is only one stage of encoding and a narrow entrance or exit slit at the opposite end of the disperser.

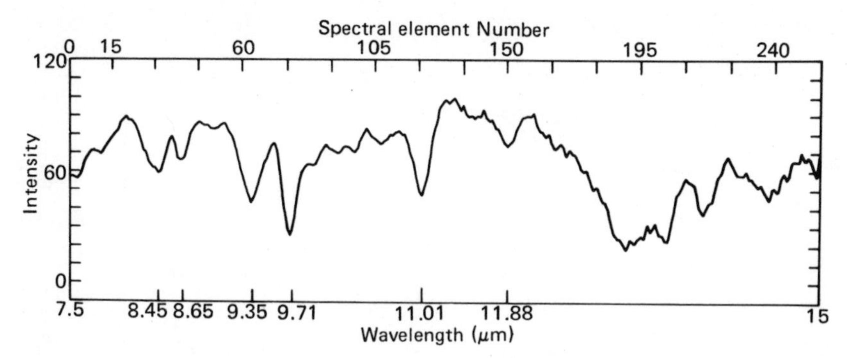

FIGURE 7.6. Infrared spectrum of a polystyrene film obtained with a commercial Hadamard transform spectrometer. (Courtesy of Dr. John A. Decker, Jr., Spectral Imaging Inc., Concord, Massachusetts.)

eter as well as a minicomputer and cathode ray tube display system to be taken to mountain top observatories. Data reduction is carried out as each data point is recorded, and the entire spectrum or set of spectra obtained can be displayed at the end of each run. In certain spectral observing runs, in which astronomical data are gathered over longer periods of time—mainly because of the faintness of astronomical sources—the mask cycle may be repeated many times, and the cumulative spectrum obtained may then be called for at the end of each cycle.

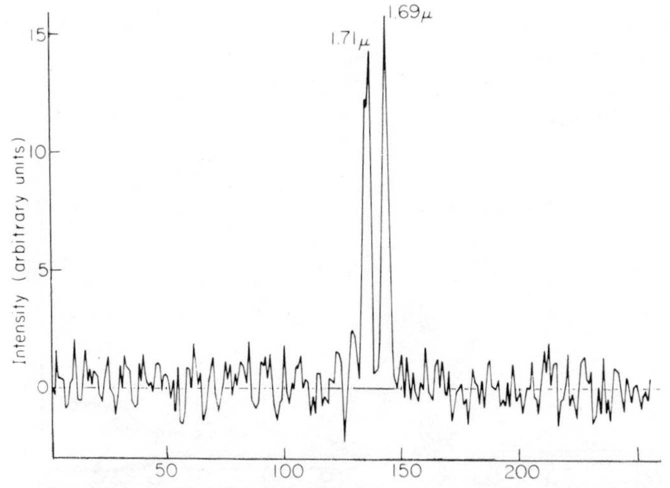

FIGURE 7.7. A 255-element laboratory spectrum of a mercury vapor lamp obtained at Cornell University with a spectrometer having 15 entrance and 255 exit mask elements. The resolution is the same as with a single entrance and a single exit slit but the signal-to-noise ratio is improved. (From *Applied Optics*,[14] with permission.)

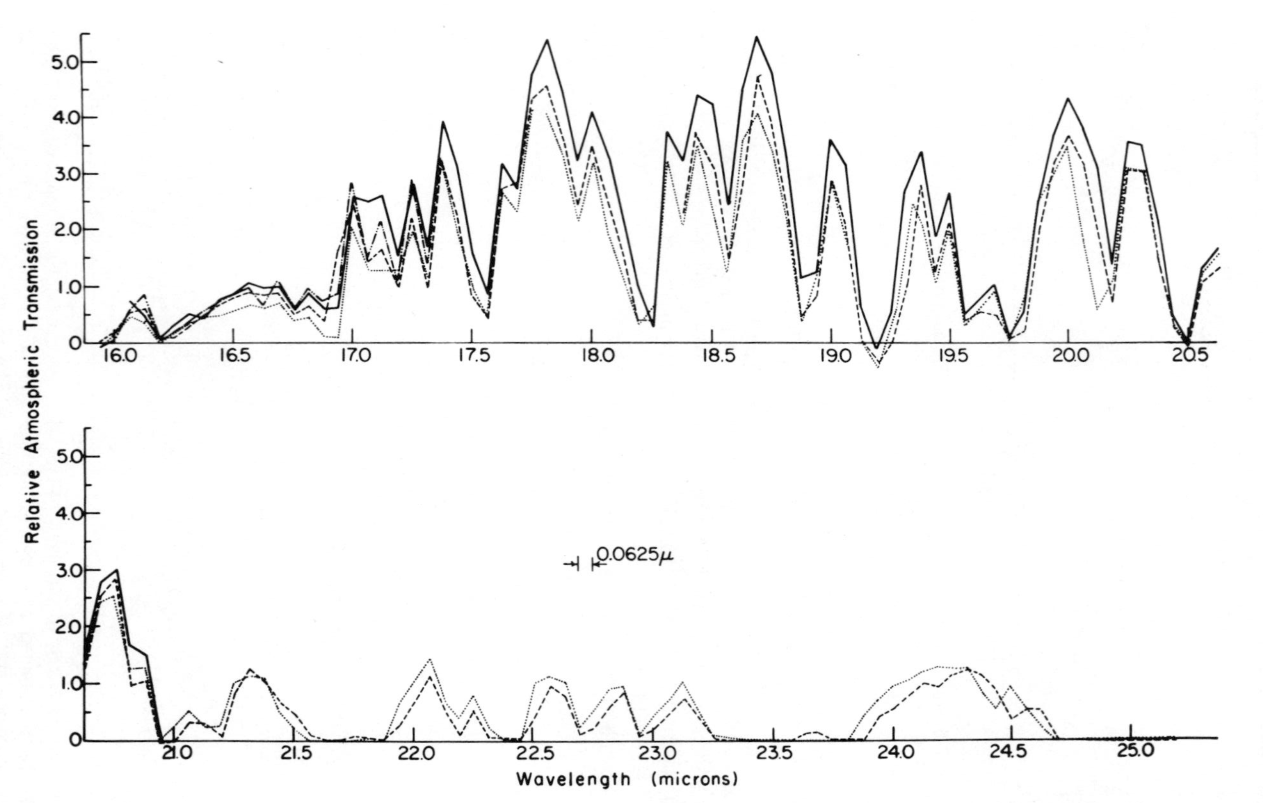

FIGURE 7.8. Atmospheric transmission spectrum at 20 μm for three days measured by P. G. Phillips and D. A. Briotta, Jr. These data were obtained through lunar observations at Mt. Lemmon, Arizona. (From *Applied Optics*,[19] with permission.)

7.6. LIMITATIONS: HTS INSTRUMENTS AND INTERFEROMETERS

A variety of limitations affect any given technique in spectrometry. Since these limitations will differ for different types of instruments, one may expect one type of apparatus to excel where another fails, and vice versa.

There have been several discussions of the relative merits of interferometers and HTS spectrometers in the literature, but the work has not been quantitative and the results are mainly qualitative.(20,21)

In summary form, however, one might be able to characterize the situation in this way. The spectral resolution of a grating instrument is ultimately limited by diffraction effects. These in turn depend on the length of the grating. Corresponding diffraction effects in interferometers depend on the limitation in pathlength differences. Available gratings generally are limited in length, to something of the order of some 20 cm, while interferometers have pathlengths that can differ by up to about 2 m. The ultimate resolution of an interferometer may therefore be some ten times greater than the resolving power of a grating instrument. This conclusion, of course, is based on currently available components.

There is a somewhat similar limitation on the number of resolution elements that can be contained in a spectrum obtained in an HTS instrument, if a fixed-grating instrument is used. Here, the ultimate limitation is either the diffractive limit or the limitation produced by off-axis aberrations, whichever happens to be larger. These effects tend to diffuse the image produced by a point source placed in the spectrometer's entrance focal plane, and therefore determine the minimum size of the mask opening required for a practical spectrometer. Typically these limitations imply that HTS instruments can produce spectra containing no more than a few hundred spectral elements, for any particular grating position. Since many HTS instruments use the grating mainly in one fixed position, this might be considered an inconvenience, but when the grating can be stepped to alter the wavelength intervals displayed across the mask, this limitation is shown not to be severe.

A further limitation of HTS instruments is their ineffectiveness at very long wavelengths, where diffraction effects would make large slit sizes necessary in the masks, and where correspondingly few spectral elements would be resolved. A similar limitation, however, also affects interferometers at long wavelengths.

A final disadvantage of the HTS instruments lies in their inability to obtain the entire high-throughput advantage obtained by, say, the Michelson interferometers. The need for a second mask placed at the entrance aperture of the spectrometer implies a second stage of encoding. In general this second coding stage is not quite as efficient as the first, and effectively the S/N increase goes only as the square root of the number of entrance slits, rather than in direct proportion to that number.

However, one may expect that this limitation will not normally be very serious. Practically used interferometer plates frequently have apertures no larger than an inch or two in diameter, and only a fraction of this aperture may actually be put to use. Moreover, this clear aperture limits the beam at a location where it has already been collimated. This implies correspondence to a rather small entrance aperture on a dispersive spectrometer, and the practical consequences of this modulation limitation of HTS instruments seem quite small. In many HTS devices the total effective energy throughput may actually be larger than that in interferometers.

One definite advantage of HTS instruments lies in the use of binary digital codes. The required computing facilities for inverting the spectrum are made quite modest by these codes. Moreover, the fast Hadamard transform is faster than the fast Fourier transform, for the same reasons. An HTS spectrometer can therefore be set up with more modestly priced minicomputers than a corresponding interferometer would probably require.

Mask motion in the HTS systems needs to be less carefully controlled than the mirror drive motion in a Michelson interferometer. Typically, the width of a mask slot is 0.1 mm, and it is sufficient to move the mask roughly with an accuracy of one-tenth that width. This is considerably less stringent than the control required in an interferometer.

At the moment, it appears that HTS instruments may find their greatest use in relatively low-cost applications where medium resolution spectra, say, with 200–300 resolution elements and roughly 1% accuracy are required. Michelson interferometers are likely to compete more effectively in very high-resolution applications, in applications requiring the display of very large numbers of resolved data points, and conditions where extreme S/N requirements apply. Most of these applications are rather highly specialized.

In HTS instruments special use may be made of blocking apertures that are potentially able to remove undesired portions of the spectrum. A strong emission line right next to some faint emission feature may therefore be blocked out, or uninteresting stretches of a spectrum may be effectively blocked so that the detector responds only to the data of actual interest. Interferometers, on the other hand, are virtually required to pass all available data whether interesting or not, and in some situations that added wealth of data may become a burden.

From this qualitative discussion it becomes clear that both classes of instruments have advantages and weaknesses, and that both classes may be expected to make their marks in different applications in the next few years.

7.7. IMAGERS AND SPECTROMETRIC IMAGERS

We have already mentioned Gottlieb's discussion of an imager that could use a folded (Simplex) S-coded pattern to obtain two-dimensional pictures by multiplexing the radiation incident on a single detector.

Such an instrument has recently been constructed by Swift *et al.*[22] It divides a scene into 1023 spatial elements, in a rectangular array of 31 × 33 elements. Figure 7.9 shows the actual appearance of the instrument. Figure 7.10 shows the appearance of the two-dimensional mask that modulates the image as it is rotated by a motor. Figure 7.11 shows the infrared emission of a human hand, as recorded in a fraction of a second.

A further patented device invented by Harwit[23] has been called a spectrometric imager. In one form of this device, a two-dimensional en-

FIGURE 7.9. Hadamard transform imager. (Courtesy of American Science and Engineering, Inc., Cambridge, Massachusetts, and Spectral Imaging, Inc., Concord, Massachusetts; from *Applied Optics,*[22] with permission.)

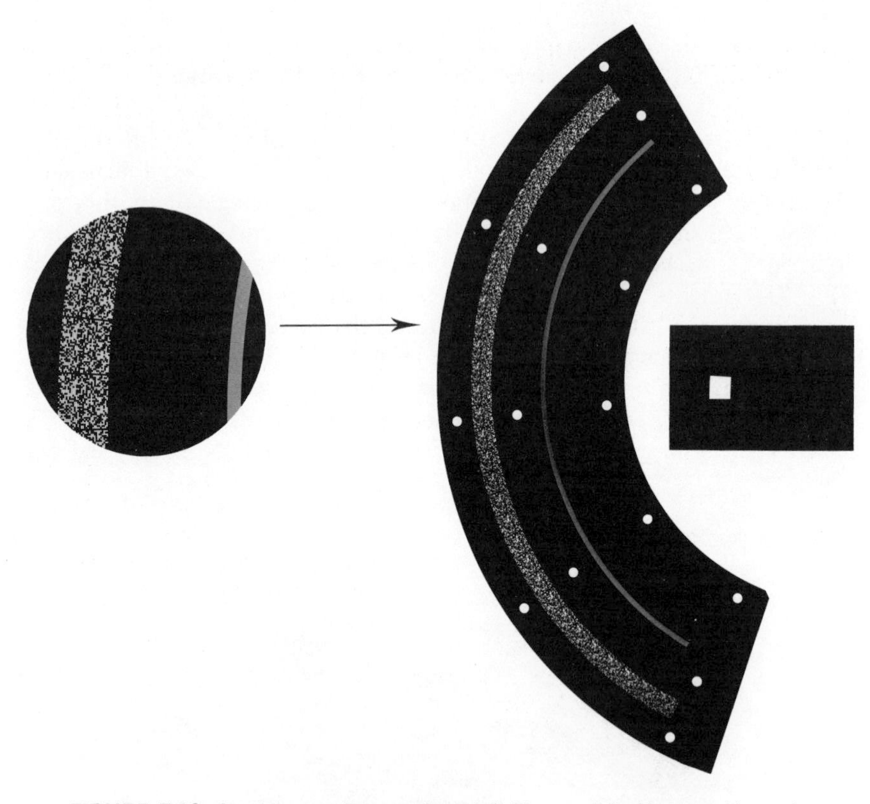

FIGURE 7.10. Spatial encoding mask and field stop (blocking mask) used both for the Hadamard transform imager and spectrometric imager. (Courtesy of Spectral Imaging, Inc., Concord, Massachusetts, and American Science and Engineering, Inc., Cambridge, Massachusetts; from *Applied Optics*,[22] with permission.)

coding mask is placed in the focal plane of a camera or telescope, and spatially encodes the image of a scene. The radiation then passes through a spectrometer where it is encoded again. This spectrometer could be an interferometer, or else an HTS instrument with a second, spectral encoding mask in the instrument's exit focal plane. A variety of combinations has been suggested. Swift *et al.* have constructed a two-mask system in which a 31×33 spatial mask works in conjunction with a spectral mask that resolves 63 spectral elements. The optical layout is shown in Figure 7.12. Some 64,000 data points are obtained and transformed into a corresponding number of spatiospectral intensities.

An interesting question involves the display of that much data. Figure 7.13 shows a cathode ray display of 12 frames that show a flame stretching from lower left to upper right, with a point source blackbody emitting above

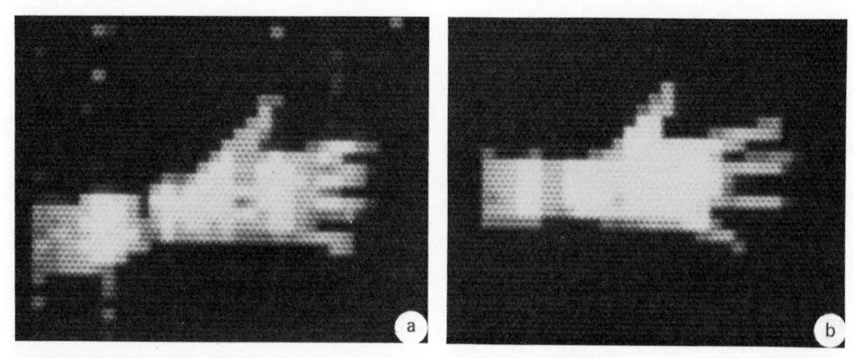

FIGURE 7.11. Thermal emission of a hand reconstructed by a Hadamard transform imager. These data were obtained in the 8–14 μm region with a cooled (Hg, Cd)Te detector. (a) Picture obtained in a single frame time of 25 ms. (b) Image integrated over 16 frame times, each frame lasting 25 ms. Note that along the arm one can distinguish shirt sleeve, bare arm, wristwatch, and hand. The palm of the hand is warmer than the fringes. (Courtesy of American Science and Engineering, Inc., Cambridge, Massachusetts, and Spectral Imaging, Inc., Concord, Massachusetts; from *Applied Optics*,[22] with permission.)

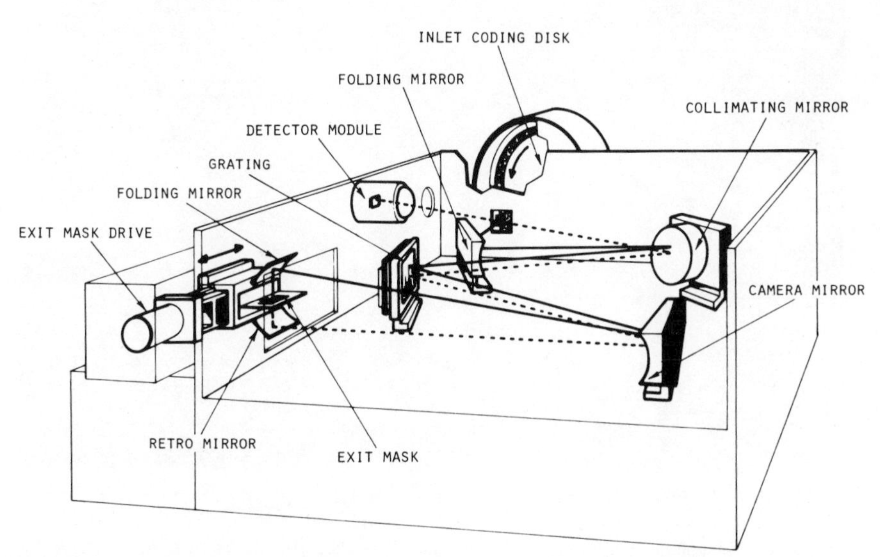

FIGURE 7.12. Optical path through a Hadamard transform spectrometric imager. (Courtesy of Spectral Imaging, Inc., Concord, Massachusetts, and American Science and Engineering, Inc., Cambridge, Massachusetts; from *Applied Optics*,[22] with permission.)

the flame and to the right of center. Each of the 12 frames shows the appearance of this scene at a slightly different wavelength in the vicinity of the 4.3-μm carbon dioxide band. The flame brightens as the center of the band is reached. In comparison the blackbody weakens as its radiation is increasingly absorbed by CO_2 along the atmospheric path between source and spectrometer. Figure 7.14 shows the same scene, but the cathode ray display now has a different format. A cursor spot (bright point) on the flame shows one point (out of the available 1023) for which a detailed spectrum is sought. The spectrum is displayed between 3.06 and 6.33 μm, to the right of the image. It clearly shows the CO_2 emission emanating from that point. A gray scale is shown just below the image on the left. When the cursor is placed so as to select the blackbody point source, the displayed spectrum is a broadly peaked continuum with a sharp absorption line at 4.3 μm. That display is not shown here for lack of space.

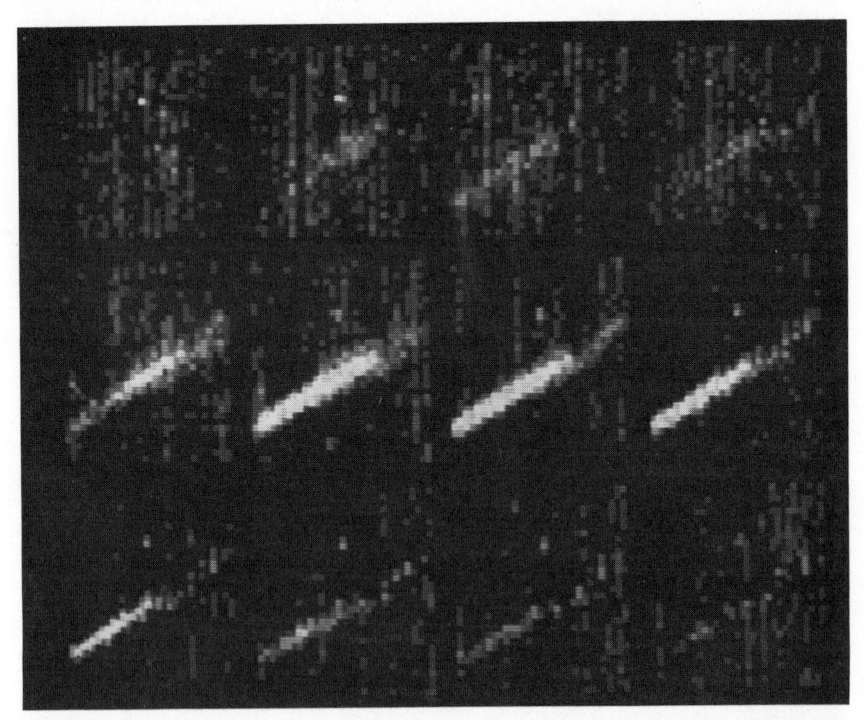

FIGURE 7.13. Twelve frames showing a scene containing a flame and a small (nearly point source) blackbody (see bright spot in upper left frame). The frames represent successive wavelength increments around the 4.3-μm carbon dioxide band. Outside the band, the flame emits very little light. Within the band its emission dominates that of the blackbody. (Courtesy of American Science and Engineering, Inc., Cambridge, Massachusetts, and Spectral Imaging, Inc., Concord, Massachusetts; from *Applied Optics,*[22] with permission.)

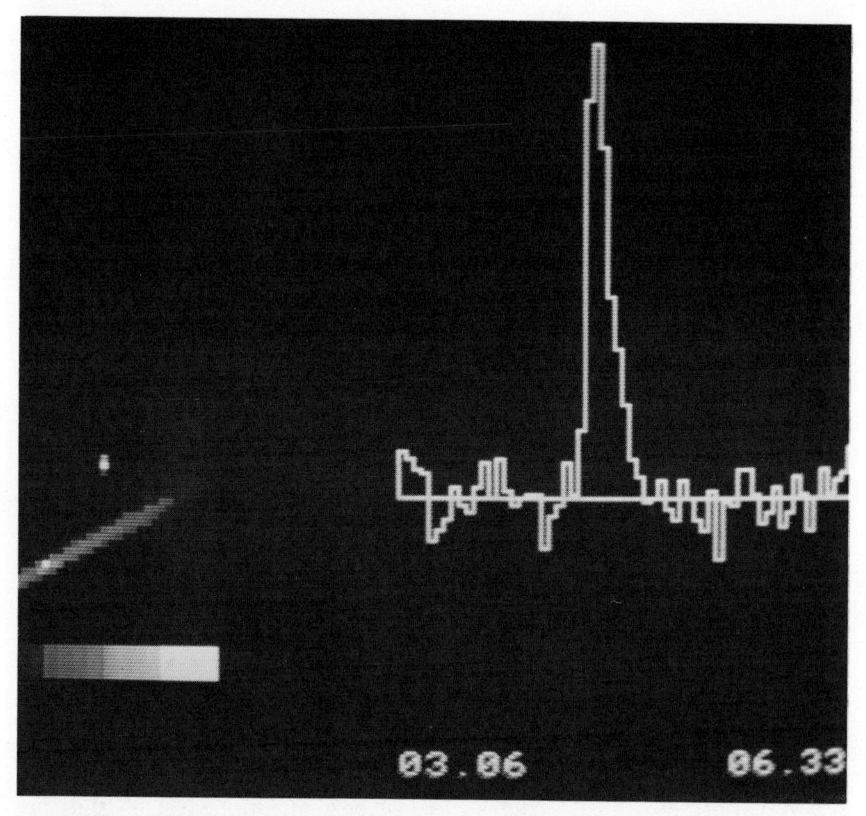

FIGURE 7.14. Display showing the infrared image of the scene shown in Figure 7.13, together with a gray scale and the spectrum of a point on the flame. The selected point is indicated by the cursor spot on the image. The wavelength extremes are 3.06 and 6.33 μm. (Courtesy of Spectral Imaging, Inc., Concord, Massachusetts, and American Science and Engineering, Inc., Cambridge, Massachusetts; from *Applied Optics*,[22] with permission.)

7.8. SIGNAL AND NOISE LIMITATIONS

A wide variety of different sources of noise may be encountered in different spectral or imaging applications. The primary noise source for which multiplexing tends to provide an advantage is intrinsic detector noise. However, sometimes the prime source of noise is photon noise from the detector housing. Other times it is photon noise in the radiation beam from the source. This latter type of noise is irreducible and, in some situations, makes multiplexing ineffective. In various astronomical, atmospheric, or pollution studies, variable atmospheric transmission or emission may be a problem. For many types of detectors, the detector size is an important element that adds noise. If the optical method requires a large detector,

TABLE 7.1. Different Sources of Noise

N_1	Photon noise in radiation source
N_2	Photon noise in field of view, but other than from source to be analyzed
N_3	Variable (atmospheric) transmission between source and detection system
N_4	Variable (atmospheric) emission from foreground field of view
N_5	Size-independent detector noise and amplifier-noise-limited systems
N_6	Size-dependent noise (D^* detector)
N_7	Slow detector—rapid scanning not possible
N_8	Minimum detector size larger than finest resolution obtained from masks
N_9	Thermal photon noise from detector housing

the detector noise may increase to the point where multiplexing is no longer effective.

In some applications the advantage of multiplexing may simply be to obtain finer spectral or spatial resolution than could be obtained if the smallest available detector were used as a limiting aperture in a system. Mask elements nowadays can be made rather finer than the smallest available detectors of some types, and this type of usage may therefore find increasing numbers of applications. Tables 7.1 and 7.2 show the way that different types of noise affect the performance of the various types of instruments discussed in previous sections.

7.9. SPECIAL OPTICAL SYSTEMS

A variety of optical systems has not been mentioned at all thus far but should be briefly discussed, at least for completeness. Girard has constructed a number of different spectrometers in the last few years, in which

TABLE 7.2. Signal-to-Noise Ratio Gain Relative to Single Element Scanning for n Spectral and m Spatial Elements[a,b]

Noise(1)	Singly encoded spectrometer(8)	Doubly encoded spectrometer(14)	Imager(7)	Spectrometric imager(23)
N_1, N_2	$1/2^{1/2}$	$1/2$	$1/2^{1/2}$	$1/2$
N_3, N_4	Depends on ability to compensate noise by means of reference beam			
N_5, N_7	$n^{1/2}/2$	$(nm)^{1/2}/4$	$m^{1/2}/2$	$(nm)^{1/2}/4$
N_6, N_9	$n^{1/2}/2$	$n^{1/2}/4$	$1/2$	$n^{1/2}/4$
N_8	$n^{1/2}/2$	$(nm')^{1/2}/4$	$(m')^{1/2}/2$	$(nm')^{1/2}/4$

[a] This assumes dedispersion for all spectral instruments.
[b] m' is the number of spatial elements that could be imaged onto the minimum size detector.

different types of modulating patterns were employed. One particular article that represents this technique[24] dates back to 1963, but there is a wide variety of publications describing different encoding procedures and discussing many interesting points. A related approach has been that of Mertz[25] in constructing his “mock interferometer.”

In the sense that the Hadamard codes can be proven to be optimum, it is likely that the use of other types of masks is not going to lead to any coding advantages. However, there are other considerations. For example, the mock interferometer manages to provide spectral data with a grating instrument having wide apertures both at the entrance and exit focal planes, and it manages to do this despite its use of only a single mask. In this procedure it potentially is able to avoid the encoding losses that accompany a second stage of encoding mentioned in Section 7.6.

Another way in which differences in mask patterns may make themselves felt is in their ability to minimize the effects of different types of aberrations. In an Ebert Fastie spectrometer, for example, one might expect to benefit from slits having a radius of curvature approximately equal to the mean distance of the mask from the optical axis. For such slits aberrations might be expected to be minimum. In such instances the main advantage to be derived from the use of different types of patterns lies in optical resolution rather than in coding fidelity and the reduction of random noise.

Special wide-aperture spectrometers may be expected to prove useful in HTS applications. The main drive to make improvements along these lines has come from Strong's group at the University of Massachusetts.[26,27] Their instrument obtained intermediate (0.2 cm^{-1}) resolution spectra in the 700-cm^{-1} region. One may also expect that holographic gratings might permit simplification of optical systems in the next few years and that this might affect the construction of HTS systems.

7.10. SOME FUTURE APPLICATIONS

Potentially the most competitive use of Hadamard transform instruments may come from their ability to yield both spatial and spectral data. Chemical differences across the surfaces of solids might be studied by the techniques that led to the types of displays shown in Figures 7.13 and 7.14. It is already clear from those figures that spectra of different portions of flames can be displayed to show combustion sequences. A variety of gas reactions might perhaps also be studied by this combined spatiospectral technique.

Mapping from spacecraft for meteorological studies can yield atmospheric temperature and humidity profiles through studies in the wings of spectral lines. With some refinements vertical wind structure might be similarly studied to obtain relatively detailed meteorological information. There is also a similar need for spatiospectral data from spacecraft in connection with various Earth resources studies. Such applications have not been sufficiently explored, to date, mainly because the instrumental capabilities did not appear to be at hand. An experiment that currently is being planned for the Space Shuttle is a downward-looking high-resolution ozone mapper. This study is under the direction of Dr. Donald F. Heath at the NASA Goddard Space Flight Center in Greenbelt, Maryland.

On the simpler, purely spectral level, one may expect substantial usage of HTS techniques in photoacoustic spectroscopy where devices are intrinsically noisy. Rosencwaig,[28] who has pioneered that technique currently is combining it with a Hadamard transform spectrometer for greater sensitivity.

Similarly one may expect that tuned laser spectroscopy, as embodied in the thermal lens technique described by Long *et al.*,[29] may be improved through the use of a spatial mask. In this technique the laser is tuned through an absorption line of a tenuous gas. As the line center is approached, the gas becomes increasingly heated along the optical path and its density decreases. This produces a defocusing effect, and the laser light, which initially was transmitted through a pinhole just in front of the detector, spreads out of the pinhole and is blocked from reaching the detector. If a spatial encoding mask were used in conjunction with this technique, light could be normally kept from falling onto the detector, and only the defocused radiation would be permitted to fall on it. Moreover, the shape of the defocused beam would be mapped and the selection of the proper laser mode ensured in that fashion. By measuring only the spilled radiation rather than the decrement in the primary beam, additional sensitivity should be attained in this already highly sensitive method.

ACKNOWLEDGMENTS

Work on Hadamard transform spectroscopy at Cornell has been kindly supported through contracts NGR 33-010-210 and NSG-1263 from the National Aeronautics and Space Administration and by contract F19628-73-C-0110 from the Air Force Cambridge Research Laboratories.

REFERENCES

1. M. Harwit and J. A. Decker, Jr., Modulation Techniques in Spectrometry, *in: Reports on Progress in Optics XII*, pp. 101–162 (E. Wolf, ed.), North-Holland, Amsterdam, 1974.
2. N. J. A. Sloane and M. Harwit, Masks for Hadamard transform optics, and weighing designs, *Appl. Opt.* **15**, 107–114 (1976).
3. F. Yates, Complex experiments, *J. Roy. Stat. Soc. Supp.* **2**, 181–247 (1935).

4. M. J. E. Gollay, Multi-slit spectrometry, *J. Opt. Soc. Am.* **39**, 437–444 (1949); Static multislit spectrometry and its application to the panoramic display of infrared spectra, *J. Opt. Soc. Am.* **41**, 468–472 (1951).
5. R. N. Ibbett, D. Aspinall, and J. F. Grainger, Real-time multiplexing of dispersed spectra in any wavelength region, *Appl. Opt.* **7**, 1089–1093 (1968).
6. J. A. Decker, Jr., and M. Harwit, Sequential encoding with multislit spectrometers, *Appl. Opt.* **7**, 2205–2209 (1968).
7. P. Gottlieb, A television scanning scheme for a detector-noise-limited system, *IEEE Trans. Inform. Theor.* **IT-14**, 428–433 (1968).
8. N. J. A. Sloane, T. Fine, P. G. Phillips, and M. Harwit, Codes for multiplex spectrometry, *Appl. Opt.* **8**, 2103–2106 (1969).
9. S. W. Golomb, ed., *Digital Communications with Space Applications,* Prentice-Hall, Englewood Cliffs, New Jersey, 1964.
10. Marshall Hall, Jr., *Combinatorial Theory,* p. 204, Blaisdell, Waltham, Massachusetts, 1967.
11. E. D. Nelson and M. L. Fredman, Hadamard spectroscopy, *J. Opt. Soc. Am.* **60**, 1664–1669 (1971).
12. H. Hotelling, Self improvements in weighing and other experimental techniques, *Ann. Math. Stat.* **15**, 297–306 (1971).
13. M. Harwit, P. G. Phillips, T. Fine, and N. J. A. Sloane, Doubly multiplexed dispersive spectrometers, *Appl. Opt.* **9**, 1149–1154 (1970).
14. M. Harwit, P. G. Phillips, L. W. King, and D. A. Briotta, Jr., Two asymmetric Hadamard transform spectrometers, *Appl. Opt.* **13**, 2669–2674 (1974).
15. P. G. Phillips and M. Harwit, Doubly multiplexing dispersive spectrometer, *Appl. Opt.* **10**, 2780–2781 (1971).
16. Ming Hing Tai, D. A. Briotta, Jr., N. Kamath, and M. Harwit, Practical multi-spectrum Hadamard transform spectrometer, *Appl. Opt.* **14**, 2533–2536 (1975).
17. John A. Decker, Jr., Experimental realization of the multiplex advantage with a Hadamard-transform spectrometer, *Appl. Opt.* **10**, 510–514 (1971).
18. John A. Decker, Jr., Private communication.
19. P. G. Phillips and D. A. Briotta, Jr., Hadamard-transform spectrometry of the atmospheres of Earth and Jupiter, *Appl. Opt.* **13**, 2233–2235 (1974).
20. T. Hirschfeld and G. Wintjes, Fourier transform vs. Hadamard transform spectroscopy, *Appl. Opt.* **12**, 2876–2880 (1973); Fourier transform vs. Hadamard transform spectroscopy: Author's reply to comments, *Appl. Opt.* **13**, 1740–1741 (1974).
21. J. A. Decker, Jr., Comments on: Fourier transform vs. Hadamard transform spectrometry, *Appl. Opt.* **13**, 1296–1297 (1974).
22. R. D. Swift, R. B. Wattson, J. A. Decker, Jr., R. Paganetti, and M. Harwit, Hadamard transform imager and imaging spectrometer, *Appl. Opt.* **15**, 1595–1609 (1976).
23. M. Harwit, Spectrometric imager, *Appl. Opt.* **10**, 1415–1421 (1971); Spectrometric imager, Part 2, *Appl. Opt.* **12**, 285–288 (1973); covered by U.S. Patent No. 3,720,469.
24. A. Girard, Spectromètre à grilles, *Appl. Opt.* **2**, 79–87 (1963).
25. L. Mertz, *Transformations in Optics,* John Wiley and Sons, New York, 1965.
26. J. Strong, Achromatic doublet lenses for infrared radiation, *Appl. Opt.* **10**, 1439–1443 (1971).
27. P. Hansen and J. Strong, High resolution Hadamard transform spectrometer, *Appl. Opt.* **11**, 502–506 (1972).
28. A. Rosencwaig, Photoacoustic spectroscopy: A new tool for investigation of solids, *Anal. Chem.* **47**, 592A–604A (1975).
29. M. E. Long, R. L. Swofford, and A. C. Albrecht, Thermal lens technique: A new method of absorption spectroscopy, *Science* **191**, 183–184 (1976).

Chapter 8

Pulsed and Fourier Transform NMR Spectroscopy

Thomas C. Farrar

8.1. INTRODUCTION

Modern Fourier transform (FT) NMR spectrometers have been available commercially for only about six years, but in this short period of time many advances and improvements in instrumentation have been made. The first FT–NMR instruments were conventional, continuous wave (CW) spectrometers to which a pulsed RF transmitter, a modified receiver–detector system, a signal-averaging device, and a small minicomputer were attached. It is not too surprising that all of the early FT instruments were quite expensive ($100,000 to $200,000), difficult to operate, fraught with numerous operating problems, and not very reliable.

Extremely rapid advances have been made in modern electronics. These advances have brought about almost unbelievable changes, improvements, and cost reductions in minicomputers and RF and digital electronic devices; they also have revolutionized FT–NMR instruments. Today's FT–NMR system is relatively inexpensive ($65,000 and up); in terms of 1965 dollars a modern FT–NMR unit costs no more than a 1965 vintage A-60 type proton NMR system did. The FT unit has, however, about 1000 times greater sensitivity, and capabilities undreamed of in an A-60 type instrument. In addition, the new models are very easy to operate, very reliable, and capable of operating in a number of automatic and semiauto-

Thomas C. Farrar • National Science Foundation, Chemistry Division, Washington, D.C. 20550

matic modes. It is quite common for these instruments to be operated 24 hours a day, 7 days a week. Rapid development is still in progress and many additional improvements in sensitivity and sophistication are continually being made.

In this chapter we will give (a) a brief description of FT–NMR spectroscopy, (b) the requirements and design of a modern FT–NMR spectrometer, (c) some standard applications of the method, (d) a brief summary of some of the current developments in progress.

8.2. BASIC CONCEPTS OF FT–NMR

We shall give here only a brief account of the theory of FT–NMR, since many excellent accounts are readily available.[1-4] It is now rather common knowledge that the free-induction decay (FID) signal following an RF pulse and the CW spectrum that is obtained by conventional frequency or field sweep methods are Fourier transforms of one another. It is also widely recognized that FT–NMR has a very large sensitivity advantage over the CW method. This arises primarily from the manner in which the data are generated.

In a conventional spectrometer the field or frequency is changed slowly and the intensity of each resonance line is recorded as the resonance condition is traversed. If the total spectral range to be observed is Δ Hz and the resolution desired is r(Hz), then the observation time t_{CW} required is given by

$$t_{CW} \simeq \Delta/r \tag{8.1}$$

In the FT–NMR experiment a short (about 10 μs) RF pulse is used to excite *all* of the nuclear spins simultaneously; we shall see how this is done in our discussion of FT–NMR instrumentation. This pulse is then followed by the nuclear "free induction decay" (FID) signal, which contains all of the information present in a normal NMR spectrum. The time required, t_{FT}, to obtain a spectrum with a resolution of r Hz is

$$t_{FT} = 1/r \tag{8.2}$$

The time and sensitivity enhancement factors e_t and e_s, respectively, are given by

$$e_t = t_{CW}/t_{FT} = \Delta \tag{8.3a}$$

$$e_s = e_t^{1/2} = \Delta^{1/2} \tag{8.3b}$$

Since $\Delta \simeq 1000$ for ^{1}H in a 23 kG field and $\Delta \simeq 5000$ for ^{13}C, one expects (and observes) a sensitivity enhancement of about 30 for ^{1}H and of about 70 for ^{13}C.

The price one pays for this large gain in sensitivity is that the FID signal is almost impossible to interpret in all but the very simplest of samples.

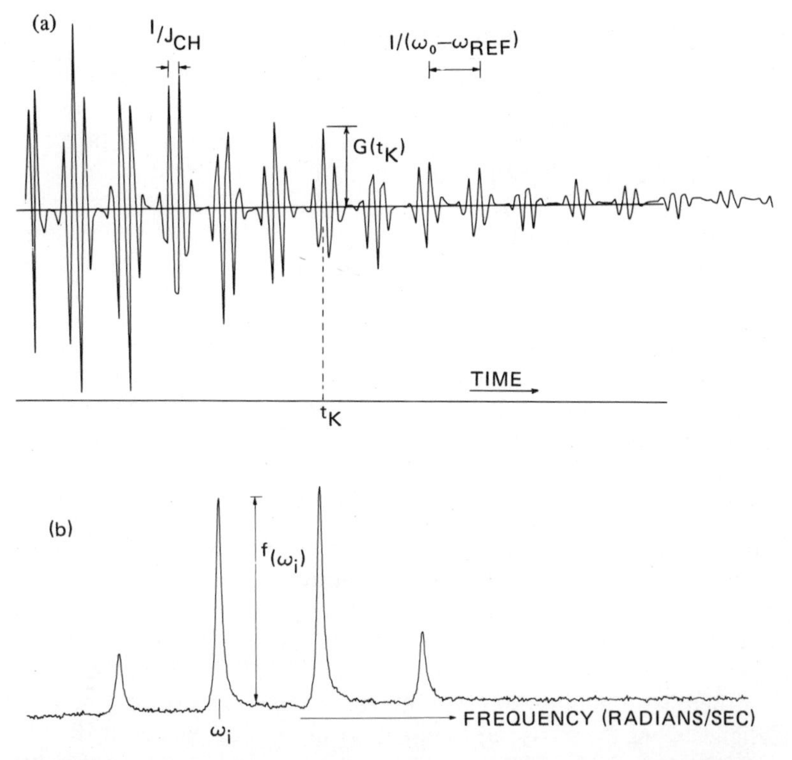

FIGURE 8.1. (a) The FID signal of a sample of CH_3OH. The nucleus observed is ^{13}C; the system is spin-coupled to the three chemically equivalent protons. The two beat frequencies are related to the chemical shift $(\omega_0 - \omega_{ref})$ and the $^{13}C-^{1}H$ spin-coupling constant $J_{C\text{-}H}$. (b) The Fourier transform of the FID shown in (a).

It is therefore necessary to convert the information to a format that is more generally useful to the average chemist. This conversion is done via the Fourier transformation.

Figures 8.1a and 8.1b show the FID (or time domain) signal and the frequency domain spectrum. A detailed discussion of how one obtains one from the other is given in Chapters 2 and 4. For the sake of clarity and continuity in this chapter we shall state only that the transformation from the time domain to the frequency domain is executed according to the following prescription:

$$f(\omega_j) = \sum_k G(t_k) \exp(i\omega_j t_k)\, \Delta t \tag{8.4a}$$

and from the frequency domain to the time domain via

$$G(t_j) = \sum_k f(\omega_k) \exp(i\omega_k t_j)\, \Delta\omega \tag{8.4b}$$

where the symbols are defined as follows:

$f(\omega_j)$ the intensity in the frequency domain at the frequency value ω_j
ω_j the frequency value (in rad/sec) at the jth point in the spectrum
$G(t_j)$ the amplitude of the signal in the time domain at time t_j
t_j the time elapsed at the jth point in the time domain
i $(-1)^{1/2}$, i.e., the data are imaginary or complex
Δt the time interval between adjacent points in the time domain
$\Delta\omega$ the frequency interval between adjacent points in the frequency domain

In order to discuss pulsed NMR experiments it is most convenient to use a frame of reference that is rotating about the z axis at the rate ω_0, at which the nuclei are precessing. Recall that this precessional frequency depends upon the magnetic field strength of the spectrometer, such that

$$\omega_0 = \gamma H_0 \tag{8.5}$$

where ω_0 is the nuclear precession frequency expressed in rad/sec (this frequency is sometimes referred to as the Larmor frequency), γ the magnetogyric ratio of the nucleus (it is a fundamental property of the nucleus), and H_0 the intensity of the DC magnetic field. For a value of H_0 of about 23,000 G (or 2.3 T), ω_0 is about 6×10^8 rad/sec (or about 100 MHz). In this rotating frame there is, at equilibrium, a stationary net macroscopic magnetization M, aligned along the z axis (see Figure 8.2a). In a typical pulsed NMR experiment this magnetization M is driven to some nonequilibrium position by an RF pulse and either the FID signal is recorded or the return of M to equilibrium is monitored. The RF field used to drive M about in the desired manner is usually designated by the symbol H_1 and has units of gauss; it is also often expressed as $(\gamma/2\pi) H_1$, in which case the units are hertz. In most experiments H_1 lies along the $+x$, $-x$, $+y$, or $-y$ axes in the rotating frame of reference. This RF field is generated in the transmitter unit of the spectrometer and coupled to the sample via the sample coil. All nuclei have angular momentum and if a torque is applied to them (in this case via H_1 and their magnetic moments) they will precess about an axis that is perpendicular to the plane defined by the torque vector H_1 and the nuclear angular momentum vector, which is coincident with the magnetic moment vector M (see Figure 8.2a). In a time t_p seconds M will precess through an angle θ given by

$$\theta = \gamma H_1 t_p \tag{8.6}$$

In order to designate the axis along which H_1 is applied one usually writes $H_1|_{+x}$, $H_1|_{+y}$, $H_1|_{-x}$, or $H_1|_{-y}$. The terms 90° RF phase shift or 180° RF phase shift are often used in discussing pulse experiments. Phase shifts of 0, 90, 180, and 270° give H_1 fields directed along $+x$, $+y$, $-x$, and $-y$, respectively. A 90° pulse directed along the $+x$ axis, that is, $H_1|_{+x}$, will rotate

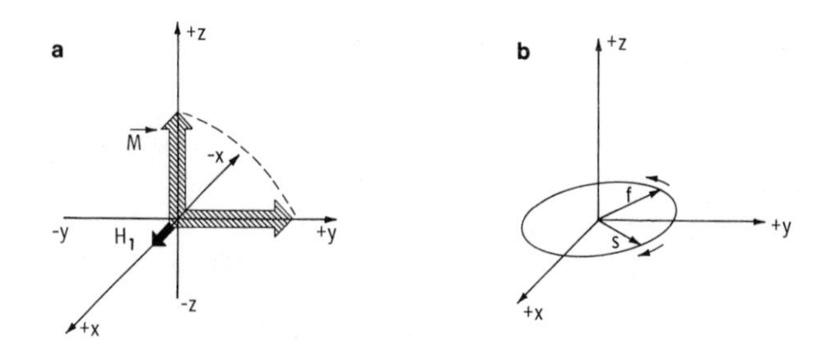

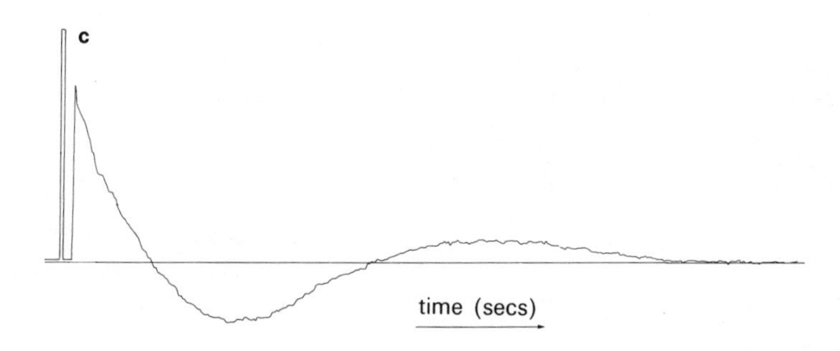

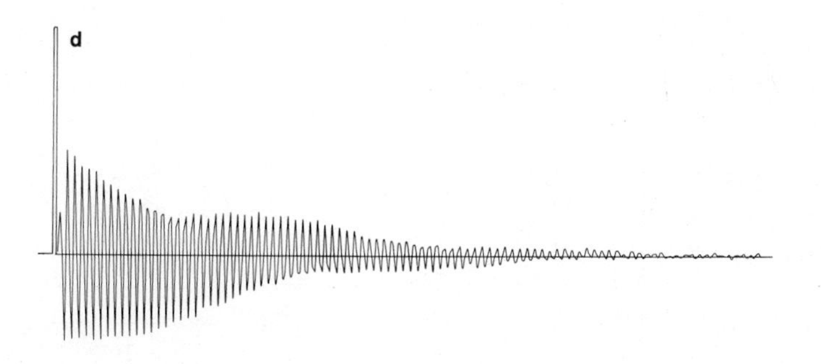

FIGURE 8.2. (a) The torque exerted by the radio frequency field H_1 on the net macroscopic magnetization vector M causes it to precess about the x axis in the rotating reference frame. (b) The dephasing of M due to magnetic field inhomogeneities. The f nuclei precess faster than average because they are in a higher-than-average magnetic field. The s nuclei precess slower than average because they are in a lower-than-average magnetic field. (c) The FID signal for a sample of water that is almost exactly "on-resonance," i.e., $\omega_0 = \omega_{\text{ref}}$. The beat frequency is about 0.5 Hz. (d) The FID signal for a water sample that is "off-resonance" by about 100 Hz.

M 90° about the $+x$ axis. If M was originally directed along the $+z$ axis, it will, after the pulse, lie along the $+y$ axis (see Figure 8.2a).

If M is rotated from the $+z$ to the $+y$ axis and the H_1 field turned off, M will remain along the $+y$ axis in the xy plane. Now a stationary M along the $+y$ axis in the rotating frame is a rotating M in the laboratory frame. It is this rotating magnetization in the xy plane of the laboratory frame of reference that generates the nuclear RF signal that is picked up by the sample coil in the NMR probe, amplified by a factor of about 10^6, detected, converted to a digital signal, and stored in the computer. Because of relaxation processes and magnetic field inhomogeneities this net magnetization M becomes dephased and the amplitude of the nuclear RF signal decays to zero (Figure 8.2b). What one actually observes on an oscilloscope is shown in Figure 8.2c. The process is initiated by a 90° RF pulse $H_1|_{+x}$, which is followed by the nuclear RF signal, i.e., the FID signal, which then decays to zero. The trace depicted in Figure 8.2c is for a sample of water "on resonance," that is, the nuclear precession frequency ω_0 and the spectrometer reference oscillator frequency f_0 are exactly the same. If the value of ω_0 is changed by changing the magnetic field strength H_0, then the signal observed is as shown in Figure 8.2d. This "beat" frequency is equal to the difference $\omega_0 - f_0$. The same sort of beat pattern is obtained by keeping the field constant and using different samples with different values of the chemical shift. The sequence of events in a typical experiment is as follows:

(a) Apply a small-angle RF pulse to the sample.

(b) Record the FID signal.

(c) If necessary repeat (b) many times and add each new FID to the previous sum.

(d) Execute a Fourier transformation to obtain the frequency domain spectrum.

The optimum value of the angle through which M should be rotated for the best signal-to-noise (S/N) ratio in a time-averaged FT–NMR experiment depends upon the values of the spin–lattice relaxation time T_1, the spin–spin relaxation time T_2, and the time between the pulses. A reasonable rule of thumb is that if a 90° pulse is used, the time between pulses should be approximately equal to T_1. In most molecules of interest there is a distribution of relaxation times for the various chemically different nuclei. For example, ^{13}C nuclei in a CH_3 group typically have relaxation times of 5 to 15 sec, whereas quaternary or carbonyl carbon, which have no directly bonded protons, have relaxation times of 10 to 100 sec or more. This often requires some considerable compromise to be made. This is all discussed in great detail in the paper by Ernst.[6]

In the CW experiment the RF field needed should be small enough that it does not significantly perturb the equilibrium distribution of nuclear spins ($\gamma H_1 \simeq 1$ mG). In the FT–NMR experiment the requirement on the

RF field is that

$$\gamma H_1 > 2\pi\Delta \tag{8.7}$$

where Δ (in units of hertz) is the entire range of chemical shifts to be observed ($\gamma H_1 \simeq 5$ to 50 G). In order to make accurate quantitative measurements about the relative amounts of the different sorts of nuclear species present, it is important that this condition [equation (8.7)] be met. For $\gamma H_1 = 2\pi\Delta$, an intensity error of about 2% or less usually results. If equation (8.7) is satisfied and a sufficient number of data points are used, then the intensity accuracy depends primarily on the signal-to-noise ratio. This is discussed in more detail later. A convenient way to measure the amount of γH_1 irradiation *at the sample* is to measure the time, $t(\pi/2)$, required to rotate the magnetization M by an angle of 90° (i.e., the angle required to produce the maximum FID signal):

$$\frac{\gamma}{2\pi} H_1 \simeq \frac{240}{t(\pi/2)} \quad \text{(kHz)} \tag{8.8}$$

This equation is valid for a spectrometer that operates using quadrature phase detection (QPD), but not for one that uses conventional single-phase detection. A discussion of quadrature and single-phase detection is given in Chapter 9. If single-phase detection is employed, the RF transmitter power is not used very effectively and the 240 kHz in the numerator of equation (8.8) must be changed to about 80 kHz. For most general ^{13}C work at 25 MHz (in a 23,000-G magnetic field) $t(\pi/2)$ should be 50 μsec or less (with QPD) since $\Delta \simeq 5$ kHz for ^{13}C at this field. For ^1H at 100 MHz $t(\pi/2)$ should be 100 μsec or less. In most modern FT–NMR spectrometers the values for $t(\pi/2)$ for ^{13}C and ^1H are both 10 to 20 μsec or less, and so no problems should be encountered because of insufficient RF power.

The factors that determine the value of γH_1 are the RF transmitter power P (in watts), the quality factor of the sample circuitry Q, the operating frequency of the spectrometer f_0 (in hertz), the volume of the transmitter coil V_c (in cc), and the filling factor of the sample coil η (η is the ratio of the volume of the NMR sample to the volume of the receiver coil). More succinctly,

$$\gamma H_1 = k\eta \frac{(PQ)^{1/2}}{f_0 V_c} \tag{8.9}$$

where k is a constant. We will discuss in detail some of the ramifications of this equation in Section 8.3.1.

The new FT–NMR spectrometers have made a major impact on chemistry in two different ways: (1) they have dramatically increased the sensitivity of the NMR technique and (2) they have made relaxation time experiments almost routine. In addition these new systems can execute, in manual or automatic mode, a number of new experiments that provide a wealth of new information about molecular and chemical dynamics and molecular structure. This is done by measuring *n*uclear *O*verhauser *e*ffects (NOE), spin–lattice (T_1), spin–spin (T_2), and rotating-frame spin–lattice ($T_{1\rho}$) relaxation times. Descriptions of such experiments are given later in this chapter and in Chapter 10.

At this point in time it is, from our point of view, safe and accurate to state that for proton NMR there is no experiment that can be done by CW methods, that cannot be done much faster and with higher quality data using FT methods. For ^{13}C NMR it is not so easy to compare CW and FT methods since *routine* ^{13}C experiments cannot be done using CW methods. The ^{13}C NMR experiment, although now routine and mostly automatic, is a fairly complex experiment. The details are given in the next section but the important point to keep in mind is that the proton nuclear spins are almost always decoupled from the ^{13}C nuclei. Since ^{13}C has a natural abundance of only 1 % there is only a very small chance (1 % or less) that any given molecule will contain more than one ^{13}C nucleus and an even smaller chance that a second ^{13}C nucleus, were it in the molecule, would be adjacent to the first one. Consequently, the ^{13}C NMR spectra that one usually sees are first order. That is, a single sharp line is seen for each chemically different ^{13}C nucleus and no fine structure due to spin–spin interactions is present. The proton decoupling is normally accomplished by constantly irradiating the sample with RF power at the resonance frequency of the protons. A shorthand way to describe this experiment is ^{13}C–{^{1}H}. Further details are given later.

In both ^{13}C–{^{1}H} and ^{15}N–{^{1}H} spectroscopy the nuclear Overhauser effect (NOE) plays a very important role. For heteronuclear interactions the NOE causes a change in the intensity of the observed NMR signal I_o, due to the irradiation of a second, spin-coupled nucleus I_r, with which it interacts. The maximum signal gain or enhancement E that one obtains is given by

$$E = 1 + \gamma_r/2\gamma_0 \tag{8.10}$$

For ^{13}C observation with proton decoupling where γ_r/γ_0 is equal to γ_H/γ_C = 3.97, E = 2.99, that is, with a maximum NOE the ^{13}C NMR signal is three times as great as without the enhancement. Since the proton decoupling also collapses fine structure into a single resonance line this enhances the S/N values of the remaining lines further. For a methyl carbon with proton decoupling and a maximum NOE, the resulting single line has an intensity 24 times greater than the outer lines in the 1:3:3:1 quartet that would be observed without the decoupling. The maximum NOE is obtained only when the relaxation is entirely due to the dipole–dipole interaction between the observed and the irradiated nucleus. If paramagnetic ions are present, if the molecular correlation times become too long, or the spin–rotation interaction contributes significantly to the relaxation then the enhancement can be greatly reduced or entirely eliminated. For ^{15}N (and a few other nuclei) the NOE is negative, and $\gamma_H/\gamma_N = -9.87$; thus $E = -3.93$. If nondipolar relaxation processes are present then in principle the value of E may be zero and the signal will disappear. This does, in fact, happen and great care should be exercised when doing ^{15}N experiments. More details of the NOE and its applications can be found in the recent excellent book by Noggle and Schirmer.[7]

Before presenting some applications of FT–NMR, we shall first discuss the basic NMR spectrometer.

8.3. BASIC INSTRUMENTATION

8.3.1. The Spectrometer

A block diagram of a typical FT–NMR spectrometer is shown in Figure 8.3a. For convenience we shall consider a typical example, ^{13}C FT–NMR. For ^{13}C NMR three separate frequencies are required: 25 MHz for the ^{13}C

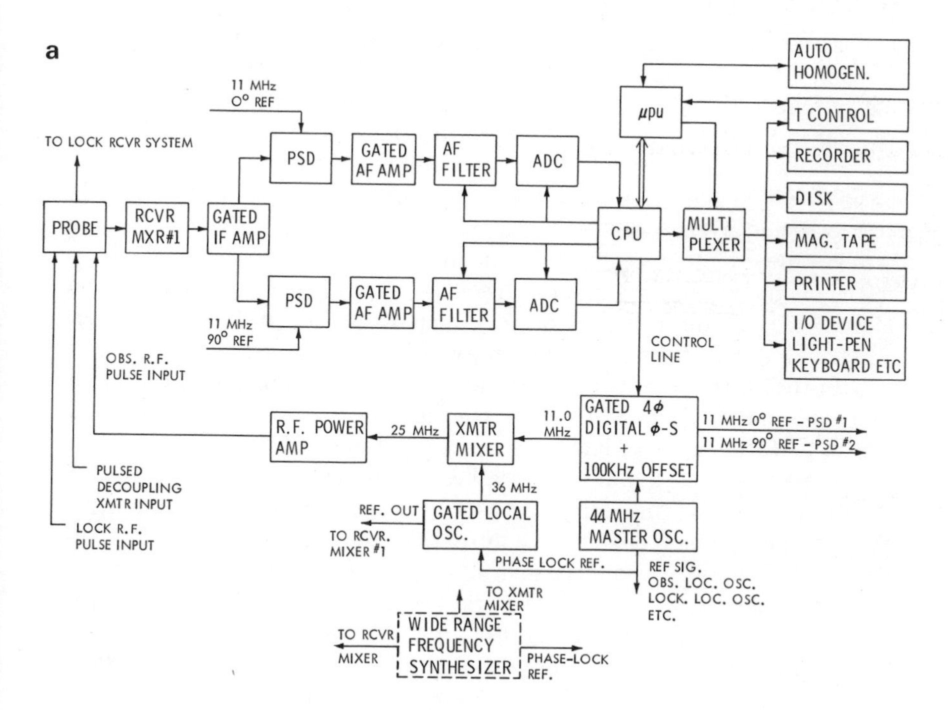

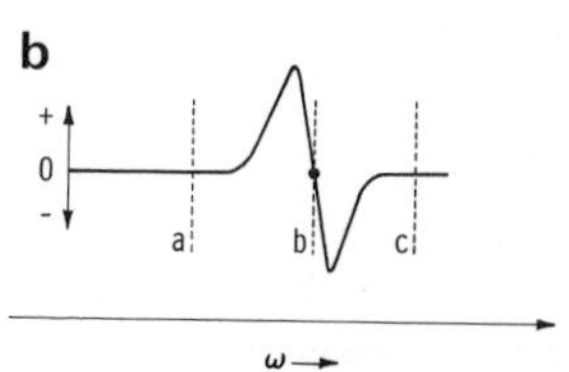

FIGURE 8.3. (a) A block diagram for a modern FT–NMR spectrometer. See text for details. (b) Lock-frequency amplitude vs. "error" frequency, $\Delta\omega$, between Larmor frequency ω_0 and Larmor field H_0.

observation channel, 100 MHz for the ^1H decoupling channel, and 15 MHz for the ^2H internal or external lock channel. The details are shown only for the ^{13}C observation channel; the ^2H lock circuitry is almost identical to the ^{13}C observation circuitry for both the receiver and the transmitter. The ^1H decoupling channel does not require a receiver and so it is omitted. In the interest of economy the transmitter of the lock channel is sometimes designed to be operated at a fixed frequency. This requires making slight changes in the value of the magnetic field when the solvent is changed.

The lock channel for the spectrometer is necessary to stabilize the system so that long-term signal averaging can be carried out. Its operation is basically rather simple. The solvents used for most ^{13}C NMR studies contain deuterium. Typical solvents are D_2O, d_6-acetone, $CDCl_3$, and d_6-dimethyl sulfoxide. They share a common property, in that they all have single deuterium resonance signals. By proper adjustment of the spectrometer a "spectrum" similar to that shown in Figure 8.3b is obtained as the deuterium RF is swept. If the RF is adjusted so that this dispersion signal is at point (a) and the sweep then turned off, the signal will remain at (a). Any small variations in either the magnetic field or the spectrometer frequency will cause the system to deviate from point (a) and a positive or negative signal will register in the lock channel; the sign of the DC signal tells one whether the system has drifted up or down in field (or frequency). This DC signal is used as an "error signal" and can correct very minute changes (up to 1 part in 10^9 or better) in the stability of the spectrometer system. The error signal is incorporated into a feedback system and compensates for any instabilities in the magnetic field or any one of the three radio frequencies. This system allows one to routinely obtain very high resolution (0.1 Hz or better) spectra even with time averaging up to 24 hours or more.

In the early FT–NMR systems the role of the computer was a minor one and primarily limited to executing the Fourier transform on the FID signal, performing some data reduction operations and recording the spectrum on an *x-y* recorder. It is now rather common for the computer to play a more central role. In addition to executing the transform and displaying the frequency domain data it can control the frequency of the lock, observation, and decoupling channels, and in some cases the homogeneity of the magnetic field and the sample temperature. It is now common practice for a magnetic tape cassette, a floppy-disk, or a large disk unit to be an integral part of the spectrometer system. This increases the flexibility of the system considerably since a large number of software programs can be stored, as well as large amounts of data. If the computer has the capability of controlling the spectrometer, a large number of difficult experiments can be run in the automatic mode. Computer control also means that as new methods and techniques are developed the spectrometer's capability can be expanded, in many instances by writing new software, without the need to buy new hard-wired accessories.

In all modern spectrometers it is essential that the multiple RF phase capability mentioned earlier be included. This is required for the systematic noise reduction described in Chapter 9 as well as for the T_1, T_2, and $T_{1\rho}$ experiments described later. It is also absolutely essential for the new techniques being developed to execute high-resolution NMR studies in solids.[8-15] In most cases the accuracy and stability of the relative phases is quite important. For operation in the quadrature detection mode (see Chapter 9) accurate adjustment of the 0° and 90° phases is important and it is essential that the phases be drift free. It is true that by using the techniques described in Chapter 9 it is possible to correct many of the spectral anomalies that arise from inaccurate phase settings in the phase detectors, but for these methods to work well it is important that the RF voltages to the transmitter and receiver be phase-stable. One way to generate accurate, stable phases is to use digital techniques. By using a pair of matched, digital binary dividers enclosed in the same integrated circuit (IC) device it is possible to convert a single-phase digital RF signal into four separate digital signals that differ one from another in that their phase angles are 0, 90, 180, and 270° (measured relative to a standard frequency). The frequencies of these four signals are the same. The details are shown in Figure 8.4. A binary digital divider or "flip–flop" is a device that changes its output state from high (+5 V) to low (0 V) whenever it experiences a sharp rise at its input; no change in output is made when the input changes from a high state to a low state.

These devices are usually constructed such that two outputs are simultaneously available; the phases at these outputs differ by 180°. Thus when one output is high, the other one is low and vice versa. As can be seen in Figure 8.4, if one starts with a digital frequency of $4f$ MHz, the final result is a four-channel output of frequency f MHz, with relative phases 0°, 90°,

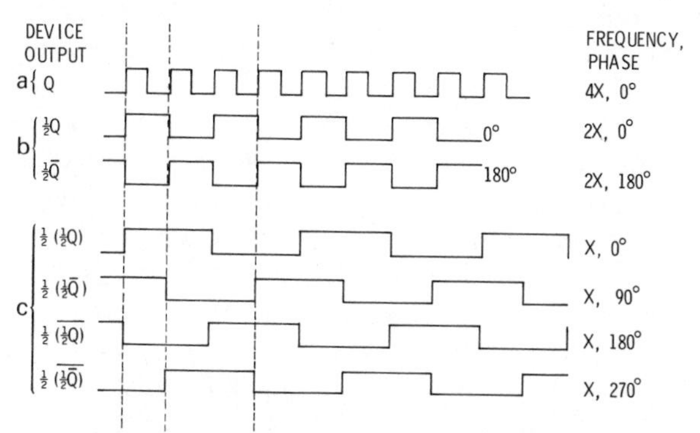

FIGURE 8.4. The diagram of the generation of a radio-frequency source with four stable, accurate outputs whose phases differ by 0, 90, 180, and 270°. (a) Waveform for the fundamental signal, (b) outputs of the first flip–flop, and (c) outputs of the final two flip–flops.

180°, and 270°. (These correspond to an H_1 field vector along the $+x$, $+y$, $-x$, and $-y$ directions in the rotating frame; see Figure 8.2a). In order to ensure a high phase accuracy it is desirable to work at a reasonably low frequency; this is because the phase accuracy depends upon the ratio of the inverse frequency f^{-1} of the square wave to the rise time of the pulse. Since at 11 MHz f^{-1} is about 10^{-7} sec and typical rise times of digital dividers are about 10^{-9} sec the uncertainty in the phase angle is 1% or less. By using trimmer capacitors the accuracy may be adjusted to 0.1% or better.

By incorporating the RF phase shifting and the frequency offsetting at the IF (intermediate frequency) range it is relatively easy and inexpensive to operate at any frequency between a few megahertz and several hundred megahertz [required when superconducting magnets (SCM) are used] with the same accurate phase capability and offset capability. The spectrometer system shown in Figure 8.3 uses gated mixing and has the advantage that the gating requirements are now much easier to meet.

Most early FT–NMR requirements used frequency sources that operated at the Larmor frequency. If even a small amount of this frequency found its way to the receiver coil or the preamplifier via ground loops or other means, a number of unhappy consequences ensued. In the gated mixing scheme it is only required that in the *off* state the RF signals be 50 mV or less; this condition is very easy to meet. If the voltages in the mixer are below 50 mV it remains a linear device and no mixing occurs. Leakage of 11 and 35 MHz (for ^{13}C) can and does occur, but this does no harm since they are not amplified by the high-gain, high-Q-preamplifier circuits. Since the gating (or switching) is all controlled by the computer a large variety of experiments can be done. It should perhaps be noted that the gating is not directly controlled by the computer since for many experiments direct computer control introduces too much uncertainty or "jitter" in the duration of the pulse widths and their timing. Instead the computer is used to program a programmable, read-only memory whose timing is controlled by a 10-MHz crystal oscillator (see, for example, reference 9). This method has the advantage of great flexibility without the disadvantage of timing jitter. The gating feature is also important in the decoupling channel and examples of gated decoupling will be given later in this chapter as well as in Chapter 9.

Gating techniques are also used in the receiver circuitry of most modern systems. This is quite easy to do since most current, wide-band, RF amplifier chips provide an automatic gain control capability. It is important to turn the signal amplifiers off when the transmitter pulse is on in order to prevent overload and ringing in the detection and filter circuitry.

Another significant development is the development of single side-band (SSB) (crystal filter) and quadrature phase detection methods. Both methods are discussed in more detail in Chapter 9. Although for routine work dealing with a single nucleus the result is about the same (a 40% increase in sensitivity and a consequent doubling of throughput), the increased flexibility of the

QPD technique makes it clearly the superior method. QPD also has the often overlooked, but important advantage of a fourfold increase in the efficiency of the RF transmitter power. This is quite important for samples involving metal complexes, shift reagents, and solid samples, or for nuclei with very large chemical shifts such as ^{19}F. Furthermore, the sampling requirements of the ADC units using QPD are half what they would be for single phase detection.

Since two low-frequency ADCs cost less than a single ADC with twice the sampling rate capability, it is actually cheaper to build a spectrometer with QPD than with SSB detection, *if* it is done in the initial production process. The cost to add QPD later is much more expensive, especially if the original spectrometer is not equipped with accurate, stable, multiple-phase capability.

It is becoming more common now to have the audio-frequency filters controlled by the computer. In many cases this is only a convenience feature, but for some of the more sophisticated automatic modes of operation, this capability can shorten significantly the data acquisition times needed.

One distinct advantage of the sort of spectrometer shown in Figure 8.3 is that it can be used at any frequency from a few MHz up to 600 MHz or even higher. This is done simply by replacing the gated local oscillator by a general-purpose frequency synthesizer. Since all the phase-shifting is done at the intermediate frequency no additional electronics for this purpose is required when the synthesizer is added. Since gated mixing is used, isolation and shielding problems of the synthesizer are greatly reduced. For operation at frequencies above 100 MHz, double-frequency conversion methods make the problems of eliminating unwanted side bands much simpler. (It is most difficult for this author to understand why most manufacturers use different spectrometers for different field strengths when a single well-designed system can do the job at any field equally well and at an overall reduced cost.) Similar spectrometer designs or conversions have been published by several other workers.[9,10,16,17]

As can be seen, the computer plays an important role in the overall operation of the spectrometer. Another important development in the NMR spectrometers is the use of two computers (a minicomputer or microcomputer) or, alternatively, the use of a hardwired signal averager plus a computer. The sensitivity of modern FT–NMR machines is such that in many cases 15 minutes or less is all that is required to obtain high quality spectra. It can easily take another 15 minutes or more to reduce the data and/or make several records of the data (e.g., one standard spectrum plus several spectra plotted with the x or y axis expanded, etc.). If signal averaging of a new sample can take place simultaneously with the data reduction of the previous sample, the sample throughput of the spectrometer can often be doubled. This mode of operation is sometimes referred to as foreground/background operation.

8.3.2. The Sample Probe

Before closing our discussion of instrumentation details at least a few words should be said about the sample probe itself, since it will determine in large part the sensitivity of the system, the quality of the spectra obtained, and in some cases the ease of operation of the entire system. Optimization of the probe operating parameters is especially important when signal averaging is required, since an N-fold increase in the sensitivity results in an N^2-fold increase in the number of samples that can be run in a given time period. For the optimum performance where the best resolution, line shape, and sensitivity are all desirable and/or necessary, it is best to use a fixed frequency probe that is carefully adjusted for a single nucleus and then readjusted only when really needed; even changing inserts can lead to a gradual, but significant degradation in the sensitivity.

There are, however, meaningful and useful compromises that can be made. If one is interested in observing many different nuclei for which sensitivity is not a major consideration it is often useful to use a tunable probe that can cover a reasonably wide range of observation frequencies. Such probes often have an external lock, which is normally adequate and greatly simplifies the operation and adjustment of the probe. Such systems have been described in the literature.[16,17]

Some of the more recent versions of FT–NMR spectrometers are designed in such a way that high-quality spectra of either ^{13}C or ^{1}H can be obtained without the need to change sample probes or receiver inserts. This can be done without any significant loss for ^{13}C, but some compromise is necessary in regard to the ^{1}H sensitivity and line shape. However, for experiments containing enough sample to obtain natural abundance ^{13}C spectra in a reasonable time, ^{1}H sensitivity is no problem whatsoever, and the ^{1}H line shape and resolution is quite adequate for all but the most demanding work.

In the view of this author, the proton disadvantages are outweighed by the ^{13}C advantages. Since both ^{13}C and ^{1}H can be done using the same sample, only a single tube must be prepared. This can be very important where the amount of sample is limited and/or sample stability is a problem. It is also important in experiments that call for selective heteronuclear decoupling. Since neither the sample nor the probe is moved and the lock is not lost or disturbed in switching from ^{1}H to ^{13}C observation, the precise frequency of each proton resonance line can be easily and reproducibly set in the irradiation (decoupling) channel. Another very distinct advantage is that it is not necessary to change probes when switching from ^{13}C to ^{1}H observation. Although this change seems at first glance to be quick and simple it is in fact not too quick or simple especially when good resolution and stability are required.

The design of the probe is very important when the amount, solubility, or concentration of the sample are limiting factors. Since the beginning development of ^{13}C NMR it has been popular to use rather large sample volumes. By using 20- or even 30-mm-diameter sample tubes, 10 to 15 ml or more of the sample can be used in the NMR sample tube. At such large tube diameters, however, great care must be used in order to maintain adequate RF power requirements. This is especially important if T_1 studies or quantitative analysis work are planned. In addition to the possible difficulties that may arise because of the probe design, there is also the problem, in the long run, of financing an experimental program that requires large amounts of costly deuterated solvents and biological samples. This is not to say there is no application for large samples. In many cases the samples in question are not very soluble or it is important to do experiments at relatively low concentrations (10^{-4} M or less). There are, however, many times when it is not possible or desirable to use such large sample tubes. The design and use of small sample tubes (1.6-mm diameter, 30-μl volume) is discussed in more detail in Section 8.4.3.

Much recent interest has been shown in two areas that promise rapid growth: fast chemical exchange and high-resolution NMR in solids. The chemical aspects of these methods are discussed in Section 8.4. The instrumentation requirements for these experiments are within the capability of some of the presently available commercial instruments. The principal requirements are

(a) Accurate, stable 0°, 90°, 180°, and 270° RF sources are essential.

(b) Short (10μsec or less), stable RF pulses are required for both ^{13}C and ^{1}H frequencies.

(c) Flexible, programmable, jitter-free pulse controllers are needed.

(d) For solid-state, high-resolution NMR work, sample spinning at about 2000 rpm is necessary.

A number of excellent recent articles are available describing the latest instrumentation available[9-12,16,17] that allows such experiments to be carried out.

8.4. RECENT INSTRUMENTAL IMPROVEMENTS

8.4.1. Coherent Broad-Band Decoupling

For both ^{13}C and ^{15}N NMR studies it is necessary to decouple the fine structure arising from the spin-coupling to the protons. It has been common practice to use proton "noise" decoupling in the standard ^{13}C or ^{15}N FT–NMR experiments.[18] This method, however, is not too effective and, especially at higher fields, it becomes very difficult to decouple all the proton-induced fine structure in the ^{13}C spectrum and at the same time keep the RF heating effects under control. A much more effective method of decoupling has been described by Grutzner and Santini[19] and by Anet.[20]

By using coherent phase modulation (with a 50% duty cycle square wave) much more efficient decoupling is achieved. For small molecules (mol. wt. 600 or less) a S/N improvement of a factor of 2 or more can be achieved. Complete details on how this can be achieved are described in references 19 and 21. This method is especially helpful over a restricted chemical shift range decoupling (e.g., $^{1}H–\{^{19}F\}$); there is still an advantage in the noise modulation methods.

Work is also progressing on "tailored excitation" methods. In these experiments the modulation of the proton decoupling power is designed such that the power appears only at those regions in the proton spectrum that contain proton resonance lines. This method requires reasonably accurate prior knowledge of proton NMR spectrum. Further details are given in references 22 and 23.

8.4.2. Gated Decoupling Methods and Quantitative Measurements

In the normal $^{13}C–\{^{1}H\}$ or $^{15}N–\{^{1}H\}$ FT–NMR experiments the proton decoupler is operated continuously. In this mode of operation, the ^{13}C NMR spectrum consists of numerous single resonance lines with no proton fine structure and with the presence of the NOE. It is often helpful or even essential, however, to observe the proton-induced fine structure or to suppress the NOE. A variety of such experiments can be executed by turning the proton decoupler on and off (i.e., by "gating" the decoupler) at the proper times. Figure 8.5 shows some examples of the sorts of experiments that are now more or less standard.

Figure 8.5a depicts the standard FT–NMR experiment, in which the ^{1}H irradiation power is on continuously, ^{13}C pulses are applied repetitively, and the corresponding FID signals are collected and added together in the computer. In Figure 8.5b the proton decoupling power is substantially reduced; the result is that the proton-induced fine structure is present, but the observed $^{13}C–^{1}H$ coupling constants are all reduced. The experiment is still very useful because it can show how many protons are bonded to each carbon atom. For example, a single ^{13}C doublet indicates one proton, a triplet shows two protons, and a quartet indicates three directly bonded protons. In the case of partial decoupling the full NOE is maintained. In Figure 8.5c the proton decoupler is turned completely off. In this case the full $^{13}C–^{1}H$ coupling is seen and there is no NOE. It should be noted that this is a very time-consuming experiment since the NOE is lost (a factor of 3 in sensitivity) and the decoupled ^{13}C singlet signals are now divided among the several lines arising from the fine structure. This sixfold or greater loss in "sensitivity" means that 36 times as much time (or more) will be required to obtain this spectrum over the time required to obtain the fully decoupled spectrum.

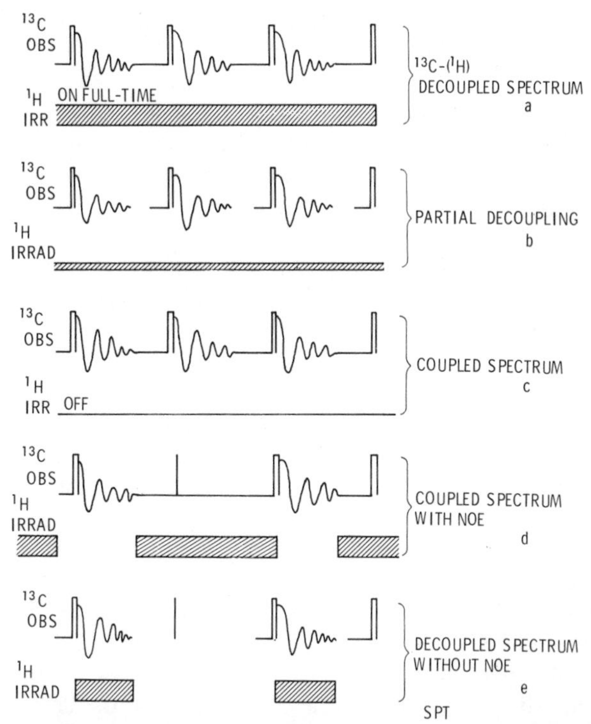

FIGURE 8.5. (a) Spectrometer operation of the "standard" ^{13}C observation with broad-band ^{1}H decoupling mode. This is denoted by ^{13}C–{^{1}H} and generally gives a full nuclear Overhauser enhancement (NOE). (b) Partial decoupling. The spin–spin splittings are all shown, but are greatly reduced in size; the full NOE is maintained. (c) Fully coupled ^{13}C spectrum, but with no NOE. (d) Same coupled spectrum as in (c), but with the NOE. (e) Decoupled spectrum as in (a), but without the NOE.

In order to minimize this loss in sensitivity it is now customary to execute partially decoupled or fully coupled ^{13}C spectra with the NOE. Figure 8.5d shows how this is done. It requires a time interval of about three times $T_1(^1\text{H})$ for the NOE to build to its full value and about three times $T_1(^1\text{H})$ for it to decay again. Typical proton spin–lattice relaxation times $[T_1(^1\text{H})]$ in small molecules (mol. wt. 600 or less) are 2–5 sec and the time required to obtain a ^{13}C FT–NMR spectrum is usually about 1 sec (for a resolution of 1 Hz). In this case the decoupler is turned on for 5 to 10 sec to allow build-up of the NOE. The decoupler is then turned off and a ^{13}C pulse applied in order to obtain the coupled ^{13}C spectrum. The decoupler is then turned on again for a few seconds to revitalize the NOE and the process is then repeated. In this way fully or partially proton-coupled ^{13}C spectra can be obtained without losing the threefold sensitivity enhancement of the NOE.

The converse of this experiment is shown in Figure 8.5e, a decoupled ^{13}C spectrum without NOE. Suppression of the NOE is done primarily for two reasons. In one case ^{13}C spectra are recorded both with and without NOE (Figures 8.5a and 8.5e, respectively) in order to measure the actual NOE. If the ^{13}C nuclides are relaxed *entirely* by the dipole–dipole interaction with the protons (see references 4 and 7) then a sensitivity enhancement of 2.9 will be observed. If this is not observed it indicates that "other" relaxation mechanisms are contributing. In most cases the mechanisms are spin–rotation (for small molecules) or paramagnetic impurities (such as dissolved oxygen, traces of iron, or copper ions).

It is now becoming more common to make quantitative ^{13}C measurements in order to determine the relative amounts of the different sorts of carbon atoms present. This is sometimes a tricky task because of the differences in spin–lattice relaxation times (T_1) and NOE values for the various carbon nuclides. The T_1 distribution problem can be solved either by using a viscous solvent or by lowering the temperature to the point where the ^{13}C T_1 values are all reasonably short and about equal (about one second). In any event one must wait about five times the T_1 value of the largest T_1 between pulses. The NOE distribution problem can be solved by performing the experiment shown in Figure 8.5e. The decoupler is turned on for only the one second during which spectra are recorded. One then waits several seconds, with the decoupler turned off (in order for any small initial NOE buildup to decay), and the experiment is then repeated. When it is possible, both the T_1 and NOE distribution problems can be circumvented by adding "shiftless" relaxation agents such as $Cr(AcAc)_3$ to the solution. This shortens the relaxation times to about one second and suppresses the NOE. The loss in NOE enhancement is usually more than compensated for by the fact that the relaxation times are now all relatively short and large ^{13}C pulse angles can be used.

If one executes the experiment in such a way that the T_1 and NOE differences are eliminated or accounted for, then it is rather routine to carry out quite accurate quantitative measurements, with results that agree very well with other analytical methods.* It is very important to maintain a high S/N and to use an adequate number of data points for quantitative work. Sampling theory requires that if a spectral range of Δ Hz is to be observed then a sampling rate of 2Δ Hz must be used. For ^{13}C observation at 25 MHz, Δ is typically 5 kHz. For a resolution of 1 Hz it is necessary to sample for one

* Dr. J. Schoolery of Varian Associates and Dr. E. Brame of DuPont have done a great deal of quantitative ^{13}C studies with results that agree well with gas chromatographic results and spectroscopic results.

second. This means that 10,000 data points are required but because of the 2^q requirement imposed by the Cooley–Tukey algorithm one usually samples for a shorter time in order to limit the number of data points to 8k ($8 \times 1024 = 8192$). After execution of the Fourier transform one is left with 4k data points for the absorption spectrum and 4k data points for the dispersion spectrum. This means less than one data point per hertz. For quantitative work this is clearly not adequate, unless rather broad lines are involved such as in viscous polymers or heavy oils. For molecules with molecular weights up to about 600 or 700 the ^{13}C line widths are usually 1 Hz or less, and since four or more data points per line are needed for accurate intensity measurements one must consider 32k transforms or use partial spectra. It should be mentioned that really accurate T_1 measurements also require accurate total line intensity information.

Two methods that are currently used to increase the number of data points per spectral line are exponential filtering and zero filling. If the resonance lines in the spectrum are not closely spaced then a large amount of exponential filtering can be added by the computer. This will greatly enhance the S/N and broaden the lines appreciably. The result is that without adding to the total number of data points it will greatly increase the number of data points per line with a concomitant increase in the accuracy of T_1 measurements and quantitative measurements. An alternate method is to use zero filling. If x data are taken originally one may add an additional x data points, which are all set to zero. For example, if 8k zeros are added to 8k original data points one can now execute a 16k transform that results in 8k real (absorption) and 8k imaginary (dispersion) data points in the frequency domain. Thus, zero filling has the advantage of doubling the number of data points per line. It also has the distinct advantage of increasing the resolution by a factor of two. It has been argued that this is not possible since adding zeros does not add new information and consequently cannot improve the resolution, and that even the gain in spectral definition (as opposed to resolution) is a bit of a mathematical boondoggle. This point of view, however, is somewhat extreme; it could be reasoned that the Fourier transform itself adds no new information. This is in fact true since all the information is contained in the FID, but it is not in a very usable form. Adding zeros does not add new information but it does allow one to make full use of the information contained in the FID. As a result of the Fourier transformation, half of the resolution information contained in the FID is lost; adding the additional zeros allows one to regain this misplaced information. Adding more than x zeros to the original x data points serves no useful purpose, other than giving more interpolated data points between each of the original data points. Another way to improve the resolution is to record 16k rather than 8k data points. This improves the resolution twofold, but doubles the scan time and degrades the S/N by a factor of $2^{1/2}$ for an equal number of scans.

8.4.3. Microsample Techniques

In many instances the amount of sample available for investigation is rather limited. In this case the best results are achieved by using as small a sample volume *and as small a receiver coil* as possible. The S/N of an NMR spectrometer may be expressed as[1]:

$$S/N = \frac{g}{n}\left\{ f\chi_0 N \left[\frac{vQ\chi_0 H_0^2 V_c}{(\Delta v)\, kT} \right] \right\} \tag{8.11}$$

where g is a composite of several constants; $f = V_s/V_c$, the filling factor; V_s is the sample volume in cm^3; V_c is the receiver coil volume in cm^3; Q is the quality factor of the receiver coil network; v is the operating frequency of the spectrometer; Δv is the spectrometer band width; χ_0 is the magnetic susceptibility of the sample; H_0 is the DC magnetic field value in gauss; k is the Boltzmann constant; n is the noise factor for the sample circuitry; N is the number of nuclei in the sample. For the purpose of comparing the S/N as a function of the diameter of the sample tube we will in the following discussion hold the values of n, v, Δv, H_0, and T constant. We may then write equation (8.11) in the simpler form.

$$S/N = qf(QV_c)^{1/2}\, N \tag{8.12}$$

where q is a new constant. We are especially interested in three particular experimental situations:

Case A. The receiver coil volume is constant. This is the general situation encountered using commercial spectrometers with a single, fixed probe. For this case we may write

$$S/N = qfNQ^{1/2} \tag{8.13}$$

For the maximum S/N value the filling factor should be as close to unity as possible and the concentration (or number of resonant nuclei available) should be as high as possible. Note that from equation (8.13) we see that the larger the sample volume the greater the S/N, hence the general idea that more is better. There are upper bounds, however, on this philosophy that if a little bit does a little bit of good, a whole lot will do a whole lot of good. In addition to the constraints imposed by the general availability and cost of 10 to 20 cm^3 of sample to obtain a spectrum, there are also electronic constraints involving the Q of the sample circuitry, the maximum allowable inductance of the sample coil, the homogeneity of H_0 over the entire sample volume, and the decrease in γH_2 decoupling power as V_c increases.

Case B. The sample volume is constant. For this situation equation (8.12) becomes

$$S/N = q(Q/V_c)^{1/2}\, C \tag{8.14}$$

where C is the concentration in moles/liter. It is intuitively obvious that the concentration should be kept as high as possible to maximize the total number of nuclei in the sample. Another important factor is that, in general, the viscosity of the sample increases as the concentration increases. As the viscosity increases T_1 decreases and larger pulse angles can be used. This leads to an additional appreciable increase in the S/N. In this case the volume of the receiver coil V_c should be as small as possible.

Case C. The total amount of sample is fixed. Here the expression for the S/N is given by

$$S/N = q(Q/V_c)^{1/2} \tag{8.15}$$

This shows clearly that the volume of the receiver coil is the critical item; it should be as small as possible. As usual, a number of compromises must be made. Since it is not possible to vary the receiver coil volume easily, some small fixed value must be used. Volumes less than about 50 to 100 μl are not practical since high Q values are hard to maintain at such small V_c values, and the filling factor degrades because of the greater importance of the wall thickness of the sample tubes and the receiver inserts for such small volumes. A reasonable design is one that uses a melting-point capillary of about 2.0 mm internal diameter and is filled to a height of about 10 to 12 mm. Such a sample tube has a volume of about 35μl, is easy to fill and to work with, and is inexpensive. For a typical molecule (for example, cholesterol, which has a molecular weight of 387) 5 mg dissolved in 35 μl of $CDCl_3$ gives a 0.37 M solution. A good quality ^{13}C NMR spectrum on a 1 mg sample (0.07 M) can be obtained in less than an hour. The corresponding ^1H spectrum at such high concentrations requires one second to obtain. A high-quality ^1H spectrum of 50 μg of cholesterol (3.7 mM) can be obtained in one or two minutes and an acceptable quality ^1H spectrum of a 5 μg (0.37 mM) sample can be obtained in about 10 to 15 minutes.

A better idea of what this means can be obtained by looking at some actual spectra. The spectra shown in Figure 8.6 are all for a *constant concentration* sample of 50% ethylbenzene in 50% $CDCl_3$. Figure 8.6a,b,c represents 10.0-, 5.0-, and 2.0-mm sample tubes, respectively. The 10-, 5-, and 2-mm tubes hold about 1.2, 0.3, and 0.03 ml of sample. It is very clear that for a fixed concentration situation, the greater the amount of solution used, the better the S/N (this is true up to a limit of about a 25-mm-diameter sample tube, *if* a large-volume, homogeneous magnetic field is available). Figure 8.7 shows spectra for samples that contain a *fixed amount* of sample. In this case 5 mg of cholesterol was dissolved in 1.2, 0.3, and 0.03 volumes of $CDCl_3$, respectively. The times required to collect the data for the 10-, 5-, and 2.0-mm samples were 10, 5, and 1 hours.

Thus for small amounts (5 mg or less) of sample, it is generally about ten times quicker to obtain spectra using a 2-mm sample tube than a 10-mm

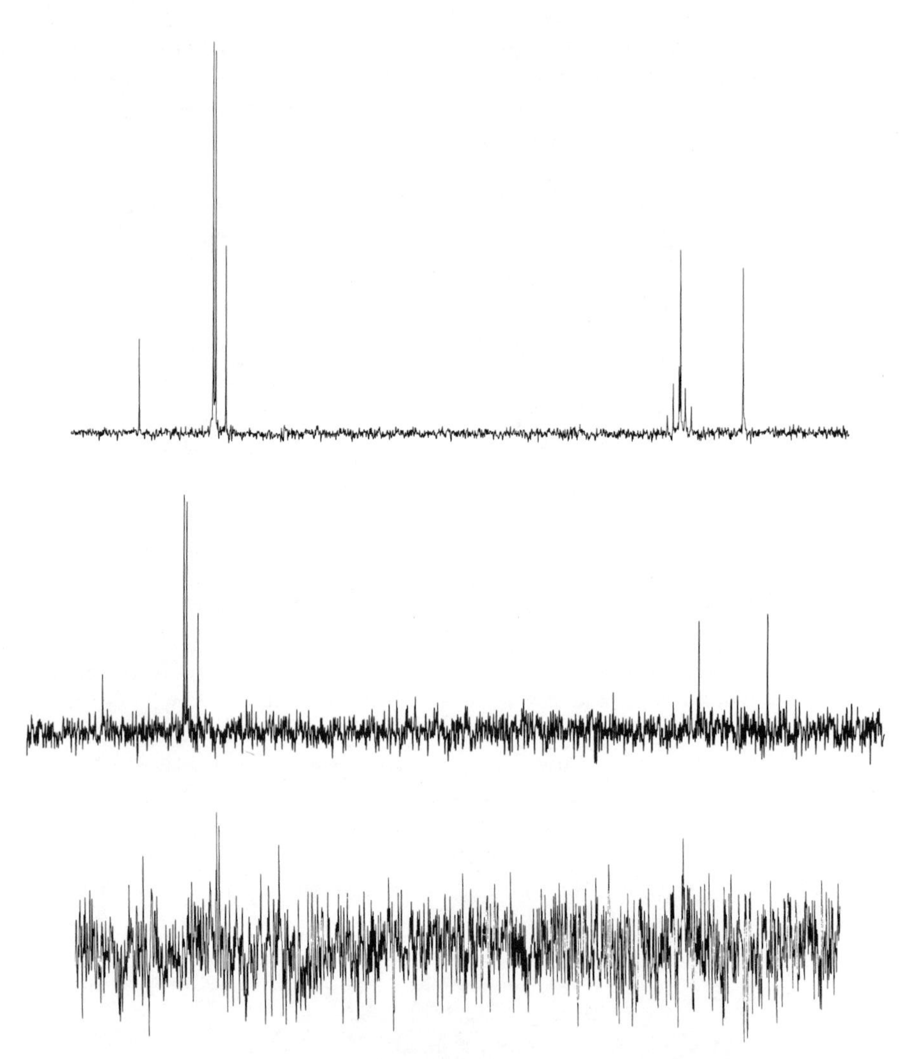

FIGURE 8.6. These $^{13}C-\{^{1}H\}$ spectra are all single-scan recordings of a solution of constant concentration: 50% ethylbenzene in $CDCl_3$. The upper spectrum is for a 1.2-cm^3 sample in a 10-mm o.d. sample tube. The middle spectrum is for a 0.25-cm^3 sample in a 5-mm o.d. sample tube. The lower spectrum is for a 0.03-cm^3 sample in a 1.6-mm o.d. sample tube (actually a melting-point capillary tube).

sample. It should be mentioned that these data are all preliminary and considerable improvements can still be made. The 2-mm sample coils currently in use have rather low Q-values and very poor filling factors. With improvements that are now in progress with regard to these and other factors, it should be possible to routinely obtain ^1H spectra 100-ng sample quantities

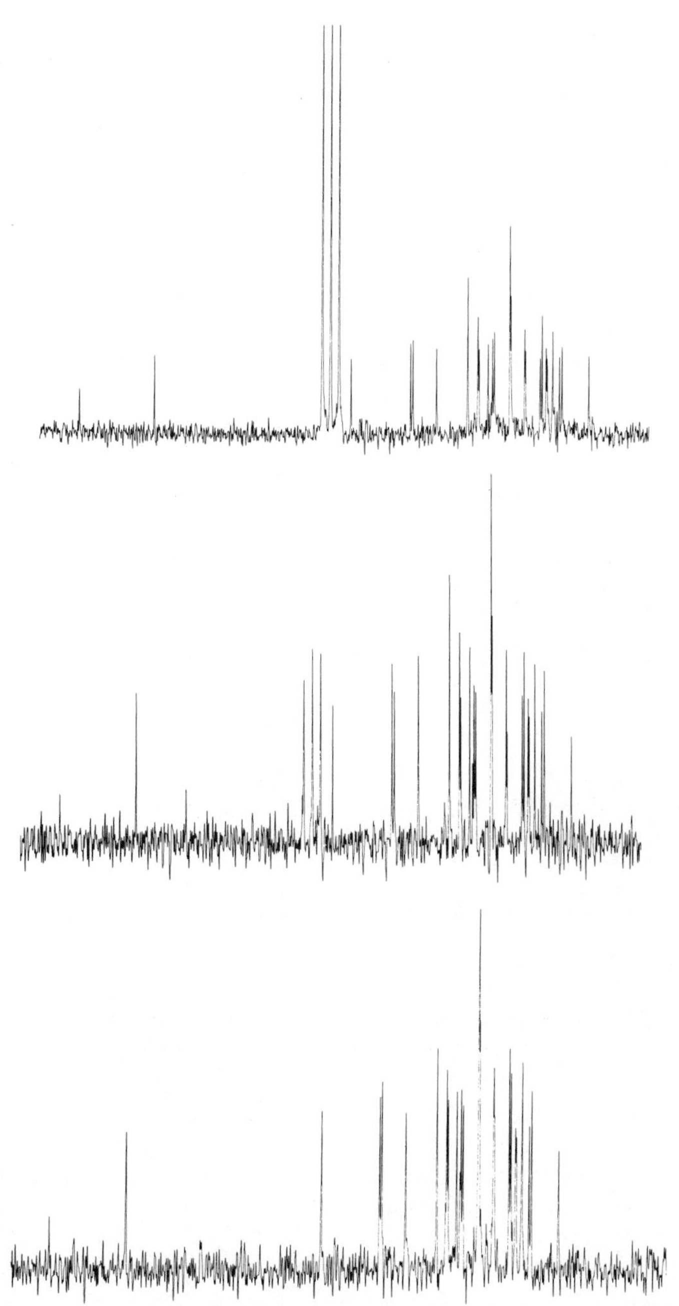

FIGURE 8.7. These are $^{13}C-\{^{1}H\}$ spectra of a fixed amount (5.0 mg) of cholesterol in $CDCl_3$ solution. The *S/N* in the three spectra is about the same, but the data accumulation times are quite different. The upper spectrum of a 1.2-cm^3 solution required 10 hours of time averaging (in a 10-mm sample tube); note the intense triplet arising from the $^{13}CDCl_3$ solvent. The middle spectrum required 5 hours and is for a 0.25-cm^3 sample in a 5-mm tube. The bottom spectrum required only one hour of time averaging and is for 5 mg of material in 0.03 cm^3 of $CDCl_3$ in a 1.6-mm o.d. sample tube; note the absence of the solvent peaks.

and to obtain ^{13}C spectra on 50μg quantities. Of course, if ^{13}C-enriched samples are used, then ^{13}C spectra of 1-μg quantities or less could be obtained. At these levels great care must be used in preparing the sample tubes and in handling the samples and the solvents.

8.4.4. Selective Population Transfer

It has been recently shown[24-26] that in many cases the same sort of information that is available from CW INDOR experiments can be obtained using FT gated double-resonance methods. Such selective population transfer (SPT) experiments can give information about the assignments and relative signs of ^{13}C–^{1}H (or ^{13}C–^{31}P, etc.) spin-coupling constants. The experiment is relatively straightforward. A very weak pulse is applied to a particular line in the ^{1}H spectrum; the amplitude, $\gamma H_2/2\pi$, should be about equal to the width of the proton line of interest (usually 0.1 to 0.5 Hz). The duration τ of the pulse is such that the proton magnetization of this particular line is inverted, $\gamma H_2 \tau = \pi$ (usually τ is between 1 and 5 sec). This will cause complete inversion of the proton line provided that $\tau < T_1$, where T_1 is the shortest spin–lattice relaxation time for all connected transitions. Immediately following the inverting γH_2 pulse to the proton line, a 90° pulse is applied to all the ^{13}C resonances. Those ^{13}C lines that are spin-coupled to the ^{1}H line that was inverted will show a large change (enhancement and/or inversion). For further details the reader is referred to references 24–26.

8.4.5. Studies of Chemical Dynamics

All of the standard methods that were used with CW spectrometers to measure activation energies and exchange rates[27-30] are applicable to FT–NMR spectrometers. The double-resonance CW methods developed by Forsén and Hoffman[29,30] allow rate constant measurements in the range of $k = 0.01$ to about 1 sec to be measured. CW line shape studies cover the range from $k = 1$ to 100 sec. Both of these methods depend upon the magnitude of the chemical shift difference between the two exchange sites. Since very accurate line shape measurements can also be done using FT methods the same sort of information is available. Using pulsed NMR methods the range of systems that can be studied can be increased enormously since many of the constraints that apply to the CW methods no longer apply to the pulse techniques.[31,32] Measurement of the spin–spin relaxation times (T_2) or the spin–lattice relaxation time in the rotating frame ($T_1\rho$) cover the range $k = 10$ to 10^6.[33] Although T_2 studies afford information over a much wider dynamic range, the analysis of the data is rather complex[31] and the experiment is experimentally extremely difficult.[32]

Much the same sort of information is available from $T_1\rho$ studies.[4,33] The $T_1\rho$ data are much simpler to interpret and the experiments are easier to execute. Both the T_2 and T_1 experiments require accurate timing in the pulse sequences and very accurate and stable adjustment of the different RF phases (which are absolutely essential for both methods).

If exchange rates above 10^6 are involved, variable-field T_1 studies can be used to determine rate constants in the range $k = 10^7$ to 10^{13}.

8.4.6. High-Resolution ^{13}C NMR in Solid Materials

One of the more exciting recent developments in FT–NMR spectroscopy is high-resolution NMR in solids. Ordinarily one does not observe high-resolution proton spectra in solids because the relatively strong dipole–dipole interaction does not average to zero in a solid, as it does in a liquid. Consequently, the chemical shift and spin-coupling information is all obscured. The possibility exists, however, to obtain high-resolution ^{13}C NMR spectra in solids since at natural abundance levels the ^{13}C spins are very dilute (1 %) and consequently there are no significant ^{13}C–^{13}C dipole–dipole interactions. There is a strong ^{13}C–^{1}H dipole–dipole interaction, but this can be removed by very strong decoupling.[34] A further increase in resolution for solid samples can be obtained by spinning the sample about an axis that is about 54° to the axis of the magnetic field, the so-called magic angle. Although these experiments are at the present time difficult to execute,[8-12] the wealth and importance of the data are so great that the present rapid development will undoubtedly continue. This opens the possibility of obtaining accurate qualitative and quantitative information about the nature and the amounts of the chemically different carbon atoms in a solid sample.

8.4.7. FT–NMR at High Fields

Recent advances in superconducting magnet (SCM) technology have been very rapid, with the result that SCMs are now as easy and reliable to operate as the older conventional iron permanent magnets and electromagnets. SCM FT–NMR systems offer a number of distinct advantages: (a) they have a much higher sensitivity, (b) the spectra obtained are much easier to interpret, and (c) they are much cheaper to operate.

The increased sensitivity arises simply from the Boltzmann factor. It can be shown that in general S/N is proportional to $v_0^{3/2}$ where v_0 is the operating frequency of the spectrometer. Thus a 450-MHz spectrometer has about 20 times greater sensitivity than a unit that operates at 60 MHz. Consequently proton spectra of samples containing about 1 μg of material can be obtained in about 10 minutes, and by running for longer times (e.g., overnight) samples down to 50 ng or less may be run. For proton NMR an equally, if not more, important consideration is the fact that since the chemical shift σ is linearly dependent on the magnetic field, the high-field spectra are much simpler. This is shown clearly in Figure 8.8. The proton spectrum shown is for 1,3,4,6-tetra-O-acetyl-2-(N-acetyl-benzamido)-2-deoxy-β-D-glucopyranose in $CDCl_3$(35),* at 60, 100, 220, and 450 MHz, respectively. A sketch of the molecule is shown in Figure 8.9. At 450 MHz one can clearly see the six chemically different proton resonances and since the spectrum is practically first order,† the interpretation is rather simple and straightforward. Although the greatest impact of the SCM systems will probably be in the proton NMR area it should also be of great

* I would like to thank Dr. Robert Highet of NIH for calculating the above spectra.

† One usually obtains such simple first-order spectra whenever the chemical shifts σ are large compared to the coupling constants J. That is, $(\sigma/J) \geq 3$. See references 1 and 36 for further information.

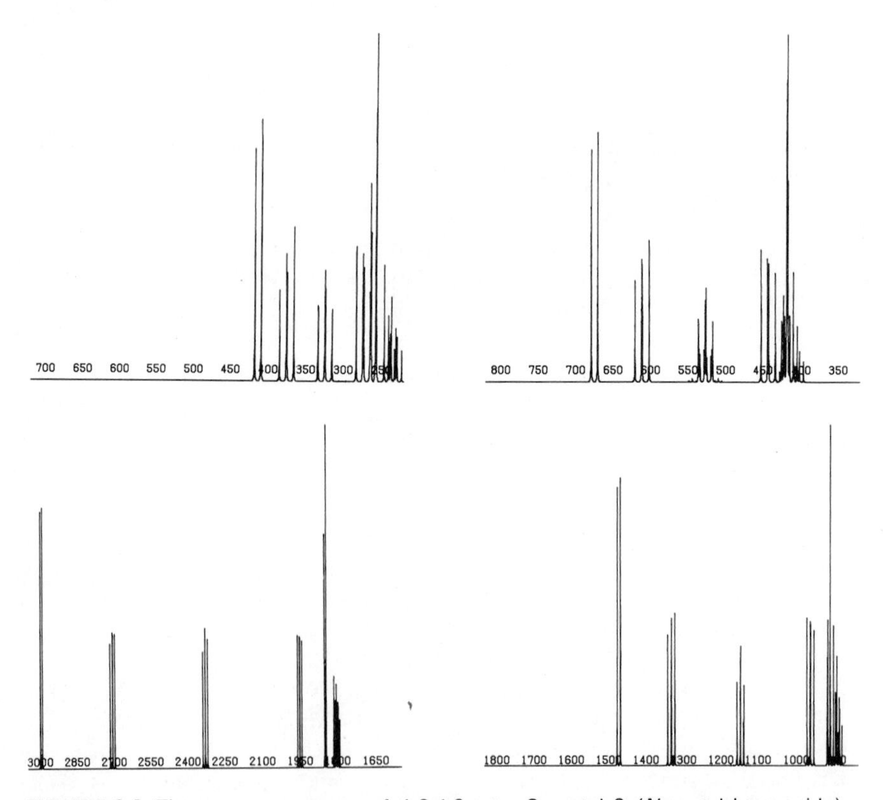

FIGURE 8.8. The proton spectrum of 1,3,4,6-tetra-O-acetyl-2-(N-acetyl-benzamido)-2-deoxy-β-D-glucopyranose at 60, 100, 220, and 450 MHz, arranged clockwise and starting at the upper left with 60 MHz (the acetyl part of the spectrum has been deleted in all cases). At 450 MHz the six chemically different protons are clearly separated. Even at 220 MHz considerable second-order interactions are clearly present.

use in ^{13}C NMR of moderately large molecules. At the present rate of development it is probably safe to predict that within 5 to 10 years at proton NMR frequencies of 100 MHz and above, electromagnets will no longer be used, since it is now cheaper to build and to operate SCMs; the operating cost of an SCM is about $4,000 per year, that of an electromagnet is $10,000 to $20,000 per year. At this time 11-T (about 470 MHz for ^1H) fields represent state-of-the-art capability for homogeneous, persistent high-resolution SCM units. Several research-type instruments operating at this frequency are now being built.

As can be seen, NMR spectroscopy is still in a state of rapid progress and development. As the sensitivity becomes even greater and the spectra even simpler to interpret we can expect it to become an increasingly powerful tool for chemists of all persuasions.

REFERENCES

1. A. Abragam, *The Principles of Nuclear Magnetism,* Oxford University Press, London and New York, 1961.
2. C. P. Slichter, *Principles of Magnetic Resonance,* Harper & Row, New York, 1963.
3. E. D. Becker, *High Resolution NMR*, Academic Press, New York, 1969.
4. T. C. Farrar and E. D. Becker, *Pulse and Fourier Transform NMR*, Academic Press, New York, 1971.
5. J. W. Cooley and J. W. Tukey, *Math. Comput.* **19**, 297 (1965).
6. R. R. Ernst and W. A. Anderson, *Rev. Sci. Instrum.* **37**, 93 (1966).
7. J. H. Noggle and R. E. Schirmer, *The Nuclear Overhauser Effect*, Academic Press, New York, 1971.
8. A. Pines, M. G. Gibby, and J. S. Waugh, *J. Chem. Phys.* **59**, 569 (1973).
9. J. D. Ellett, Jr., M. G. Gibby, U. Haeberlen, L. M. Huber, M. Mehring, A. Pines, and J. S. Waugh, *Advan. Mag. Res.* **5** (1971).
10. A. G. Redfield and R. K. Gupta, *Adv. Mag. Res.* **5** (1971); A. G. Redfield and S. D. Kunz, *J. Mag. Res.* **19**, 250 (1975).
11. R. W. Vaughan, D. D. Ellerman, L. M. Stacy, W.-K. Rhim, and J. W. Lee, *Rev. Sci. Instrum.* **43**, 1356 (1972).
12. E. O. Stejskal and J. Schaefer, *J. Mag. Res.* **18**, 560 (1975).
13. I. J. Lowe, *Phys. Rev. Lett.* **2**, 285 (1959).
14. H. Kessemeier and R. E. Norburg, *Phys. Rev.* **155**, 321 (1967).
15. E. R. Andrew, *Progr. NMR Spectrosc.* **8**, 1 (1971).
16. D. D. Traficante, J. A. Sims, and M. Mulcay, *J. Mag. Res.* **15**, 484 (1974).
17. H. C. Dorn, L. Simeral, J. J. Natterstad, and G. E. Maciel, *J. Mag. Res.* **18**, 1 (1975).
18. R. R. Ernst and W. A. Anderson, *Rev. Sci. Instrum.* **37**, 93 (1966). See also other references cited here.
19. J. B. Grutzner and R. E. Santini, *J. Mag. Res.* **19**, 173 (1975).
20. F. A. L. Anet, *Topics C-13 NMR Spectrosc.* **1** (1974).
21. G. C. Levy, I. R. Peat, R. Rosanske, and S. Parks, *J. Mag. Res.* **18**, 205 (1975).
22. B. L. Tomlinson and H. D. W. Hill, *J. Chem. Phys.* **59**, 1775 (1973).
23. R. Freeman and H. D. W. Hill, *J. Mag. Res.* **4**, 366 (1971).
24. K. G. R. Pachler and P. L. Wessels, *J. Mag. Res.* **12**, 337 (1973).
25. S. Sorenson, R. S. Hansen, and H. J. Jakobsen, *J. Mag. Res.* **14**, 243 (1974).
26. H. J. Jakobsen, S. A. Linde, and S. Sorensen, *J. Mag. Res.* **15**, 385 (1974).

27. H. S. Gutowsky, D. M. McCall, and C. P. Slichter, *J. Chem. Phys.* **21**, 279 (1953).
28. C. S. Johnson, *Advan. Mag. Res.* **1**, 33 (1965).
29. R. A. Hoffman, *Advan. Mag. Res.* **4**, 87 (1970).
30. S. Forsén and R. A. Hoffman, *J. Chem. Phys.* **39**, 2892 (1963).
31. H. S. Gutowsky, R. L. Vold, and E. J. Wells, *J. Chem. Phys.* **43**, 4107 (1965).
32. R. L. Vold, R. R. Vold, and H. E. Simon, *J. Mag. Res.* **11**, 283 (1973).
33. T. K. Leipert, J. H. Noggle, W. J. Freeman, and D. L. Dalrymple, *J. Mag. Res.* **19**, 208 (1975).
34. A. Pines, M. G. Gibby, and J. S. Waugh, *J. Chem. Phys.* **59**, 569 (1973).
35. T. D. Inch, *J. Am. Chem. Soc.* **31**, 1825 (1966).
36. J. A. Pople, W. G. Schneider, and H. J. Bernstein, *High Resolution Nuclear Magnetic Resonance*, McGraw-Hill, New York, 1959.

Chapter 9

Advanced Techniques in Fourier Transform NMR

James W. Cooper

9.1. INTRODUCTION

In this chapter, we shall be concerned with various techniques for the rapid acquisition of NMR data. We shall discuss their advantages and their limitations as well as some of the relevant experimental details. Since there are so many methods being published regularly in so many subfields of NMR, we shall limit ourselves here to the more common newer techniques and for the most part to techniques in which the Fourier transform plays some important part.

We first discuss methods of getting the most out of the data system by various manipulations to reduce the coherent noise during signal averaging, and then discuss techniques that make use of this feature, notably the increasingly popular two-input quadrature detection approach to NMR. Following this we discuss some other nonpulsed methods of acquiring NMR data rapidly including stochastic resonance, Hadamard transform NMR, and tailored excitation. Finally, we summarize several methods for suppressing strong solvent peaks.

James W. Cooper • Department of Chemistry, Tufts University, Medford, Massachusetts 02155

9.2. SYSTEMATIC NOISE REDUCTION

9.2.1. Noise Reduction Methods

One important and easy to implement modification in the method of data acquisition is some method of systematic noise reduction. There are several sources of systematic noise that can cause coherent addition of unwanted information during signal averaging: (a) noise in the analog-to-digital converter (ADC), (b) noise pickup by input leads, (c) DC bias in the spectrometer amplifiers, (d) spin–echo formation by too-rapid pulsing. All of these noise contributions can cause memory to fill more rapidly than desired, leading to less signal averaging and less enhancement of the signal-to-noise ratio, as well as spurious data. These can all, however, be suppressed by simple input and spectrometer modifications, many of which are now standard in commercial spectrometers.

Noise in the ADC is perhaps the easiest to eliminate. Every analog-to-digital converter, no matter what conversion method it uses, will have some coherent noise associated with the conversion process, arising primarily from the switching during the conversion itself. This noise is coherent and will be of approximately the same size throughout the spectrum, regardless of the nature of the input data. It can be easily eliminated, however, by the simple expedient of inverting the analog input on alternate scans and then adding or subtracting these alternate scans into the running average. This method is most easily understood if we consider the simple chart for one point during a six-scan signal average of the spectral data as shown in Table 9.1.

After six scans the average stored in that memory location has the value of 60. Clearly, this is a straightforward case of signal averaging of input data having a value of about 10 V per scan. Now let us consider what would

TABLE 9.1. Contents of a Single Data Point during Six Scans

Scan	Input data	Average
1	11	11 + 0 = 11
2	9	11 + 9 = 20
3	12	20 + 12 = 60
4	8	32 + 8 = 40
5	9	40 + 9 = 49
6	11	49 + 11 = 60

TABLE 9.2. Contents of a Single Point during Six Scans with Coherent ADC Noise

Scan	Input	ADC	Converted	Average
1	11	+1	12	12 + 0 = 12
2	9	+1	10	12 + 10 = 22
3	12	+1	13	22 + 13 = 35
4	8	+1	9	35 + 9 = 44
5	9	+1	10	44 + 10 = 54
6	11	+1	12	54 + 12 = 66

happen to our average if the analog-to-digital converter introduced a coherent noise value of one count per scan. This is shown in Table 9.2. Here a coherent value of 1 has been added to each scan by the noise in the ADC. We can see that if the maximum memory contents were 64, that memory would have overflowed, making the signal averaging impossible because of the coherent noise term. Now, let us consider Table 9.3, where we remove this conversion noise by the simple application of a circuit that inverts the analog input on alternate scans and on these same scans *subtracts* the converted value from the average rather than adding it. Clearly, this alternate inversion–subtraction technique will indeed cancel out the systematic noise generated by the ADC.

Actually, the ADC noise is very much smaller than this, and will appear only if hundreds of thousands of scans are taken without the noise reduction feature. However, this description serves as a model for more complex noise reduction methods where all of the system noise is canceled out in alternate scans.

Suppose that instead of inverting the analog input to the ADC, we perturb the nuclei in reverse fashion so that a negative signal is generated. This can be accomplished by changing the phase of the RF of the excitation pulse

TABLE 9.3. Contents of a Single Point during Six Scans with Coherent ADC Noise and Noise Reduction

Scan	Input	Inversion	ADC	Converted	Average (+/−)
1	11	—	+1	12	12 + 0 = 12
2	9	−9	+1	−8	12 − (−8) = 20
3	12	—	+1	13	20 + 13 = 33
4	8	−8	+1	−7	33 − (−7) = 40
5	9	—	+1	10	40 + 10 = 50
6	11	−11	+1	−10	50 − (−10) = 60

by 180°. This will result in a negative signal of equal amplitude, and if at the same time the converted data is subtracted from memory, this will on alternate scans cancel out the entire noise of the spectrometer system from the input excitation pulse all the way through to the data system. Such common problems as RF pick-up from cable leads and small DC biases in the spectrometer amplifiers will be canceled, leaving only the random noise from the spin system under study. This random noise, of course, averages out in the usual fashion as discussed in Chapter 4.

9.2.2. Relaxation Times and Spin Echoes

It is often suggested that the pulse repetition rate in a pulsed NMR experiment be approximately three to five times the longest T_1 of spin–lattice relaxation time in the molecule.[1] This suggestion carries the implicit assumption that the RF pulse tips the magnetization by 90° each time. If a shorter RF pulse is used, the nuclei will take less time to relax back to their equilibrium condition but will produce less signal information per pulse. This trade-off of more rapid acquisition versus less signal per scan is seldom analyzed in detail. Instead, the pulse repetition rate is selected more or less empirically.

However, while it may take less time for nuclei to relax back to the z axis by the spin–lattice relaxation mechanism, the faster repetition rates are also faster than the spin–spin relaxation time T_2 and this means that some phase coherence may be built up in the individual nuclear magnetic vectors of each nucleus in the sample. When this happens, the vectors will under some conditions become coherent at a time later than the time of the application of the RF pulse. This leads to *echo* formation, or the formation of an unexpected maximum in the free-induction decay, usally near the end of the FID.

Clearly the formation of such echoes leads to essentially spurious information in the transformed spectrum, which may obscure the data of interest. However, these echoes can be suppressed by the use of a phase-alternating pulse sequence much like that described for systematic noise reduction. If the RF phase is kept the same for *two* successive pulses and then inverted for two more pulses and the cycle repeated, the result is an echo buildup for two pulses, which is then canceled by a similar, negative echo buildup for the two remaining pulses. This process then cancels out not only the echo buildup but also all of the coherent noise of the system every four scans. This technique is known as PAPS[2] or phase-alternating pulse sequence by one manufacturer but is commonly in use by a number of workers with other equipment. Another method of reducing such echoes has been described by Freeman and Hill[3] and amounts to a small variation in the delay time between pulses.

9.3. SIDEBAND FILTERS AND QUADRATURE DETECTION NMR

9.3.1. The Crystal Sideband Filter

The pulsed experiment in its simplest form assumes that the RF carrier is placed at one end of the spectrum so that data will be obtained from only one side of the carrier. If the RF carrier were placed in the center of the spectrum, peaks from both sides of the carrier would fold back into the spectrum causing a complex and uninterpretable result. It was assumed by many that there would be no foldback of information if the carrier were placed at one end of the spectrum, but Allerhand[4] has pointed out that random noise will indeed be folding back from the other side of the carrier. He therefore proposed that a crystal sideband filter be installed in the spectrometer, which not only limits the bandwidth but filters out all data from the other side of the carrier. This method effectively increases the signal-to-noise ratio per scan by a factor of 1.4 at a cost of only a few hundred dollars. This disadvantage to this technique, however, is that these filters, made for the electronics industry, do not allow variation of the selected bandwidth, and one must be purchased for each bandwidth to be observed. For a 90- or 100-MHz proton frequency spectrometer a crystal sideband filter of 1500 Hz is used for ^{1}H NMR and one of 5000–6000 Hz is used for ^{13}C. When other frequency ranges are observed either no filter is used or the larger one is left in place.

9.3.2. Quadrature Detection Spectroscopy

Another method for getting more information per unit time is by quadrature detection spectroscopy.[5,6] Here the RF carrier is placed in the center of the spectrum, and two phase detectors set 90° apart are used to detect the resulting free-induction decay. The data from these two detectors are sampled simultaneously by a two-input ADC and stored in adjacent blocks of the computer's memory as shown in Figure 9.1. These two FIDs are used as the real and imaginary parts of a *complex* transform that produces the NMR spectrum shown in Figure 9.2a. If conventional single-phase detector NMR had been used, at the same carrier position, the data would transform to a mixture of real and folded-back lines, as shown in Figure 9.2b.

Now the complex transform is actually simpler than one for transforming real data, as we saw in Chapter 4. However, it does necessitate reprogramming the computer system for the new experiment. This can be avoided by using a technique described by Kunz and Redfield[7] in which the outputs of the two detectors are read not simultaneously, but sequentially. Here the inputs to the computer having but a single ADC become $a, b, -a, -b,$

FIGURE 9.1. Dual free-induction decays as detected by two-input ADC connected to two phase detectors for quadrature FT–NMR.

$a, b, -a, \ldots$, etc. Data acquired in this fashion can be transformed with a conventional real transform. Thus a programming modification is traded for a hardware modification involving two switches that switch between detectors *a* and *b* and between positive and negative data.

One of the principal problems in quadrature detection is the matching of the two phase detectors and associated amplifiers needed for the experiment. If they are both provided in a commercial spectrometer system, this matching will probably have been done during manufacture, but if a second phase detector system is added to an existing spectrometer, the matching may be difficult. Stejskal and Schaefer[6] and Hoult[8] have suggested ways of avoiding this problem by some simple hardware modifications. Let us divide the data in the spectrum into two halves, called *A* and *B*, where *A* is the data below and *B* the data above the carrier frequency. Then, let us call the computer inputs to the dual ADC *a* and *b*. Now we need only develop a simple switching arrangement so that on the first scan, data *A* go through phase detector 1 and into ADC channel *a* while at the same time data *B* go througn 2 and into *b*. Then on the next scan, data *A* go through 2 and into *a* and data *B* go through 1 and into *b*. Thus, both phase detectors are used by both channels on alternate scans equalizing the weight of both detectors.

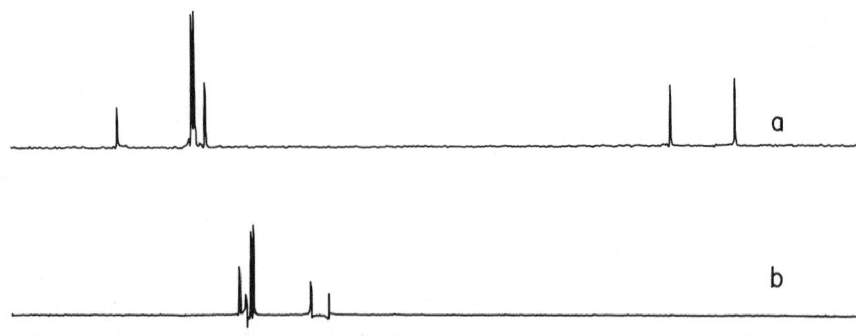

FIGURE 9.2. (a) Ethylbenzene as detected by quadrature FT–NMR. (b) Ethylbenzene detected at the same carrier position as in (a), by single-detector methods. This illustrates the foldback of the information on both sides of the carrier.

TABLE 9.4. Two-Pulse Swapping Sequence

Scan	Data	Phase detector	Input	X memory data
Odd	*A*	1	*a*	*y*
	B	2	*b*	*z*
Even	*A*	2	*b*	*y*
	B	1	*a*	*z*

This sounds quite simple and indeed can be implemented quite readily. Levy[9] has reported that there is an image formation of perhaps 8% using unmatched detectors in quadrature detection, but that this is totally removed by the two-pulse swapping technique.

Implementation of this two-pulse swapping technique can be somewhat simplified by allowing the connections between phase detector and computer input *a* and detector 2 and input *b* to remain constant. Then the data accumulating in memory can be swapped by software between scans. This amounts to negating the two blocks of memory where data are accumulating *y* and *z* and swapping them between scans. The result is shown schematically in Table 9.4.

This technique of memory and input swapping can be extended further to allow for systematic noise reduction by using a phase-alternating pulse sequence, or a total of four pulses in a cycle as shown in Table 9.5. Note that the data are inverted by the RF phase on scans 2 and 4 and then subtracted from memory by adding to memory after it is negated. Memory is then renegated after these scans to make it positive again.

Now it turns out that this entire process of switching and swapping can be simplified both from the point of view of software and hardware by realizing that if the phase of the irradiating RF is changed by 90° then the data that each phase detector "sees" will change since the detectors are set 90° apart.

TABLE 9.5. Four-Pulse Swapping Sequence

Scan	Data	Phase	Phase Detector	Input	Memory data
1	*A*	0°	1	*a*	*y*
	B		2	*b*	*z*
2	*A*	180°	1	*a*	$-y$
	B		2	*b*	$-z$
3	*A*	0°	2	*b*	*y*
	B		1	*a*	*z*
4	*A*	180°	2	*b*	$-y$
	B		1	*a*	$-z$

TABLE 9.6. Four-Pulse Swapping Sequence with 90° RF Phase Increments

Scan	Data	Phase	Phase detector	Input	Memory data	Next manipulation
1	A	0°	1	a	y	
	B		2	b	z	Swap y and z
2	A	90°	1	a	z	
	B		2	b	y	Swap y and z and negate
3	A	180°	1	a	$-y$	
	B		2	b	$-z$	Swap y and z
4	A	270°	1	a	$-z$	
	B		2	b	$-y$	Swap y and z and negate
5(1)	A	0°	1	a	y	
	B		2	b	z	

Thus, there is no need to actually physically switch the inputs from $A1$–$B2$ to $A2$–$B1$; instead the RF phase is changed by 90°. Since we are already changing the RF phase by 180° to arrive at the phase-alternating pulse sequence we might as well simply advance the RF by 90° four times in succession instead to accomplish both the detector switching and the noise reduction. Then the problem reduces itself to memory swapping between scans along with RF phase changes between scans. This is shown in Table 9.6. Clearly, to use the sequence described in Table 9.6, it is necessary to ensure that the RF phase and the data swapping stay in sync even if data collection is interrupted and restarted.

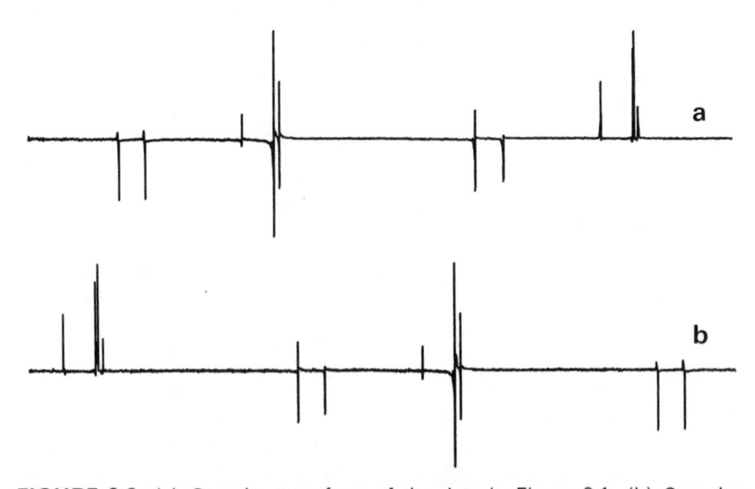

FIGURE 9.3. (a) Complex transform of the data in Figure 9.1. (b) Complex transform followed by the swapping of quadrants 1 with 4 and 2 with 3.

Other pitfalls in processing quadrature data include the fact that the straightforward complex transform of the data from quadrature detection will lead to a spectrum such as shown in Figure 9.3a. It can be divided into four parts or quadrants, which we call 1, 2, 3, and 4. Because of the fact that the data are detected relative to a frequency in the center of the spectrum, these quadrants require some swapping before phase correction to the final form. Quadrants 1 and 4 must be swapped as must quadrants 2 and 3. The result before phase correction is shown in Figure 9.3b.

9.3.3. Operational Details in Quadrature NMR

Since in quadrature detection FT–NMR, it is possible to place the carrier anywhere at all, it is possible that it will turn out to have been sitting directly on a peak of interest. If it was sitting on such a peak the frequency in the free-induction decay would be essentially 0 Hz or DC, and the baseline correction usually used before a Fourier transform would remove the peak from the spectrum. This is illustrated in Figure 9.4. Johnson[10] has suggested that this problem can be avoided by simply not baseline correcting the two FIDs before the transform, but Bradley[11] has pointed out that the DC component due to the peak at 0 Hz will have died out by the end of the free-induction decay like any other NMR signal, and suggests that the baseline correction factor be determined from the last 25% of the FID rather than from the entire FID. These are usually determined by making the integral of the FID equal to zero by subtracting an appropriate constant from the entire FID. This latter approach avoids the introduction of spurious spikes in the spectrum from actual DC errors in the spectrum.

Operationally, quadrature detection NMR is almost exactly like single-detector FT–NMR. The only significant difference is that in examining the

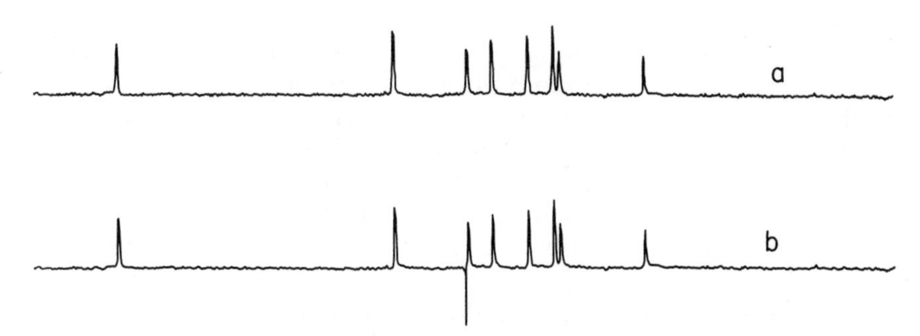

FIGURE 9.4. (a) Complex transform of quadrature data after baseline correction using only the data in the last quarter of the spectrum to determine the correction factor. (b) Complex transform of quadrature data after baseline correction using entire FID to determine the correction factor. This reveals that a peak was almost at the carrier frequency. Spike is generated because of spurious DC offset produced by erroneous baseline correction.

spectrum display (or plot) one must be cognizant of the fact that the carrier is in the *middle* of the spectrum when deciding how to change the carrier position or the spectrum width. Further, the entered spectrum width must be divided by 2 by the computer program or by the operator, since half of this width is detected on each side of the carrier and only half the expected sampling rate is needed. These problems are usually taken care of in commercial programs.

9.3.4. Comparison between Crystal Sideband Filter and Quadrature Detection

Both the crystal sideband filter approach and the quadrature detection technique have the advantage of a factor of 1.4 increase in sensitivity over the single-detector unfiltered approach. The crystal sideband filter filters out the unwanted noise on the other side of the carrier and the quadrature detection approach utilizes data on both sides of the carrier. Since quadrature detection spectroscopy is significantly more expensive to implement, why should it be considered at all? We will discuss the reasons below.

The advantages of the crystal sideband filter are primarily cost. It can be purchased and installed for one hundred dollars or so and requires no change in the software or operating procedures. On the other hand, quadrature detection is not limited to one or two bandwidths, but allows a continuous variation of the bandwidth of the spectrum. Quadrature detection requires less of the spectrometer and data system, moreover, because the RF carrier is placed in the middle of the spectrum rather than at the end. For example, only half the sampling rate need be used to cover the same bandwidth as in single-detector NMR, thus leading to wider possible spectral widths with the same computer and spectrometer limitations. Second, the amount of pulse power needed to cover the spectral width is halved as well, so that fall-off of pulse power over a wide bandwidth becomes less of a problem. Stejskal and Schaefer[12] have also pointed out that sizable distortions can occur in transformed spectra obtained by single-detector methods when the spectrum contains extremely broad Lorentzian lines. These are less troublesome in quadrature FT–NMR.

One further persuasive argument for quadrature detection NMR has to do with the maximum dynamic range of the transformed spectrum. The dynamic range, as we saw in Chapter 4, depends on the length of the computer word, the size of the array, and the number of memory locations that are full at the outset of the transform. If in quadrature detection NMR we can place the carrier very near the large solvent peak in the spectrum, the number of memory channels that are full or almost full is much smaller than if the solvent peak is near the Nyquist frequency. Thus, the dynamic range of the transform will be increased by the fact that most solvent peaks are in the middle of a spectrum in which foldback cannot be tolerated.

9.4. RAPID-SCAN (CORRELATION) NMR

9.4.1. General Description

The rapid-scan NMR technique is not a pulsed technique at all, but rather a method for the rapid gathering of data by a fast frequency or field sweep through the data. The scan is so rapid, usually only a few seconds, that all of the peaks ring together into a somewhat confusing mass, such as shown in Figure 9.5a. It turns out, however, that these data can be converted into a conventional NMR spectrum by cross-correlating them with a spectrum obtained by scanning through a single line under the same sweep conditions. The spectrum of a single line is shown in Figure 9.5b and the result of the correlation process in Figure 9.5c.

The obvious advantages of correlation NMR are that it does not re-

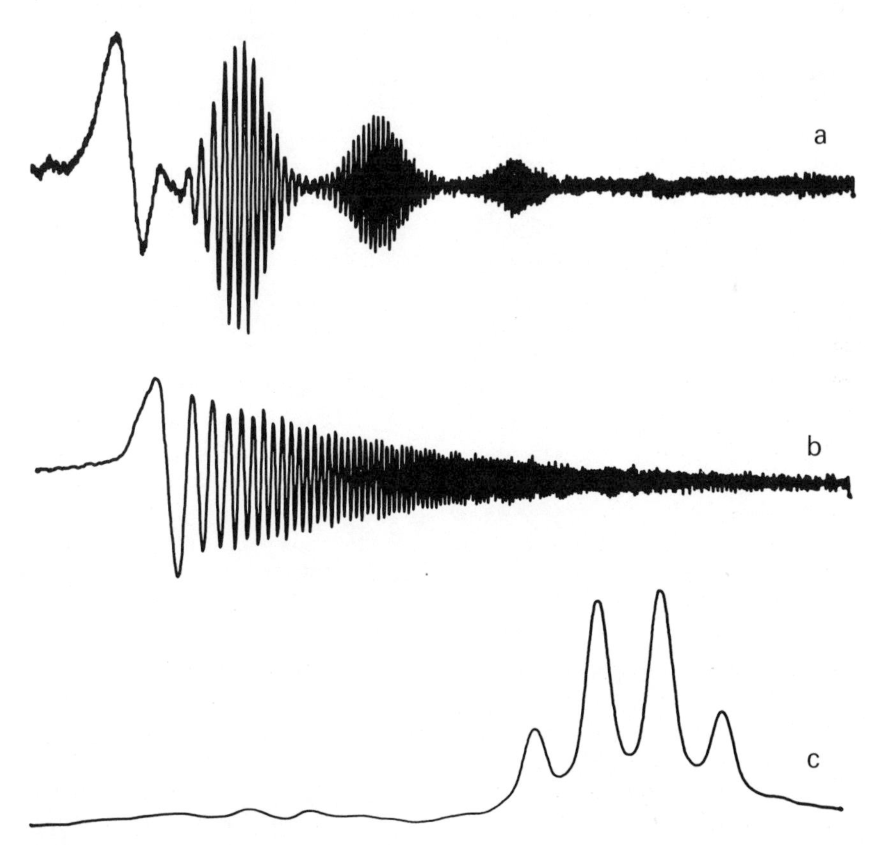

FIGURE 9.5. (a) Single rapid scan through the quartet of ethylbenzene. (b) Single rapid scan through TMS singlet. (c) Cross-correlation of (a) and (b).

quire high-powered pulse amplifiers and can thus often be carried out on older second-generation spectrometers, which would be difficult to adapt for FT work. Further, unlike FT–NMR, the user has the option of scanning any desired portion of the spectrum at any rate, so that large solvent peaks can be ignored if they might impose undesirable dynamic range requirements on the spectrum. Further, the rapid-scanning process means that data can be acquired almost as fast as in the pulsed-FT experiment.

In both of the papers published on the subject,[13,14] the authors have actually acquired or calculated a reference line with which to perform the cross correlation. Now this requires that twice as much memory be used for the correlation process as might be needed for the FT experiment, since a 4096-point scan will require a 4096-point reference to perform the cross-correlation.

It should be noted, however, that the correlation reference can be calculated point by point by the correlation routine, and while this is much slower, it allows a halving of the required memory. The actual correlation process has been described by Gupta *et al.*[13] Their technique is to generate the complex reference line by calculating the function

$$R_t = \exp(-ibt^2/2) \tag{9.1}$$

where t is the time of that data point and b the sweep rate in rad/sec.[2] This can be reduced to

$$R_n = \exp(i\pi n^2/\mathrm{ST}\cdot\mathrm{SW}) \tag{9.2}$$

where n is the index of that data point, ST the sweep time, and SW the sweep width. Then, since this is actually a complex multiplication, the values C and S are determined:

$$C = \cos(\pi n^2/\mathrm{ST}\cdot\mathrm{SW}) \tag{9.3}$$

$$S = \sin(\pi n^2/\mathrm{ST}\cdot\mathrm{SW}) \tag{9.4}$$

and the real and imaginary parts of the correlogram determined by

$$R_n = R_n C + I_n S \tag{9.5}$$

$$I_n = -R_n S + I_n C \tag{9.6}$$

Thus, correlation amounts to complex multiplication of data, according to equations (9.5) and (9.6).

9.4.2. Data Processing Methods

There are two ways of developing the data for correlation from the rapid-scan data. One method, devised by Gupta *et al.*, produces a real and imaginary part just as in FT–NMR, which can be readily phase corrected.

The other method devised by Patt[14] produces all real points, which must be phase-corrected in the complex domain. The advantages of the first method are ease of phase correction, and the advantages of Patt's method that greater resolution is attainable in the same amount of memory.

The data processing method of Gupta, Becker, and Ferretti (GBF) includes the following steps:

1. Inverse transform of data, assuming data are complex and that real and imaginary points alternate. Leaves real and imaginary in first and second halves of data.
2. Correlation performed in complex domain. Second half of data is *zeroed*.
3. Exponential multiplication or other window functions.
4. Forward complex Fourier transform assuming real in first half, imaginary in second half.
5. Phase correction on resulting real and imaginary frequency domain data.

The processing method developed by S. L. Patt (SLP) is as follows:

1. Forward real transform of data, assuming alternate points real and imaginary. Leaves real and imaginary in first and second halves of data.
2. Correlation performed in complex domain. No data are zeroed.
3. Exponential multiplication of resulting data.
4. Phase correction performed here in the "complex domain."
5. Inverse real transform, assuming real and imaginary in first and second half but shuffling to alternate points at end of transform.
6. Leaves all real points. Phase correction only possible by successive approximation by returning to step 4 by forward transform and then re-transforming to real domain.

The GBF method clearly has the advantage in terms of phase correction since it is performed exactly as in FT–NMR with a zero- and first-order term in the frequency domain. The SLP method, on the other hand, produces all real points after the last transform, so that there is inherently more resolution in the final spectrum. However, the SLP method also requires that phase correction be done "flying blind" by estimating the needed correction in the frequency domain, and then going back by forward real transform to the "complex domain" where these estimated correction parameters are applied. If several iterations are required to correct the spectrum properly, this can be somewhat time consuming. Further, as we have seen, such multiple transformations reduce the accuracy of the resulting data. Finally, the SLP method requires a real forward and a real inverse transform, which requires somewhat more memory than the corresponding forward and inverse complex routines. Both methods are illustrated in Figures 9.6 and 9.7.

The principal reason for carrying out the correlation NMR experiment is usually to acquire data rapidly in a small region without sweeping through a spectral region containing a large peak that will cause a dynamic range

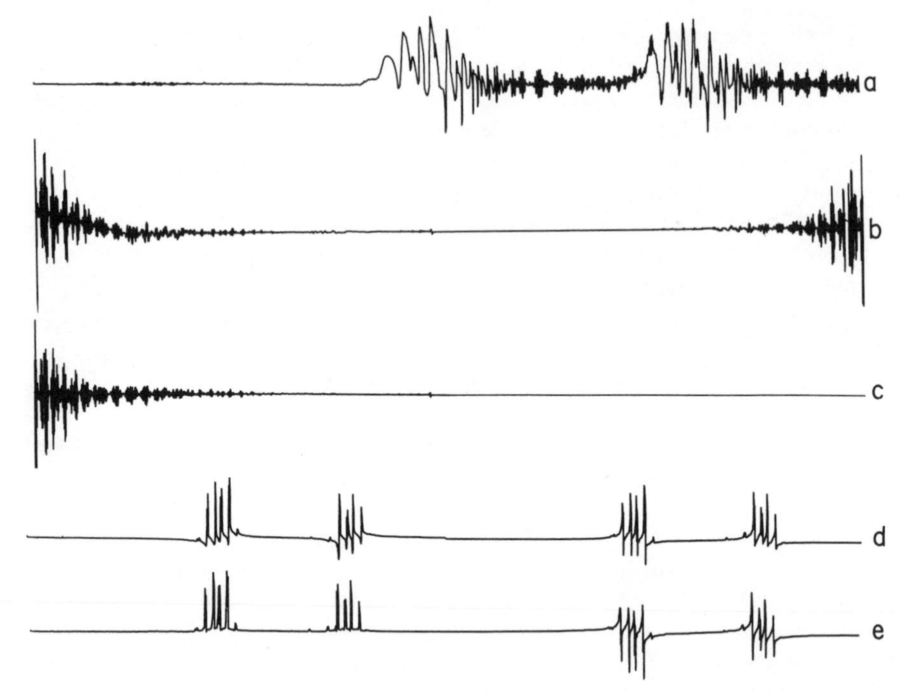

FIGURE 9.6. Processing of correlation NMR data by the GBF method. (a) Original scan. (b) After inverse transform. (c) After correlation and zeroing of right half. (d) After forward complex transform. (e) After phase correction.

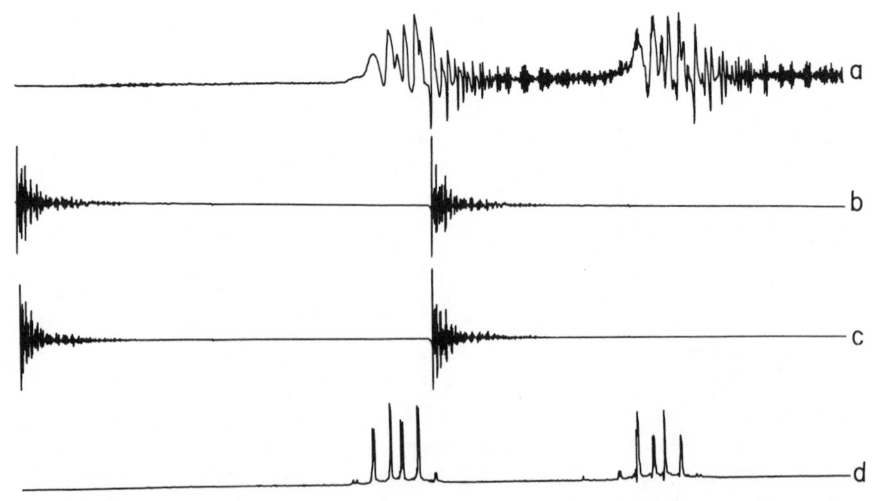

FIGURE 9.7. Processing of correlation NMR data by the SLP method. (a) Original scan. (b) After forward real transform. (c) After correlation. (d) After inverse real transform, phase correction not shown.

problem during signal averaging. The other reason is that correlation spectroscopy is somewhat simpler to retrofit to older spectrometers than pulsed-Fourier techniques.

Operationally, correlation NMR is somewhat more difficult to use than are standard pulsed-FT techniques, because the sampling theorem can be so easily violated. The sweep rate is empirically determined by the response of the nuclei and of the receiver in most swept NMR experiments. However, in correlation NMR, the ringing of the lines increases in frequency as the observation frequency goes farther and farther from the resonance frequency. In fact, the farther (in hertz) that one sweeps beyond the resonance frequency, the higher the ringing frequency of that particular line. However, the lines do not persist indefinitely but relax in the usual fashion so that a line near the beginning of the spectrum may very well have died out before the end of the spectrum through natural relaxation processes, thus preventing violation of the sampling theorem. The interrelation of the sweep rate and the sampling frequency and the ringing rates is thus quite complex and can usually only be determined by careful experiment. Selection of the appropriate filter bandwidth is likewise related to these parameters.

For example, if we are observing a rapid-scan spectrum having a bandwidth of 1000 Hz, and lines near the beginning of the spectrum are quite sharp and thus will persist through the sweep, the sampling frequency must be at least 2000 Hz so that the frequency of the ringing of these lines will not be above the Nyquist frequency. This further implies a 1000-Hz filter setting and if, say, 8192 points are to be taken, a sweep time of

$$\mathrm{ST} = 8192\,(1/2000\ \mathrm{Hz}) = 4.096 \quad \mathrm{sec}$$

However, if only 500 Hz is to be observed, then the maximum sampling rate is 1000 Hz, or one point every millisecond, so that the sweep time is

$$\mathrm{ST} = 8192\,(1/1000\ \mathrm{Hz}) = 8.192 \quad \mathrm{sec}$$

9.5. NOISE EXCITATION METHODS

9.5.1. Stochastic Resonance Spectroscopy

The stochastic resonance spectroscopy technique[16-18] amounts to the modulation of a single-frequency RF carrier by a pseudorandom number sequence, which in turn produces a frequency distribution that excites the entire spectrum region of interest. The response of the spin system to this excitation can be stored in a computer memory and signal averaged, where the start of each new "scan" corresponds to the restarting of the pseudorandom number generator. This response can then be Fourier transformed to produce a frequency domain spectrum, but has the disadvantage that the phase of each point must be corrected separately through a rather time-consuming calculation.

Further, the data acquired cannot be Fourier transformed using the standard Cooley–Tukey algorithm, since pseudorandom number sequences generated by shift registers always produce $2^n - 1$ points, and if one data point is taken for each point of the shift register excitation, there will not be the requisite 2^n points for a Fourier transform. Strictly speaking, one cannot simply add an extra point to the signal-averaged data to allow fast Fourier transformation, but in practice, the errors generated are rather small.

9.5.2. Hadamard Transform NMR

A second and more elaborate excitation method has been described by Kaiser[19] and Ziessow,[20] in which the nuclei are excited by a pseudorandom modulation of the RF carrier and then the values of the modulation, which are all ones and zeros, are used instead of a sine look-up table to perform a *Hadamard* transform on the accumulated data. The Hadamard transform is very fast, since there are no multiplications in it, and it produces a free induction decay, which can be processed using standard NMR programs. This method eliminates the problem of the $2^n - 1$ excitation points versus the 2^n points required for the transform, since the Hadamard transform performs this conversion. The phase correction necessary in Hadamard spectroscopy is now a simple zero- and first-order correction, since the Hadamard transform unscrambles the phase in converting the response to an FID.

At one time it was thought that the Hadamard transform method would reduce the problem of computer detection of dynamic range, but the argument by which this was advanced has since been shown to be incorrect.[21] The computer system must be able to detect the response of the largest and smallest peaks simultaneously, so that the dynamic range is exactly the same as in pulsed FT–NMR.

9.5.3. Tailored Excitation

Recently Tomlinson and Hill[22] reported an interesting technique for the generation of an excitation function containing or not containing any desired frequencies. This method requires the generation of a frequency domain spectrum containing the frequencies to be included as excitation frequencies as peaks in the spectrum and conversely, those frequencies not to be excited are omitted from this spectrum. This spectrum is then inverse Fourier transformed to produce a time domain spectrum that can be used to modulate the RF carrier. This modulation effectively produces irradiation whose spectral characteristics are described by the original synthesized frequency spectrum. Therefore, the excitation can be "tailored" to whatever

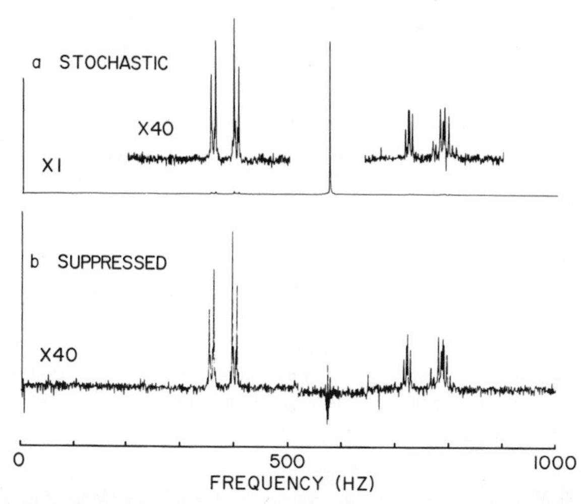

FIGURE 9.8. Tyrosine in D_2O, showing suppression of an unwanted solvent line. (a) Spectrum obtained by exciting sample with flat-power, random-phase excitation. (b) Spectrum obtained by exciting with a function having zero amplitude for the frequencies around the HDO resonance. Intensity of the HDO line in the top spectrum is approximately 150 times the intensity of the same line in the lower spectrum. (Courtesy of the Journal of Chemical Physics.)

experiment is desired. In Figure 9.8 the stochastic excitation spectrum of tyrosine in D_2O is shown at an amplitude 40 times that of the center HDO peak. The lower trace shows the same spectrum excited with a tailored excitation, which did not include the HDO frequency. The HDO peak is barely visible even at the 40× amplitude setting.

The tailored excitation method clearly can be used to suppress strong solvent peaks or decouple any number of frequencies. It has the disadvantage that, like standard stochastic resonance, the phase correction process is rather complex and time consuming. Further, in setting up the modulation scheme and keeping it under control there are significant instrumental problems that keep it out of the realm of the routine experiment at present. However, it does represent one more useful technique to use when needed.

9.6. MEASUREMENT OF THE SPIN–LATTICE RELAXATION TIME T_1

9.6.1. General Description

When a RF pulse excites a spin system in a magnetic field, it tips the magnetization into the *x-y* plane where the various magnetic vectors from

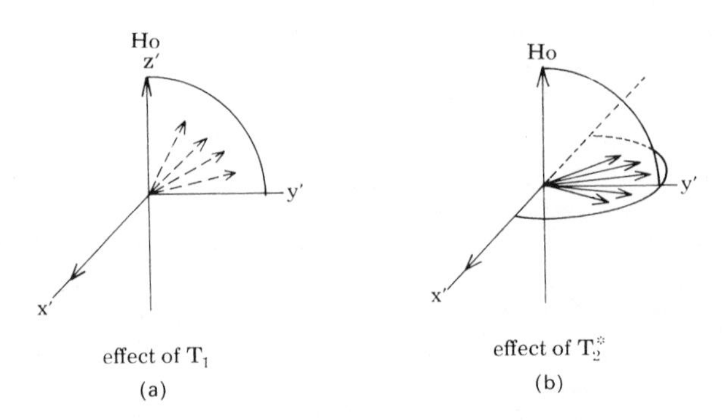

FIGURE 9.9. (a) Spin–lattice relaxation, as measured by T_1. (b) Spin–spin relaxation and inhomogeneity effects, as measured by T_2^*.

each of the nuclei precess at their respective Larmor frequencies. The rate at which these magnetization vectors tend to return to their unperturbed states precessing close to the applied field H_0 is described by a spin–lattice relaxation time T_1. This relaxation process is illustrated in Figure 9.9a. At the same time the nuclei can relax through spin–spin interaction and through interaction with the inhomogeneities of the magnetic field. The spin–spin relaxation time is usually called T_2 and the relaxation process is represented as a dephasing of the various nuclear vectors, as shown in Figure 9.9b. Now, it turns out that the magnetic field inhomogeneities cause exactly the same dephasing of the vectors so that the effective relaxation is a combination of spin–spin relaxation and that caused by the magnetic field inhomogeneities. These are lumped together and called T_2^* and are together the dominant relaxation process leading to the decay of the NMR signal following a pulse in liquids.

While measurement of the spin–spin relaxation time T_2 is thus complicated by magnetic field inhomogeneities and is thus somewhat difficult to obtain, the spin–lattice relaxation time T_1 is much easier to measure by a variety of techniques. These include inversion–recovery, progressive saturation, and homospoil–recovery. Now the T_1's of a molecule are interesting because they can provide significant structural information. The types of information available have been described by Levy and Nelson[23] and by Levy and Lyerla.[24] We summarize them briefly here.

9.6.2. Reasons for Measuring T_1

In order to see how the T_1's relate to structural parameters, it is necessary to have some appreciation of the relaxation mechanisms that dominate in most common organic molecules. They are, in decreasing order of importance, (1) dipole–dipole interaction (DD), (2) spin rotation (SR), (3) chemical shift anisotropy (CSA), (4) scalar relaxation (SC). Only the first two are of any importance in most common organic systems in ^{13}C NMR. Dipolar relaxation is simply the interaction of the various magnetic nuclei with small local magnetic fields of other nuclei. Spin rotation relaxation is uncommon except in very small molecules and can usually be neglected.

In relatively rigid molecules that tumble equally in all directions or *isotropically*, the T_1's of the various carbons are related to the number of hydrogens that are attached to them. The T_1's for nonprotonated carbons are 10–20 times those for carbons having a single hydrogen substituent. Further, in such molecules, the CH_2 T_1's are half those of CH carbons. In principle, the CH_3 carbons should have T_1's that are $\frac{1}{3}$ those of the CH carbons, but *only* if they do not rotate freely. If some rotation occurs, as is true except in very hindered molecules, the T_1's of the CH_3 groups will be somewhat longer than $\frac{1}{3}$ the CH T_1.

More interesting structurally, however, is the fact that many compounds do not tumble isotropically and have preferential tumbling modes due to their geometry. For example, the T_1's of monosubstituted benzenes are often cited, where it is found that the carbon *para* to the substituent has the shortest T_1 since the molecule presumably is rotating around the 1–4 axis. The C_1 T_1, of course, are quite long since no proton is attached to improve DD relaxation.

Measurements of T_1 are also of great interest in determining the rate of molecular tumbling in larger biologically interesting molecules. While various rotation and tumbling rates cannot always be separated, the average time for a molecule to rotate through one radian can often be determined. This is called the *rotational correlation time* τ_c and it can be related directly to the spin–lattice relaxation time T_1. Assuming that the relaxation mechanism in these larger molecules is dominantly dipolar as it almost always is, the effective rotational correlation time is related to the dipolar T_1 by

$$\frac{1}{T_1^{DD}} = \frac{n_H \gamma_H^2 \gamma_C^2 \hbar^2 \tau_c^{eff}}{r_{CH}^6} \tag{9.7}$$

where n_H is the number of hydrogens attached to that carbon, and γ_C and γ_H are the magnetogyric ratios of carbon and hydrogen.(25)

Allerhand(26) has also shown that measurement of the T_1 by the inversion–recovery technique can assist the spectroscopist in uncovering lines buried under other lines that have a different T_1.

9.6.3. Methods of Measuring T_1

One of the most popular methods of T_1 measurement is the *inversion–recovery* method originally reported by Vold *et al.*(27) It has also been referred to by Allerhand(28) as *partially relaxed Fourier transform* (PRFT) spectroscopy.

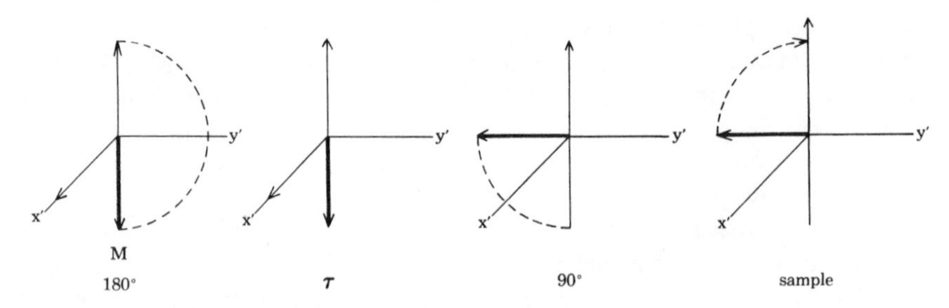

FIGURE 9.10. Inversion–recovery experiment when $\tau \ll T_1$.

In this method, the system, assumed to be at equilibrium, is excited with two RF pulses, a 180° pulse and a 90° pulse, separated by a delay time τ. For each spectrum in the set, this pulse spacing is varied. Following the 90° pulse, the free-induction decay is sampled as usual and followed by a fixed delay T, which must be at least five times the longest T_1 in the system, to allow for complete recovery of the magnetization.

The pulse sequence then is abbreviated

$$[-180 - \tau - 90 - (\text{sample}) - T -]_n$$

where n is the number of spectra to be summed.

Now let us examine what happens during this sequence. For $\tau \ll t\, T_1$ the 180° pulse rotates the magnetization M into the $-z$ axis as shown in Figure 9.10. If the delay time τ is quite short, very little relaxation will occur and the 90° pulse will cause rotation of the vector M into the x axis, from which it slowly decays back to $+z$ while sampling occurs. Since this is roughly equivalent to a 270° pulse, it is not surprising that it produces a signal 180° out of phase with that produced by a 90° pulse, or in other words, an inverted peak.

In case where $\tau < T_1$ some spin–lattice relaxation will occur during the time τ. During τ, the vector M is "shrinking" up the $-z$ axis during τ so that when the 90° pulse occurs a smaller negative signal occurs, as shown in Figure 9.11.

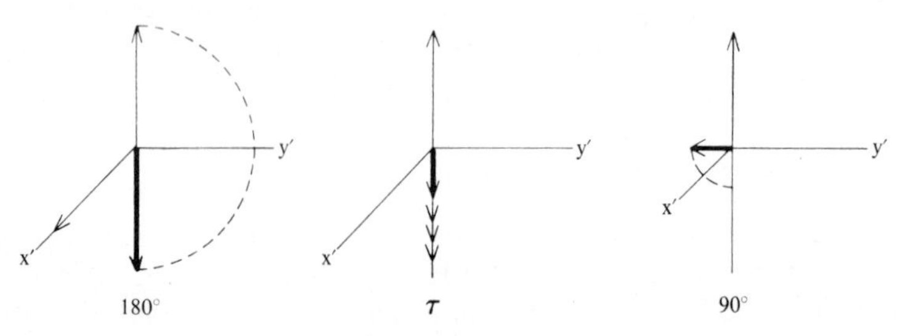

FIGURE 9.11. Inversion–recovery experiment when $\tau < T_1$.

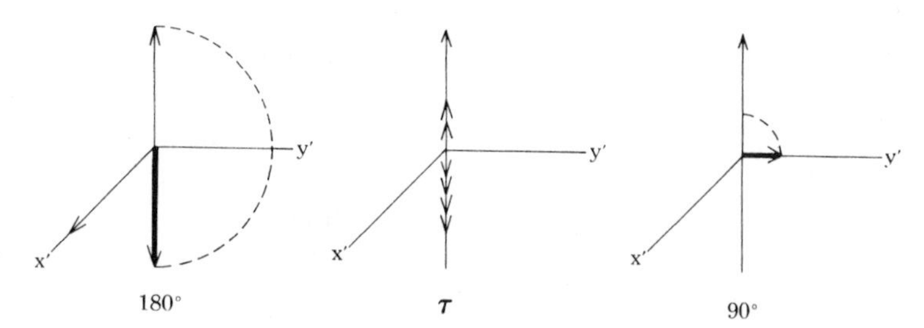

FIGURE 9.12. Inversion recovery experiment when $\tau > T_1 \ln 2$.

When $\tau \simeq T_1 \ln 2$, the magnetization will be passing through the origin of our coordinate system and a null or zero signal will be observed for this line. As τ becomes longer, the signal will become a small positive peak as the magnetization relaxes into the positive z axis, as shown in Figure 9.12.

Finally, after a time $\tau \gg T_1$, recovery from the 180° pulse will be complete and the 90° pulse will produce a full positive-intensity signal.

To take full advantage of this technique, one simply selects a range of values for the pulse interval τ and allows the data system to accumulate a number of scans at each value. These data can then be stored on disk after each scan and processed all at once at the end of the experiment. This disk storage approach saves processing time during data acquisition and allows nondestructive experimental determination of the most effective exponential and/or trapezoidal window and phase correction constants. It also allows the production of stacked offset (isometric) plots without introducing the worry of pen skipping during data readout, since any spectrum can be replotted from disk if necessary. A typical plot of such T_1 data is shown in Figure 9.13.

Now the T_1 can be calculated from the plots or from the data as stored on disk, where the intensities of the lines are given by

$$A = A_\infty[(1 - 2\exp(-\tau/T_1)] \tag{9.8}$$

where A is the intensity of the line in a given spectrum and A_∞ the intensity of the line in the spectrum where $\tau \gg T_1$. This calculation can be simplified to a linear plot or least-squares calculation by taking the log of both sides of equation (9.8), giving

$$\ln(1 - A/A_\infty) = -\tau/T_1 + \ln 2 \tag{9.9}$$

Thus, a plot of $\ln(1 - A/A_\infty)$ vs. τ gives a line with a slope of $-1/T_1$ and a theoretical intercept of 0.693. This calculation method has been commonly used in articles[28,29] and in commercial data systems for the calculation of the T_1. However, it does unduly weigh the τ_∞ value and it is sometimes desirable to take a number of spectra at the infinity τ value so that their amplitudes can be averaged in the calculation. Another method to avoid this

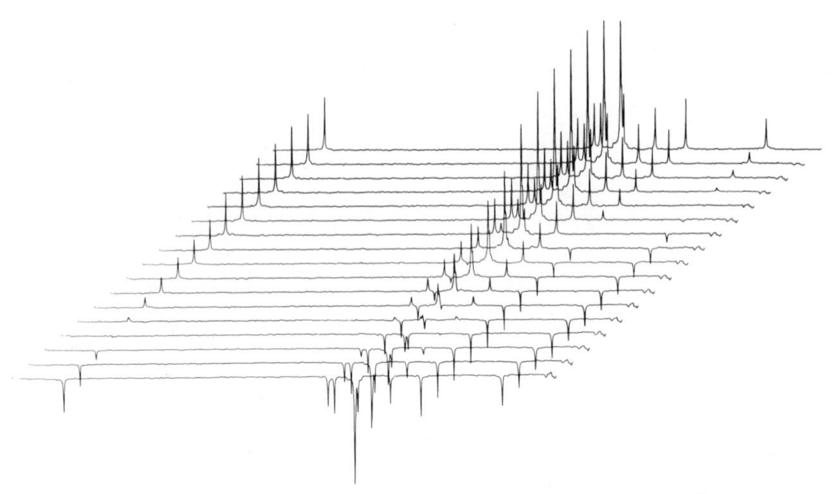

FIGURE 9.13. Typical data from an inversion–recovery T_1 experiment. The spectrum is of dodecyl alcohol in deuterobenzene, with delay times of 0.1, 0.2, 0.3, 0.4, 0.5, 0.6, 0.7, 0.8, 0.9, 1.0, 1.5, 2.0, 2.5, 3.0, 3.5, 4.0, 6.0, 8.0, and 10.0 s. (Courtesy of L. F. Johnson, Nicolet Technology, Inc.)

overweighing of τ_∞ suggested by Fagerness is to perform a nonlinear least-squares calculation on the data, but this is considerably more time consuming.

9.6.4. Progressive Saturation

The other most popular method of measuring T_1 is the progressive saturation technique developed by Freeman and Hill.[30] This method amounts to applying a series of several equally spaced 90° pulses to the data and then sampling the free-induction decay after the last of the pulses. For each spectrum a different time interval between the pulses is chosen, the spectra are plotted out, and a similar least-squares calculation is performed. The progressive saturation approach appears at first to be more desirable for samples having long T_1 because there is no long waiting time T between scans. However, the accuracy with which the 90° pulse can be determined is critical to this method and errors in the pulse width can lead to extreme errors in the determination of the T_1. Further, the total intensities of the lines in the spectra obtained by this technique are only half that obtained in the inversion–recovery technique. On the other hand, if the particular spectrometer system cannot produce an effective 180° pulse that covers the desired spectral width because of pulse power limitations, the progressive saturation technique is preferable.

9.6.5. Homospoil-T_1 Methods

Recently, methods for measuring long T_1 that do not require a lengthy $5T_1$ waiting period between pulses have been suggested by Markley *et al.*,[31] Freeman and Hill,[30] and McDonald and Leigh.[32] These are termed "homospoil-T_1" or "saturation–recovery" experiments. The sequences involve the use of a pulse to "spoil" the homogeneity of the field for a brief time. Such pulses are easily connected to most spectrometers. One such sequence is illustrated in Figure 9.14.

Initially, a 90° pulse is applied to the system, forcing the magnetization into the x-y plane. This removes all z magnetization. Here a homospoiling pulse is applied to the y gradient, causing the vectors to dephase in the x-y plane as shown in Figure 9.14b. The result is that the system has no net x-y magnetization, as shown in Figure 9.14c. The delay time τ is then used to allow z magnetization to recover according to T_1, as shown in Figure 9.14d. If τ is very short, there will be little or no recovery; if τ is long recovery will be essentially complete. After the time τ, a second 90° pulse is applied (Figure 9.14e) followed immediately by sampling of the free-induction decay (Figure 9.14f). After sampling, the magnetization will be partially recovered, having components in both the x-y planes and the z axis. The x-y magnetization is removed by an additional homospoil pulse resulting in (Figure 9.14g). Figure 9.14g is exactly analogous to Figure 9.14a and the sequence can be repeated immediately.

The sequence is thus abbreviated

$$[-90 - \text{spoil} - \tau - 90 - \text{sample} - \text{spoil} -]_n$$

It has been our experience that these sequences are most applicable to systems having long T_1's. When very short T_1's are to be measured, the shorter τ values may actually lead to sampling of the FID before the homogeneity has recovered.

Spectra produced by such sequences will start at null and move upward to fully recovered data. There are no inverted peaks since no 180° pulses are

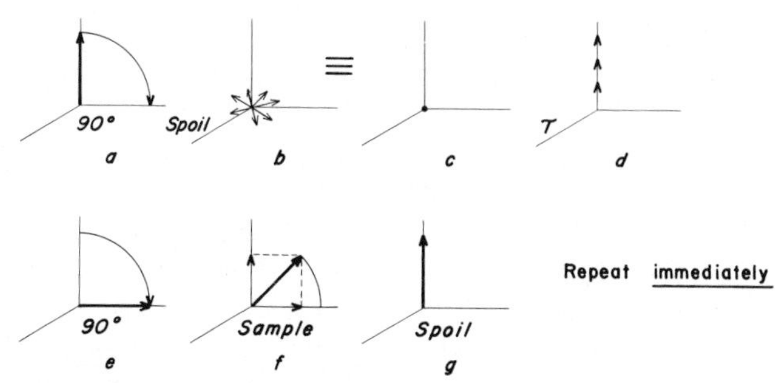

FIGURE 9.14. Homospoil-T_1 sequence as suggested by McDonald and Leigh.[32]

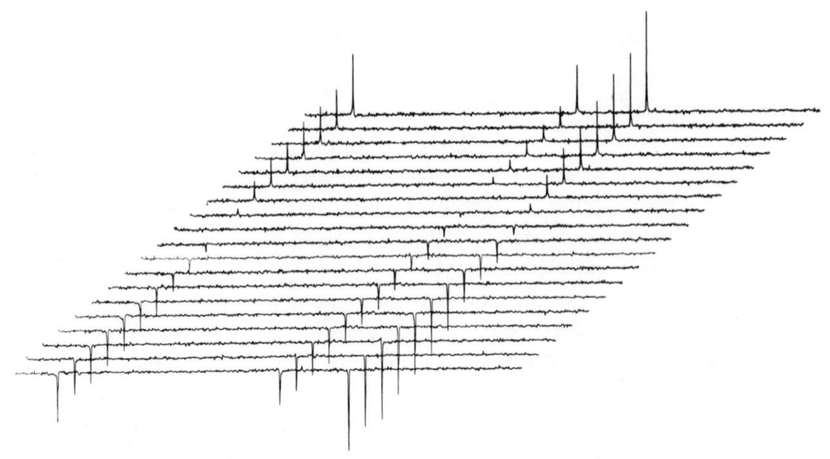

FIGURE 9.15. Isobutanol in $CDCl_3$ by inversion–recovery methods.

given. Care should be taken that the first spectrum in the set is not a set of nulled peaks, as the FT scaling factor may be calculated from the first spectrum processed during batch processing and this spectrum should have most of the peaks recovered so that the remaining spectra will be on scale. Inversion recovery and homospoil-T_1 sequence spectra of isobutanol are shown in Figures 9.15 and 9.16.

9.6.6. Experimental Techniques in the Measurement of T_1

In order to measure T_1 accurately, it is first necessary to determine the width of a 90° and a 180° pulse accurately. If one starts at a very small pulse

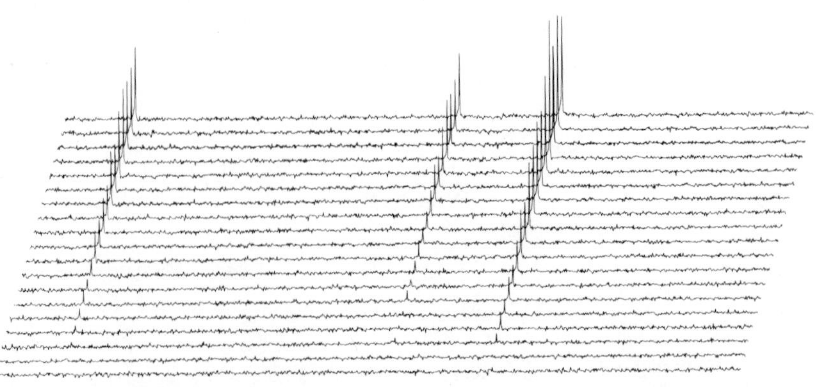

FIGURE 9.16. Isobutanol in $CDCl_3$ by homospoil-T_1 method.

width of say 1 μs or so, takes a single scan, and then does further experiments with longer pulses, the intensity of the FID is directly related to the width of the pulse. If the pulse width is slowly increased, the FID will go through a maximum as the 90° pulse is reached and begin to diminish in intensity beyond this point.

In actual experimental cases, however, the 90° pulse is difficult to determine rapidly since the user must wait around three to five times the longest T_1 in the molecule before a new pulse can be applied. Therefore, rather than using the experimental sample to determine the 90° pulse, one having very short T_1 and a very strong signal is often substituted. One common sample for this technique is benzene saturated with $Cr(acac)_3$. Another common sample is polyethylene glycol, where the T_1 are in the millisecond range.

In addition, the pulse width that is least easily determined is the 90° pulse since it is a rather difficult and subjective judgement which of several FIDs has the greatest intensity. It is somewhat simpler to determine the width of the 180° pulse, since the FID will seem to disappear completely at the 180° setting. However, a waiting time of $5T_1$ is required here between experiments. Thus, when the data system and spectrometer electronics will allow it, the 360° pulse is simplest to measure. The FID from a 360° pulse will be a null, just as for the 180° pulse, but the null is usually cleaner than for the 180°, and further, since the magnetization is returned to essentially its equilibrium position, there is no necessity for a long waiting period between pulses.

Once the 360° pulse is determined, the 180° pulse is half of that and the 90° pulse half of that. These pulse widths are then constant for all experiments performed on that nucleus unless there are marked changes in the spectrometer configuration or sample characteristics.

If the homospoil-T_1 or saturation–recovery sequence is to be used for the measurement of T_1 a pulse must be connected to the y gradient of the spectrometer, which markedly changes the homogeneity of the magnetic field. This pulse has been found[33,34] to vary from 10 μs to 10 ms in different spectrometers, and it must be determined that the pulse width used completely spoils the homogeneity quickly and that the pulse is of sufficient duration that the homogeneity returns after the pulse rapidly enough that the spectrometer lock is not lost. This can usually be determined by simply observing the characteristics of a free-induction decay, which is spoiled by applying a spoiling pulse shortly after the 90° observation pulse. There should be no return of ringing or echo once the spoiling pulse has been applied.

There are a number of sampling handling criteria that must also be considered in accurate measurement of T_1. These have been discussed by Levy and Peat[35] and references cited therein. For T_1 longer than about 50 s, molecular diffusion becomes an important source of error. This can usually be avoided by keeping the sample size small and restricted to the area covered by the coils in the probe, usually 1.5–2 cm. If the sample in solution is volatile at the temperature of the observation, the T_1 may be anomalously short because of potential interactions between the liquid and vapor phase. This is usually avoided by using specially constricted NMR tubes to keep such vaporization to a minimum. Finally, dissolved oxygen in the sample may cause the T_1's to be shortened, especially those over 5–10 sec. For short experiments, this can be avoided by simply bubbling nitrogen through the sample before the experiment and then capping it tightly. For accurate long-term measurements, vacuum degasing is necessary, followed by sealing the sample tube.

9.7. TECHNIQUES FOR THE SUPPRESSION OF STRONG SOLVENT PEAKS

9.7.1. Introduction

As we showed in Chapter 4, the large dynamic range spectrum is a particular problem in Fourier transform spectroscopy. It limits the number of scans before memory overflows, it limits the acquisition sensitivity by the resolution of the ADC, and it introduces more noise during the Fourier transform leading to an extremely inaccurate frequency domain spectrum. Some of the techniques for dealing with the strong solvent peak, usually in ^{1}H NMR have already been discussed. They are, most notably, correlation NMR and tailored excitation NMR. Further, the quadrature detection NMR experiment can also be shown to enhance dynamic range somewhat if the RF carrier can be placed near the large peak in the center of the spectrum.

The simplest solvent peak suppression technique is the notch filter. This is particularly useful when the peak is in the same position in a series of experiments such as the HDO peak in ^{1}H NMR spectra. The carrier can be easily positioned so that the filter filters out exactly the frequency range that the HDO peak covers. In fact, this positioning can often be carried out by simply viewing the free-induction decay without transformation, since there will be a marked decrease in intensity when the correct carrier position is found. Further, many spectrometers have the ability to "listen" to the FID by feeding the detected audio signal to a small speaker. The ear is quite sensitive to carrier position and the adjustment can be made even more accurately in such systems.

Another important method for solvent suppression is the WEFT (*w*ater *e*limination *F*ourier *t*ransform spectroscopy) technique reported by Patt and Sykes[36] and others.[37] It can be used whenever the peaks of interest have significantly different relaxation times from the solvent (usually water) peak. This method is actually an inversion–recovery experiment with the delay time τ between the 180° and 90° pulses chosen so that the water peak is nulled. Since the water peak is nulled it will not cause appreciable signal buildup during averaging. However, this is a somewhat dif-

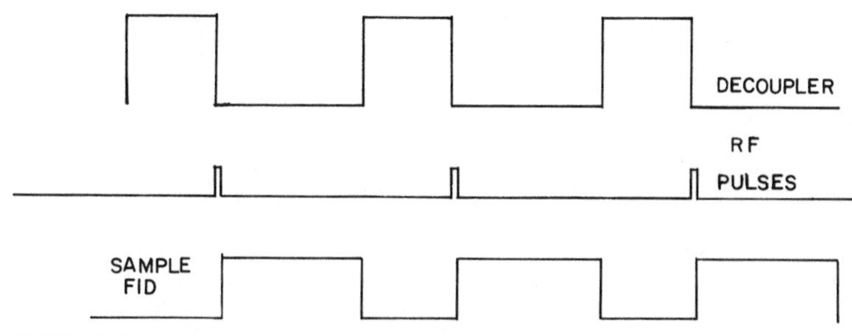

FIGURE 9.17. Pulse timing for presaturation sequence. Decoupler is turned on for a few seconds and then shut off at the time of the excitation pulse and computer trigger. Decoupler is turned on after acquisition finished and so forth.

ficult experiment to do rapidly since the water peak must be allowed to recover so that it can be inverted and nulled again for each scan.

Most of the remaining methods for solvent suppression amount to selectively saturating the strong peak in some fashion. One such method is a presaturation technique entirely analogous to the progressive saturation T_1 experiment in which a series of rapid 90° pulses is applied to the sample before the acquisition of data. As in the WEFT experiment, if the T_1 of the lines of interest are much shorter than that of the solvent peak, the solvent peak will remain saturated and the other lines will be visible.

Another presaturation technique is the turning on of the decoupler[38] at the exact resonance frequency of the solvent peak before the acquisition of data and then turning it off during acquisition. This also effectively saturates the solvent peak and allows observation of others. The pulse sequence is illustrated in Figure 9.17.

Still another method of solvent peak saturation is to apply a short low-energy decoupler pulse to the sample during acquisition but *between data points* so as not to saturate the receiver. This technique was first introduced in commercial data system by Bruker and has been described by Jesson and Meakin.[39] The pulse timing is illustrated in Figure 9.18.

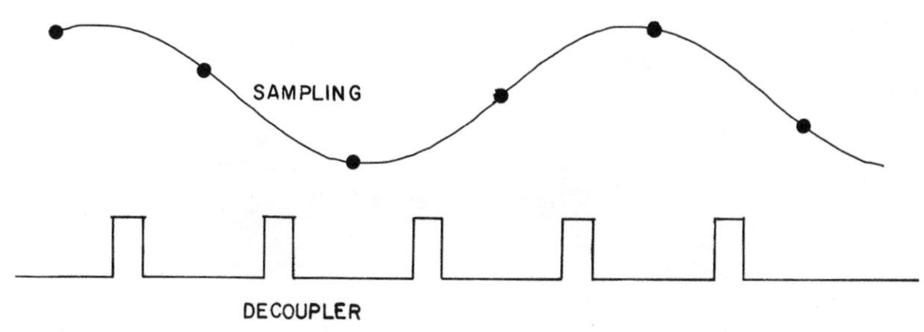

FIGURE 9.18. Timing for pulsed homonuclear decoupling experiment. Decoupler is turned on briefly between data points.

9.7.2. Block Averaging

Finally, block averaging can be used to overcome the problems of solvent peak overflow of memory, by simply accumulating scans until memory overflow is imminent and then transforming the result. This result is then stored in another region of the computer's memory or on tape or disk. Then memory is cleared and the data allowed to accumulate again until memory is again full. This new block of data is then Fourier transformed and then added in the frequency domain into the first block in another portion of memory. Thus, addition of data beyond the point where overflow will occur takes place in the frequency domain, so that the strong solvent peaks overflow memory without affecting those smaller peaks that gain in intensity by signal averaging in the frequency domain. Further, block averaging is the only convenient technique for dealing with samples containing two or more strong solvent peaks, such as when ether or THF is the solvent. We have already shown[40] however, that dynamic range cannot be improved by block averaging beyond the limitations imposed by the round-off error in the Fourier transform.

Two other computer methods that can enhance the dynamic range slightly have been recently described.[40] One is simply to place the large peak at the Nyquist frequency and to vary the phase detector setting or predelay so that the large peak is sampled at its zero crossing time as shown in Figure 4.8. The other is to place the large peak near the Nyquist frequency but not on it, so that the free-induction decay envelope looks like that shown in Figure 9.19. This is roughly equivalent to placing the large peak at a low frequency since only a few memory points are actually full, but may be experimentally more convenient in some cases. This technique works well only for very narrow lines, unlike the usual HDO resonances, and not all spectometers have a fine enough adjustment of carrier position to achieve the beat pattern shown in Figure 9.19. Both of these methods also require excellent homogeneity so that narrow lines are observed. However, preliminary experiments indicate that the dynamic range can be enhanced by this method, adding still another to the arsenal of methods available for these limiting situations.

FIGURE 9.19. Beat pattern that occurs when the carrier is placed so that the large peak is near the Nyquist frequency.

REFERENCES

1. R. R. Ernst and W. A. Anderson, *Rev. Sci. Instrum.* **37**, 93 (1966).
2. T. Keller, Bruker-Physik, private communication.
3. R. Freeman and H. D. W. Hill, *J. Mag. Res.* **4**, 366 (1971).
4. A. Allerhand, R. F. Childers, and E. Oldfield, *J. Mag. Res.* **11**, 272 (1973).
5. J. D. Ellett, M. G. Gibby, U. Haeberlein, L. M. Huber, M. Mehring, A. Pines, and J. S. Waugh, *Advan. Mag. Res.* **5**, 117 (1971).
6. E. O. Stejskal and J. Schaefer, *J. Mag. Res.* **13**, 249 (1974).
7. A. G. Redfield and S. D. Kunz, *J. Mag. Res.* **19**, 250 (1975).
8. D. I. Hoult and R. E. Richards, *Proc. Roy. Soc. London. A* **344**, 311–340 (1975).
9. G. C. Levy, private communication.
10. L. F. Johnson, *16th Exp. NMR Conf.,* Asilomar, California, 1975.
11. C. Bradley, Bruker Instruments, Inc. private communication.
12. E. O. Stejskal and J. Schaefer, *15th Exp. NMR Conf.,* Raleigh, North Carolina, 1974.
13. R. K. Gupta, E. Becker, and J. A. Ferretti, *J. Mag. Res.* **13**, 275 (1974).
14. J. Dadok and R. F. Sprecher, *J. Mag. Res.* **13**, 243 (1974).
15. S. L. Patt, Stanford Magnet Research Laboratory, private communication.
16. R. R. Ernst, *J. Mag. Res.* **3**, 10–27 (1970).
17. R. Kaiser, *J. Mag. Res.* **3**, 28–40 (1970).
18. J. W. Cooper and R. E. Addleman, *13th Exp. NMR Conf.,* Asilomar, California, 1972.
19. R. Kaiser, *J. Mag. Res.* **15**, 44 (1974).
20. D. Ziessow, *15th Exp. NMR Conf.,* Raleigh, North Carolina, 1974.
21. J. W. Cooper, *Topics in Carbon-13 NMR,* Vol. 2, p. 392–430, Wiley (Interscience), New York, 1976.
22. B. L. Tomlinson and H. D. W. Hill, *J. Chem. Phys.* **59**, 2775 (1973).
23. G. C. Levy and G. L. Nelson, *Carbon-13 Nuclear Magnetic Resonance for Organic Chemists,* Wiley (Interscience), New York, 1972.
24. G. C. Levy and J. R. Lyerla, Jr., *Topics in Carbon-13 NMR*, Volume 1, p. 81, Wiley (Interscience), New York, 1974.
25. A. Abragam, *The Principles of Nuclear Magnetism,* p. 64, Oxford University Press, London, 1961.
26. A. Allerhand and D. Doddrell, *J. Am. Chem. Soc.* **93**, 2777 (1971).
27. R. L. Vold, J. S. Waugh, M. P. Klein, and D. E. Phelps, *J. Chem. Phys.* **48**, 3831 (1968).
28. A. Allerhand, D. Doddrell, V. Glushko, V. W. Cochran, E. Wenkert, P. W. Lawson, and F. Gurd, *J. Am. Chem. Soc.* **93**, 544 (1971).
29. J. W. Cooper, *American Laboratory* **5** (9), 63, September (1973).
30. R. Freeman and H. D. W. Hill, *J. Chem. Phys.* **54**, 3367 (1971).
31. J. L. Markley, W. J. Horsley, and M. P. Klein, *J. Chem. Phys.* **55**, 3604 (1971).
32. G. G. McDonald and J. S. Leigh, Jr., *J. Mag. Res.* **9**, 358 (1973).
33. G. C. Levy, R. A. Komoroski, I. R. Peat, and P. D. Ellis, *15th Exp. NMR Conf.*, Raleigh, North Carolina, 1974.
34. G. G. McDonald, *15th Exp. NMR Conf.*, Raleigh, North Carolina, 1974.
35. G. C. Levy and I. R. Peat, *J. Mag. Res.* **18**, 500 (1975).
36. S. L. Patt and B. D. Sykes, *J. Chem. Phys.* **56**, 3182 (1972).
37. F. W. Benz, J. Feeney, and G. C. K. Roberts, *J. Mag. Res.* **13**, 243 (1974).
38. J. P. Jesson, P. Meakin, and G. Kneissel, *J. Am. Chem. Soc.* **95**, 618 (1973).
39. A. G. Redfield and R. K. Gupta, *J. Chem. Phys.* **54**, 1418 (1971).
40. J. W. Cooper, *J. Mag. Res.* **22**, 345–357 (1976).

Chapter 10

Fourier Transform Ion Cyclotron Resonance Spectroscopy

Melvin B. Comisarow

10.1. INTRODUCTION

This chapter presents a discussion of the principles of Fourier transform spectroscopy as applied to a form of mass spectroscopy called ion cyclotron resonance (ICR) spectroscopy.

As in any mass spectrometer, the ICR spectrometer has provision for ionizing a gaseous sample and then determining the masses of the ions that are formed. The ICR experiment is typically conducted at very low pressures, usually in the range 10^{-8}–10^{-4} Torr. The operating features of ICR spectrometers have been extensively reviewed[1-7] and will only be briefly discussed here.

An ion in a homogeneous magnetic field will be constrained to a circular path that is perpendicular to the magnetic field and will orbit at an angular frequency ω, called the cyclotron frequency, given by equation (10.1), where q is the ion charge, m the ion mass, and B the magnetic field strength:

$$\omega = qB/m \quad \text{(mks)} \tag{10.1}$$

According to equation (10.1), an ensemble of ions of differing masses will have a spectrum of cyclotron frequencies that is characteristic of that ensemble. For a magnetic field strength of 1 T and a mass range of 15–1500 amu, the cyclotron frequency spectrum [equation (10.1)] extends from 10 kHz

Melvin B. Comisarow • Chemistry Department, University of British Columbia, Vancouver, B. C., Canada V6T 1W5

to 1 MHz and thus falls in the radio-frequency region of the electromagnetic spectrum.

It follows from equation (10.1) that measuring the cyclotron frequencies of a sample of ions will determine the mass of the sample ions, i.e., determine the mass spectrum of that sample. The measurement of the cyclotron frequencies of a sample of ions is accomplished in the following manner. A radio-frequency electric field is applied to the sample of ions by connecting the output of a radio-frequency oscillator to the plates of a parallel-plate capacitor whose dimensions define the volume of the sample (see Figure 10.1). If the frequency of the oscillator equals the cyclotron frequency for a particular ion, that ion will absorb energy from the electric field and will follow the spiral path shown in Figure 10.1. This absorption of energy from the alternating electric field is called "ion cyclotron resonance." The increase of the ion's orbital radius is called "exciting the cyclotron motion" of the ion. If the oscillator frequency does *not* equal the cyclotron frequency for a particular ion, the cyclotron motion of that ion will not be excited.

The detection of excited cyclotron motion may be accomplished in several different ways. In the most common type of detector, the radio frequency is provided by a special type of oscillator circuit called a marginal oscillator, which provides the alternating voltage required for cyclotron excitation and in addition has the capability of measuring the power absorption from that circuit. The mass spectrum of the sample is then obtained by plotting power absorption from the marginal oscillator vs. the magnetic field strength, as the magnetic field strength is swept to successively equate the cyclotron frequencies of the ions in the sample with the fixed frequency of the marginal oscillator.[1-7] In electrometer-detection-type ICR spectrometers[8,9] the alternating electric field is used to excite the ion motion until the ions strike a collector plate that is near one or more of the plates of the capacitor in Figure 10.1. The presence of ion cyclotron resonance is in this case detected as an ion current. In Fourier transform ion cyclotron resonance (FT–ICR) spectrometers[10-14] the cyclotron motion of ions of many different masses is excited essentially simultaneously. The presence of excited cyclotron motion is then detected as the alternating signal that is induced in the plates of the capacitor (Figure 10.2) after the exciting oscillator is turned off. Thus, unlike the power-absorption-type and the electrometer-detection-type ICR spectrometers in which ion excitation and ion detection are simultaneous, ion excitation and ion detection in the FT–ICR spectrometer are temporally distinct.

From the description of the FT–ICR experiment of the previous paragraph it is clear that the FT–ICR experiment is operationally similar to the Fourier transform–nuclear magnetic resonance (FT–NMR) experiment (Chapters 8 and 9) in that many spectral positions are first excited during an excitation period, after which the excitation of the spectrum

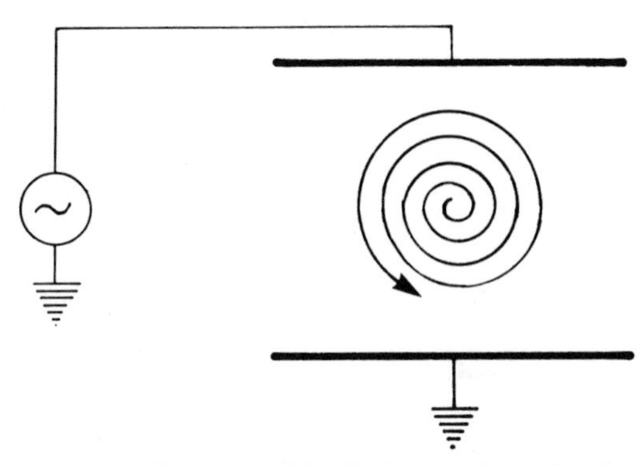

FIGURE 10.1. The cyclotron resonance principle as applied to mass spectrometers. An alternating electric field whose frequency equals the cyclotron frequency [equation (10.1)] for a particular ion mass excites the cyclotron motion of that ion. An oscillator is connected to the plates of a capacitor, whose dimensions define the sample volume, and gives rise to an alternating electric field within the capacitor. If the frequency of the oscillator equals the cyclotron frequency [equation (10.1)] of an ion located within the capacitor, the radius of the ion's cyclotron orbit will be increased (i.e., the ion cyclotron motion is excited). This phenomenon is called cyclotron resonance. The kinetic energy of the ion increases as the ion follows the spiral path shown, and the presence of cyclotron resonance is detected by measuring the power absorption by the ion from the external oscillator (power-absorption-type detectors), by exciting the ion motion until the ions strike a collector plate near the plates of the capacitor (electrometer-type detectors), or by measuring the signal that is induced in the plates of the capacitor by the excited-ion motion (Fourier transform ion cyclotron resonance spectrometers).

persists for a period of time that is long with respect to the excitation time. The excitation gives rise to a time domain signal that contains components from all spectral positions that have been excited. This time domain signal, when sampled and subjected to Fourier transformation, results in the frequency domain spectrum that is characteristic of the sample. The FT–ICR experiment is operationally unlike the Fourier transform infrared (FT–IR) experiment (Chapters 5 and 6) in which the infrared spectrum is obtained by Fourier transformation of a spatially dispersed interferogram.

Ion cyclotron resonance experiments (both conventional and Fourier) are similar to NMR experiments (both conventional and Fourier) in that a magnetic field is required for the experiment and in that the width of the spectrum is proportional to the magnetic field strength. ICR experiments differ from NMR experiments in two fundamental aspects. First, ion cyclotron resonance results from the interaction of alternating electric fields with electrically charged ions, whereas NMR involves the interaction of the alternating magnetic fields with magnetic dipoles. Second, the relaxation time of the time domain signal in the FT–NMR experiment is fixed, whereas the relaxation time of a time domain ICR signal can be made arbitrarily long.[13] For both NMR and ICR experiments, spectral linewidth is usually determined by the relaxation time of the excited sample. This is true for both conventional detection methods and Fourier detection methods. However, the relaxation time of an excited NMR sample is determined primarily by *intra*molecular couplings and thus is largely a function of the sample

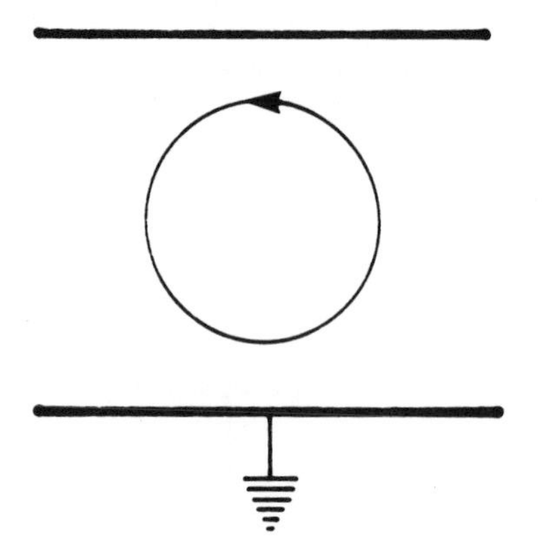

FIGURE 10.2. Ion motion during the detection period of a FT–ICR experiment. The cyclotron motion of the ions in the sample is excited along spiral paths as shown in Figure 10.1 by an oscillator, which is then turned off. The excited ions then proceed on the circular paths shown and induce a time domain signal in the plates of the capacitor that is amplified and detected. Fourier transformation of this time domain signal yields the frequency domain ICR spectrum.

molecule, thereby beyond control by the experimentalist. In ICR experiments, however, the relaxation time is determined by the rate of *inter*molecular ion–molecule collisions. Therefore, if the sample is diluted, i.e., the pressure is lowered, the ion–molecule collision rate will be lowered and the relaxation time of the excited ICR sample will become arbitrarily long.[13] As discussed in detail below, the spectral resolution obtained from an excited ICR sample of very long relaxation time will be determined by the length of the time interval used to observe that excited ICR sample. ICR resolution may thus be made arbitrarily high by using a sufficiently low sample pressure and a sufficiently long time period for observation of the ICR signal.[13]

In this chapter, the relation between mass resolution and frequency resolution for all forms of ICR spectroscopy is derived and the ICR lineshape is then derived for the Fourier transform ICR experiment for the case where there is no damping of the transient ICR signal. Representative FT–ICR line shapes are illustrated graphically. Analytical low-pressure ICR linewidth and mass resolution are then derived for the FT–ICR line shape, in terms of ionic mass, magnetic field, ionic charge, and the appropriate time period for observation. FT–ICR mass resolution is tabulated as a function of ionic mass for various typical choices for data acquisition time. Since the time domain data in actual FT–ICR experiments are stored as a digital array, convenient expressions are then derived for FT–ICR mass resolution and mass range in terms of computer size and minimum specified mass-to-charge ratio. FT–ICR mass range is tabulated as a function of computer size for representative choices of minimum specified mass-to-charge ratio. Finally, various definitions for FT–ICR spectral linewidth and mass resolution are presented, and the corresponding analytical expressions for FT–ICR frequency linewidth, mass linewidth, mass resolution, and mass range are tabulated for both absorption mode and magnitude (absolute-value) line shape. The treatment is taken from reference 14.

10.2. FUNDAMENTAL EQUATIONS FOR ICR LINEWIDTH AND RESOLUTION

Since ICR spectral line shape is conveniently computed in terms of angular frequency, while mass spectra and mass resolution are usually expressed in terms of mass units, it is desirable to relate these two approaches. Beginning from the first derivative of equation (10.1),

$$d\omega = -(qB/m^2)\,dm \tag{10.2}$$

and dividing equation (10.1) by equation (10.2), we obtain

$$\omega/d\omega = -m/dm \tag{10.3}$$

The minus sign in equations (10.2) and (10.3) results from the inverse relation between m and ω in equation (10.1) (i.e., ICR frequency increases as ionic mass decreases), and will henceforth be omitted, as in

$$\omega/d\omega = m/dm \tag{10.4}$$

By expanding ω_2 at m_2 in a Taylor series about ω_1 at m_1, it is readily shown that over any finite small mass increment, $m_2 - m_1 = \Delta m \ll m$, corresponding to the small frequency increment, $\omega_2 - \omega_1 = \Delta\omega \ll \omega$,

$$d\omega/dm \cong \Delta\omega/\Delta m, \qquad \Delta m \ll m, \quad \Delta\omega \ll \omega \tag{10.5}$$

Substituting equation (10.5) into equation (10.4) gives the general result,

$$\omega/\Delta\omega = m/\Delta m \tag{10.6}$$

which shows that resolution defined for the *frequency* scale, $\omega/\Delta\omega$, is *numerically identical* to resolution defined for the *mass* scale, $m/\Delta m$.

Substituting from equation (10.1) into equation (10.6) yields

$$\Delta m = \frac{m^2}{qB}\Delta\omega \quad \text{(mks)} \tag{10.7}$$

which gives the ICR mass increment Δm in kg, corresponding to the ICR frequency increment $\Delta\omega$ in rad/sec, for an ion of mass m kg and charge q C in a magnetic field of B T. Dividing both sides of equation (10.7) by m and inverting leads to

$$\frac{m}{\Delta m} = \frac{qB}{m\,\Delta\omega} \quad \text{(mks)} \tag{10.8}$$

which gives the ICR mass resolution at mass m as a function of m, q, B, and the frequency increment $\Delta\omega$ corresponding to the mass increment Δm. Equations (10.6)–(10.8) are valid for *any* ion cyclotron resonance spectrometer for *any* arbitrary definition of frequency or mass increment.

These equations form the basis for subsequent analytical and tabular expressions for conventional ICR and FT–ICR spectral linewidth and mass resolution. Other fundamental equations that are valid for any ICR spectrometer [equations (10.24) and (10.24a)] are derived below.

10.3. FOURIER TRANSFORM ION CYCLOTRON RESONANCE (FT–ICR) SPECTROSCOPY

In the FT–ICR experiment, the excited cyclotron motion (Figure 10.2) for ions of a particular mass-to-charge ratio is detected during an observation period T, which follows an excitation period during which the radii of the cyclotron orbits for those ions are coherently increased (Figure 10.1). In the absence of ion–molecule collisions (zero-pressure limit), the excited cyclotron motion continues indefinitely as shown in Figure 10.2. In this chapter it will be assumed that the amplitude of the signal induced in the plates of the ICR cell by the excited cyclotron motion is of the form (Figure 10.3, leftmost graph),

$$f(t) = \begin{cases} K \cos \omega t, & 0 < t < T \\ 0, & t < 0 \quad \text{or} \quad t > T \end{cases} \tag{10.9}$$

in which K is a constant that is proportional to the number of excited ions, ω is the ion cyclotron frequency of equation (10.1), and T, the data acquisi-

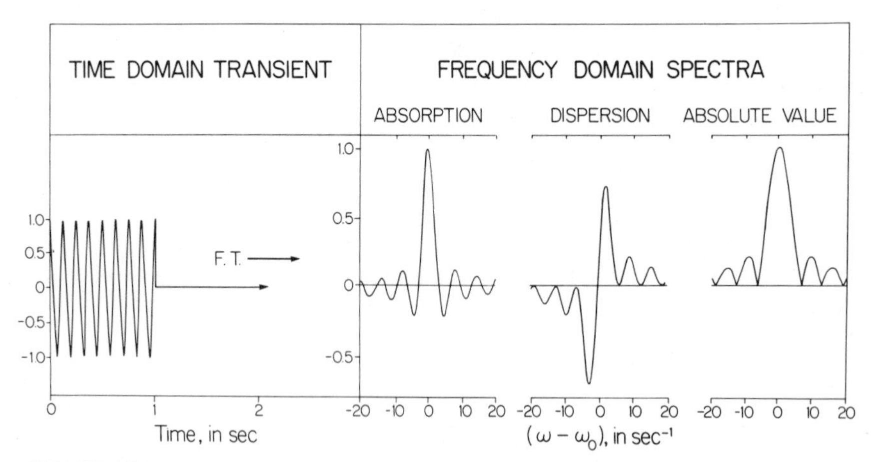

FIGURE 10.3. Cosine transform (absorption), sine transform (dispersion), and magnitude (absolute value) frequency domain ICR spectra obtained by analytical Fourier transformation of the continuous time domain signal shown at top left. These spectra were plotted from the zero-pressure FT–ICR line shape formulas listed in Table 10.1; all frequency domain spectra have the same horizontal and vertical scale.

tion time, defines the duration of the detection period. $f(t)$ is called the time domain signal. The time domain signal defined by equation (10.9) will be an accurate approximation to the actual FT–ICR signal in the low-pressure limit, $\xi \ll 1/T$, in which ξ is the reduced collision frequency for the ion.[15] In practice, for example, equation (10.9) applies for $T \leq 50$ ms for a non-reactive ion at neutral pressure of 10^{-8} Torr. For a sample containing excited ions of several different charge-to-mass ratios, $f(t)$ will be composed of a sum of sinusoids whose amplitudes and frequencies correspond to the number of ions having those respective ICR frequencies.

The amplitudes of the frequency components of $f(t)$, namely, $A(\omega')$ and $B(\omega')$ in

$$f(t) = \int_0^\infty A(\omega') \cos \omega' t \, d\omega' + \int_0^\infty B(\omega') \sin \omega' t \, d\omega' \qquad (10.10)$$

are obtained from the basic formulas for Fourier transform[16]:

$$A(\omega') = \frac{1}{\pi} \int_0^\infty f(t) \cos \omega' t \, dt = \frac{K \sin[(\omega' - \omega) T]}{2\pi(\omega' - \omega)} \qquad (10.11)$$

$$B(\omega') = \frac{1}{\pi} \int_0^\infty f(t) \sin \omega' t \, dt = \frac{K \cdot 1 - \cos[(\omega' - \omega) T]}{2\pi(\omega' - \omega)} \qquad (10.12)$$

$$C(\omega') = \{(A(\omega')]^2 + [B(\omega')]^2\}^{1/2} = \frac{K \cdot 2 \sin[(\omega' - \omega) T/2]}{2\pi(\omega' - \omega)} \qquad (10.13)$$

$A(\omega')$, $B(\omega')$, and $C(\omega')$ are called the absorption, the dispersion, and the magnitude (absolute-value) spectra, respectively. $C^2(\omega')$ is called the power spectrum. The explicit analytical evaluations of $A(\omega')$, $B(\omega')$, and $C(\omega')$ for $f(t)$ defined by equation (10.9) are collected in Table 10.1, in which the

TABLE 10.1. Summary of FT–ICR Line Shape Formulas and Their Principal Properties, Based on a Low-Pressure Limiting Time Domain Signal of the Form $f(t) = K \cos(\omega t)$ between $t = 0$ and $t = T$ and Zero Elsewhere. These Line Shapes Are Shown Graphically in Figure 10.3

Spectral display mode	Zero-pressure line shape[a]	$\lim (\omega' - \omega) \to 0$	Full linewidth, $\Delta\omega_{50\%}$, half-maximum height
Absorption	$A(\omega') = \frac{\sin[(\omega' - \omega) T]}{\omega' - \omega}$	T	$\frac{3.791}{T}$
Dispersion	$B(\omega') = \frac{1 - \cos[(\omega' - \omega) T]}{\omega' - \omega}$	$\pm \frac{0.7246^b}{T}$	$\frac{4.662^c}{T}$
Absolute value ("magnitude")	$C(\omega') = \frac{2 \sin[(\omega' - \omega) T/2]}{\omega' - \omega}$	T	$\frac{7.582}{T}$

[a] The scaling coefficient $K/2\pi$ has been omitted for simplicity.
[b] These are the extrema of $B(\omega')$, located at $\omega' - \omega = \pm(2.331/T)$.
[c] This value represents the distance between the extrema of $B(\omega')$ on either side of ω.

time domain signal is taken to have unit amplitude ($K = 1$) for simplicity in display. Under the very general condition that the amplitude of the time domain ICR signal, equation (10.9), be proportional to the magnitude of the applied excitation (i.e., the system responds linearly), the cosine [sine] transform $A(\omega')$ [$B(\omega')$] is proportional to the *instantaneous* absorption [dispersion] that would be obtained following an excitation period T in a conventional "trapped-ion" ICR spectrometer.

Representative theoretical absorption, dispersion, and magnitude (absolute-value) low-pressure line shapes are illustrated in Figure 10.3. The absorption line shape (leftmost spectrum) exhibits the small maxima on either side of the main peak characteristic of the $(\sin x)/x$ function. Expressions for peak height and for full width at half-height are listed in Table 10.1, and it may be noted that the absolute-value linewidth (rightmost spectrum) is exactly twice as large as the absorption-mode linewidth. In addition, the dispersion shape (middle spectrum) extends over a wider frequency range than the absorption spectrum. In summary, the absorption-mode display shows the narrowest line shape and is thus best suited to high-resolution spectral display.[12]

10.4. ANALYTICAL FT–ICR LINEWIDTH AND MASS RESOLUTION

Spectral linewidth, $\Delta\omega$, is commonly defined to be the full linewidth at half of the maximum peak height, $\Delta\omega_{50\%}$,

$$\Delta\omega \equiv \Delta\omega_{50\%} \tag{10.14}$$

For the FT–ICR absorption-mode line shape, equation (10.11), the frequency-scale FT–ICR full linewidth at half-maximum height, $\Delta\omega_{50\%}$, in rad/sec, is

$$\Delta\omega_{50\%} = \frac{3.791}{T} \tag{10.15}$$

where T is the acquisition time [equation (10.9)] in seconds. Application of the particular resolution criterion, equation (10.14), to the FT–ICR experiment requires combination of equation (10.14) with equations (10.7) and (10.8) to give

$$\Delta m_{50\%} = \frac{3.791 m^2}{qBT} \quad \text{(mks)} \tag{10.16}$$

$$\frac{m}{\Delta m_{50\%}} = \frac{qBT}{3.791 m} \quad \text{(mks)} \tag{10.17}$$

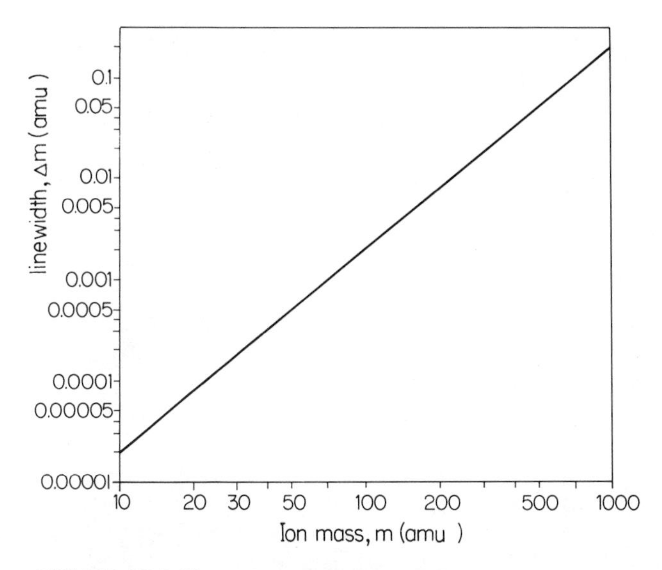

FIGURE 10.4. Theoretical zero-pressure FT–ICR linewidth Δm as a function of ion mass for a magnetic field intensity $B = 20$ kG and an acquisition time $T = 100$ ms. From equation (10.16a).

or equivalently,

$$\Delta m_{50\%} = \frac{3.929 \times 10^{-4} m^2}{qBT} \tag{10.16a}$$

$$\frac{m}{\Delta m_{50\%}} = \frac{qBT}{3.929 \times 10^{-4} m} \tag{10.17a}$$

Equations (10.16a) and (10.17a) give the zero-pressure, absorption-mode FT–ICR full mass-scale linewidth at half-maximum height $\Delta m_{50\%}$ in amu, and the corresponding mass resolution $m/\Delta m_{50\%}$, as a function of the ionic mass m in amu, ionic charge q in multiples of the electronic charge, magnetic field B in kilogauss, and acquisition time T in milliseconds. Equations (10.16) and (10.17) give the corresponding quantities in mks units. The variation of the FT–ICR linewidth with ion mass is shown graphically in Figure 10.4 for the particular linewidth criterion equation (10.14), and the particular but typical experimental values of magnetic field = 20 kG and acquisition time = 100 ms. Figure 10.5 shows the dependence of the FT–ICR resolution from equation (10.17a) upon ion mass for a magnetic field of 20 kG and for three different values of the acquisition time. Figure 10.6 shows the variation of FT–ICR resolution from equation (10.17a) with acquisition time with a magnetic field intensity of 20 kG for values of the ion mass equal to 100, 200, and 500 amu. Table 10.2 lists formulas that

TABLE 10.2. FT–ICR Linewidth and Resolution Formulas for Absorption-Mode and Magnitude (Absolute-Value) Line Shapes

	% peak height at which linewidth is defined				
	50%	25%	10%	5%	1%
	Absorption $= A(\omega') = \frac{K}{2\pi}\frac{\sin[(\omega' - \omega)T]}{\omega' - \omega}$				
Frequency-scale linewidth $\Delta\omega$ (rad/s)	$\frac{3.791}{T}$	$\frac{4.949}{T}$	$\frac{5.705}{T}$	$\frac{5.983}{T}$	$\frac{6.221}{T}$
Mass-scale linewidth[a] Δm (kg)	$\frac{3.791m^2}{qBT}$	$\frac{4.949m^2}{qBT}$	$\frac{5.705m^2}{qBT}$	$\frac{5.983m^2}{qBT}$	$\frac{6.221m^2}{qBT}$
Mass-scale linewidth[b] Δm (amu)	$\frac{3.93 \times 10^{-4}m^2}{qBT}$	$\frac{5.13 \times 10^{-4}m^2}{qBT}$	$\frac{5.91 \times 10^{-4}m^2}{qBT}$	$\frac{6.20 \times 10^{-4}m^2}{qBT}$	$\frac{6.45 \times 10^{-4}m^2}{qBT}$
Mass-scale linewidth[c] Δm	$\frac{1.207m^2}{Nm_{\min}}$	$\frac{1.575m^2}{Nm_{\min}}$	$\frac{1.816m^2}{Nm_{\min}}$	$\frac{1.904m^2}{Nm_{\min}}$	$\frac{1.980m^2}{Nm_{\min}}$
FT–ICR resolution[a] $m/\Delta m = \omega/\Delta\omega$	$\frac{0.264qBT}{m}$	$\frac{0.202qBT}{m}$	$\frac{0.175qBT}{m}$	$\frac{0.167qBT}{m}$	$\frac{0.161qBT}{m}$
FT–ICR resolution[b] $m/\Delta m = \omega/\Delta\omega$	$\frac{2.55 \times 10^3 qBT}{m}$	$\frac{1.95 \times 10^3 qBT}{m}$	$\frac{1.69 \times 10^3 qBT}{m}$	$\frac{1.61 \times 10^3 qBT}{m}$	$\frac{1.55 \times 10^3 qBT}{m}$
FR–ICR resolution[c] $m/\Delta m = \omega/\Delta\omega$	$\frac{0.829Nm_{\min}}{m}$	$\frac{0.635Nm_{\min}}{m}$	$\frac{0.551Nm_{\min}}{m}$	$\frac{0.525Nm_{\min}}{m}$	$\frac{0.505Nm_{\min}}{m}$

	Magnitude (absolute value) $= C(\omega') = \frac{K}{2\pi}\frac{2\sin[(\omega' - \omega)T/2]}{\omega' - \omega}$				
Frequency-scale linewidth $\Delta\omega$ (rad/s)	$\frac{7.582}{T}$	$\frac{9.989}{T}$	$\frac{11.41}{T}$	$\frac{11.97}{T}$	$\frac{12.44}{T}$
Mass-scale linewidth[a] Δm (kg)	$\frac{7.582m^2}{qBT}$	$\frac{9.898m^2}{qBT}$	$\frac{11.41m^2}{qBT}$	$\frac{11.97m^2}{qBT}$	$\frac{12.44m^2}{qBT}$
Mass-scale linewidth[b] Δm (amu)	$\frac{7.86 \times 10^{-4}m^2}{qBT}$	$\frac{1.03 \times 10^{-3}m^2}{qBT}$	$\frac{1.18 \times 10^{-3}m^2}{qBT}$	$\frac{1.24 \times 10^{-3}m^2}{qBT}$	$\frac{1.29 \times 10^{-3}m^2}{qBT}$
Mass-scale linewidth[c] Δm	$\frac{2.413m^2}{Nm_{min}}$	$\frac{3.150m^2}{Nm_{min}}$	$\frac{3.632m^2}{Nm_{min}}$	$\frac{3.810m^2}{Nm_{min}}$	$\frac{3.960m^2}{Nm_{min}}$
FT–ICR resolution[a] $m/\Delta m = \omega/\Delta\omega$	$\frac{0.132qBT}{m}$	$\frac{0.101qBT}{m}$	$\frac{8.76 \times 10^{-2}qBT}{m}$	$\frac{8.04 \times 10^{-2}qBT}{m}$	$\frac{8.04 \times 10^{-2}qBT}{m}$
FT–ICR resolution[b] $m/\Delta m = \omega/\Delta\omega$	$\frac{1.27 \times 10^{3}qBT}{m}$	$\frac{9.75 \times 10^{2}qBT}{m}$	$\frac{8.46 \times 10^{2}qBT}{m}$	$\frac{8.06 \times 10^{2}qBT}{m}$	$\frac{7.76 \times 10^{2}qBT}{m}$
FR–ICR resolution[c] $m/\Delta m = \omega/\Delta\omega$	$\frac{0.414Nm_{min}}{m}$	$\frac{0.317Nm_{min}}{m}$	$\frac{0.275Nm_{min}}{m}$	$\frac{0.262Nm_{min}}{m}$	$\frac{0.253Nm_{min}}{m}$

[a] In these expressions, m is ionic mass in kilograms, q is ionic charge in coulombs, B is magnetic field strength in teslas, and T is data acquisition time in seconds.

[b] In these expressions, m is ionic mass in amu, q is ionic charge in multiples of electronic charge, B is magnetic field strength in kilogauss, and T is data acquisition time in milliseconds.

[c] In these expressions, m, m_{min}, and Δm all have the same (arbitrary) units.

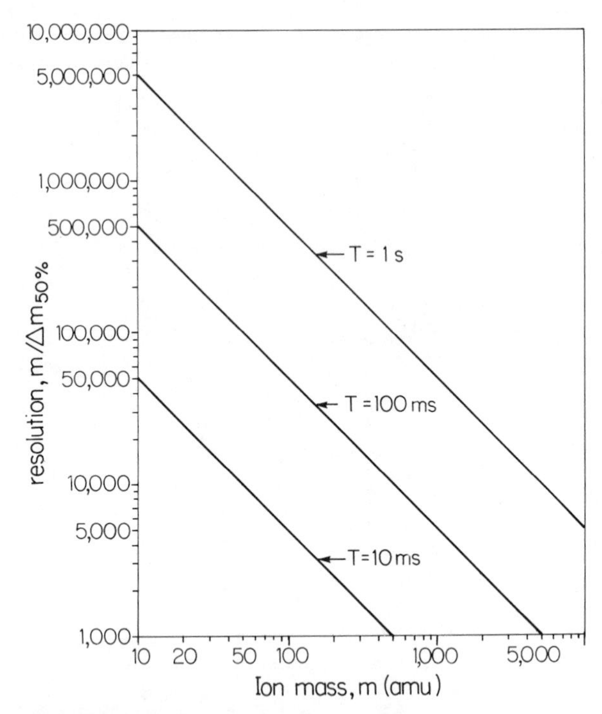

FIGURE 10.5. Theoretical zero-pressure FT–ICR mass resolution $m/\Delta m$ as a function of ion mass for a magnetic field intensity $B = 20$ kG and an acquisition time $T = 10$ ms, 100 ms, and 1 s. From equation (10.17a).

are analogous to equations (10.16)–(10.17a), for both absorption mode and magnitude (absolute-value) mode for various common resolution criteria.

10.5. FT–ICR MASS RANGE, COMPUTER DATA SIZE, AND SAMPLING RATE

Equations (10.17) and (10.17a) show that for a given ionic mass and charge and given magnetic field strength, FT–ICR mass resolution is determined by the length of observation of the excited ICR signal, namely, the data acquisition time T. Now *discrete* Fourier analysis (Chapter 4) requires that the time domain signal be sampled at (say, N) equally spaced points in the time domain, and so if the sampling rate is F points/s, then

$$T = N/F \tag{10.18}$$

in which N is the number of points in the time domain data set. Furthermore, the sampling theorem (Chapter 4) of Fourier analysis[17] requires that the sampling rate be at least twice the *highest* frequency ν_{max} (in hertz) to be analyzed. Thus, for proper Fourier analysis of a band of cyclotron frequencies corresponding to a given range of ionic masses, the sampling theorem and equation (10.1) require that

$$F = 2\nu_{max} = 2\left(\frac{\omega_{max}}{2\pi}\right) = \frac{qB}{\pi m_{min}} \quad \text{(mks)} \tag{10.19}$$

or equivalently,

$$F = 2\nu_{max} = 2\left(\frac{\omega_{max}}{2\pi}\right) = \frac{3.071 \times 10^6 qB}{m_{min}} \tag{10.19a}$$

Equation (10.19a) gives the sampling rate F in s^{-1} as a function of ω_{max}, the highest ion cyclotron frequency in rad/s in the band, corresponding to m_{min}, the minimum ionic mass in amu in the mass range, q, the ionic charge in multiples of the elementary charge, and B, the magnetic field strength in kilogauss. Equation (10.19) is the corresponding equation in mks units.

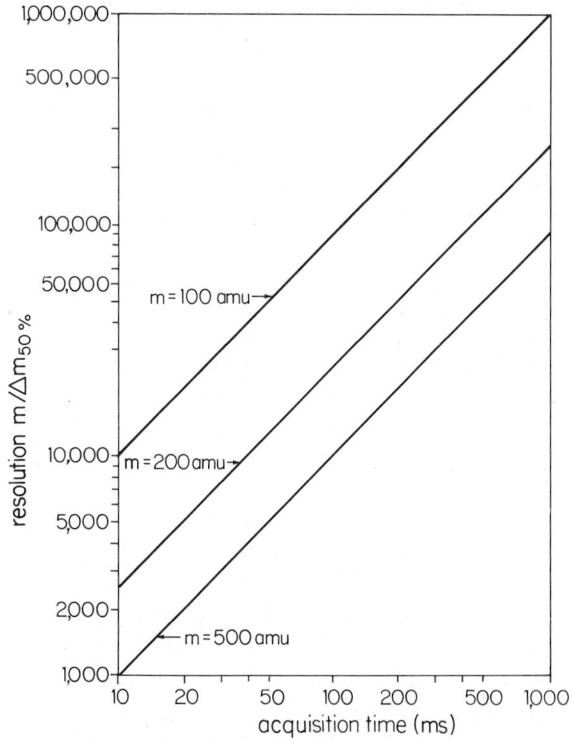

FIGURE 10.6. Theoretical zero-pressure FT–ICR mass resolution as a function of acquisition time for a magnetic field B = 20 kG and for ion mass = 100, 200, and 500 amu. From equation (10.17a).

Combination of equations (10.7), (10.18), and (10.19), shows that when a wide mass range is desired, the FT–ICR mass increment at mass m, Δm, takes the form

$$\Delta m = \frac{m^2 T}{\pi N m_{\min}} \Delta\omega \tag{10.20}$$

in which T is the data acquisition time in seconds, N the number of points in the time domain data set, $m_{\min}$ the smallest mass ion to be observed, and $\Delta\omega$ the frequency increment in rad/s [m, $m_{\min}$, and Δm have the same (arbitrary) units]. Dividing both sides of equation (10.20) by m and inverting yields the mass resolution at mass m,

$$\frac{m}{\Delta m} = \frac{\pi N m_{\min}}{m T \Delta\omega} \tag{10.21}$$

as a function of m, N, T, and $\Delta\omega$. Both equations (10.20) and (10.21) are independent of the choice of units for mass. Combining the particular resolution criterion, equation (10.14) with equations (10.15) and (10.20) gives the FT–ICR zero-pressure, absorption-mode full linewidth at half-maximum height at mass m,

$$\Delta m_{50\%} = \frac{1.207 m^2}{N m_{\min}} \tag{10.22}$$

as a function of the ionic mass m, the number of points in the time domain data set N, and the minimum ionic mass to be observed $m_{\min}$. Dividing equation (10.22) by m and inverting yields the corresponding mass resolution,

$$\frac{m}{\Delta m_{50\%}} = \frac{N m_{\min}}{1.207 m} \tag{10.23}$$

as a function of m, N, and $m_{\min}$. Equations (10.22) and (10.23) are also independent of the choice of mass units.

Table 10.2 lists several formulas analogous to equations (10.22) and (10.23) for FT–ICR mass-scale zero-pressure linewidth, for both absorption-mode and magnitude (absolute-value) line shapes, for various common resolution criteria.

It is clear from equations (10.7) and (10.20) that at fixed magnetic field strength, the ICR mass increment Δm increases quadratically with mass. Thus, for an arbitrary definition of frequency increment $\Delta\omega$, we may further *define* an ICR upper mass limit $m_{\max}$ as the mass at which ICR mass linewidth Δm has increased to some arbitrary number of mass units. Rearranging equation (10.7) then gives an expression for $m_{\max}$:

$$m_{\max} = \left(\frac{qB\,\Delta m}{\Delta\omega}\right)^{1/2} \quad \text{(mks)} \tag{10.24}$$

or equivalently,

$$m_{\max} = \left(\frac{9.649 \times 10^6 qB\,\Delta m}{\Delta\omega}\right)^{1/2} \tag{10.24a}$$

in which it is understood that a numerical value may be assigned to Δm for any definition of $\Delta\omega$. Equation (10.24a) gives the *defined* upper mass limit $m_{\max}$ in amu, for an ICR experiment conducted at a magnetic field strength of B kG, for an ion charge q multiples of the elementary charge, with a frequency increment defined as $\Delta\omega$ rad/s, and with a maximum allowed mass increment of Δm amu. Equation (10.24) is the analogous expression in mks units. Both equations (10.24) and (10.24a) are valid for any ICR spectrometer.

For example, if $\Delta\omega$ is defined to be the frequency-scale ICR full linewidth at half-maximum height [equation (10.14)] for the zero-pressure absorption mode in the FT–ICR experiment [equation (10.15)], and if the corresponding mass-scale linewidth $\Delta m_{50\%}$ (FT–ICR zero-pressure absorption-mode full linewidth at half-maximum height) is chosen to have a maximum value of 1.661×10^{-27} kg (i.e., 1.0 amu) at $m = m_{\max}$, then equations (10.24) and (10.24a) take the form

$$m_{\max} = (4.380 \times 10^{-28}\, qBT)^{1/2} \qquad (\Delta m = 1.661 \times 10^{-27}\,\text{kg}) \quad (\text{mks}) \tag{10.25}$$

or equivalently

$$m_{\max} = (2.545 \times 10^{3}\, qBT)^{1/2} \qquad (\Delta m = 1.0\,\text{amu}) \tag{10.25a}$$

Equation (10.25a) gives the FT–ICR upper mass limit in amu (i.e., the mass at which Δm has increased to 1.0 amu), as a function of the ionic charge q in multiples of the elementary charge, the magnetic field B in kilogauss, and the data acquisition time T in ms. Equation (10.25) is the analogous expression in mks units. For the specific choices of $B = 20$ kG, $q = 1$ elementary charge, and $T = 100$ ms, equation (10.25a) predicts an FT–ICR upper mass limit of 2256 amu. The upper mass limit as a *function* of acquisition time is shown in Fig. 10.7 for the above values of B and q. Table 10.3 lists several formulas analogous to equation (10.25a) for both absorption-mode and magnitude (absolute-value) FT–ICR zero-pressure line shape for various common linewidth criteria.

Equations (10.25) and (10.25a) present the FT–ICR upper mass limit as a function of (among other things) the acquisition time T. However, equation (10.18) showed that T is in turn determined by the maximum number of time domain data points N, and the sampling rate F. From equation (10.19), F is in turn determined by the smallest desired ionic mass $m_{\min}$. Therefore, the discrete nature of the actual experimental FT–ICR data reduction is introduced by combining equations (10.18), (10.19), and (10.24) to give the general expression for FT–ICR upper mass limit:

$$m_{\max} = \frac{\pi m_{\min} N}{T} \frac{\Delta m}{\Delta\omega} \tag{10.26}$$

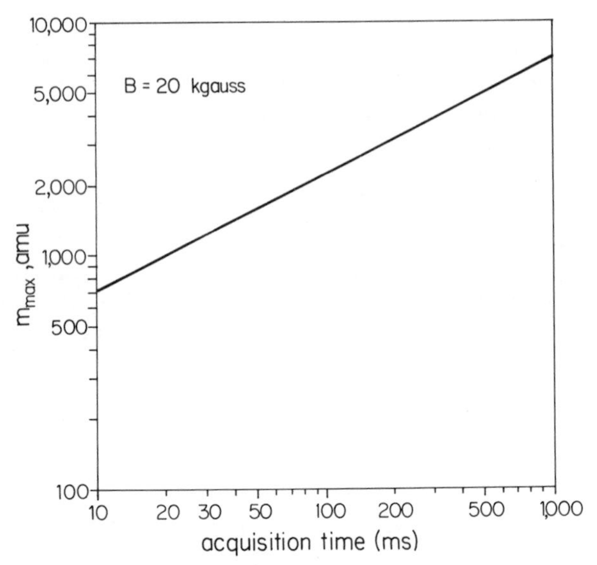

FIGURE 10.7. Theoretical FT–ICR upper mass limit as a function of acquisition time for a magnetic field intensity $B = 20$ kG. From equation (10.25a).

Equation 10.26 gives the *defined* maximum ionic mass for an FT–ICR experiment as a function of the data acquisition time T in seconds, the minimum ionic mass m_{min} in the mass range, the number of time domain data points N, and the frequency-scale increment $\Delta\omega$ in rad/s. Equation (10.26) is independent of the units for mass. For the particular resolution criterion equation (10.14) applied to the FT–ICR experiment [equation (10.15)], and the further definition of the maximum mass increment as $\Delta m = 1.0$ amu, equation (10.26) becomes

$$m_{\text{max}} = 0.8287 N m_{\text{min}} \tag{10.27}$$

Equation (10.27) gives the ionic mass in amu at which the zero-pressure absorption-mode full linewidth at half-maximum height in an FT–ICR experiment has become 1.0 amu, as a function of the number of time domain data points N, and the minimum ionic mass m_{min} in amu, in the mass range. For example, when the number of time domain data points $N = 16{,}384$, and the lower mass limit is chosen to be $m_{\text{min}} = 60$ amu, then the upper mass limit defined by equation (10.27) becomes $m_{\text{max}} = 903$ amu. Table 10.3 lists several formulas analogous to equation (10.27) for both absorption-mode and magnitude (absolute-value) FT–ICR zero-pressure line shape, for various common resolution criteria.

The principal features of the various equations derived above will now be discussed, and their practical use illustrated by representative

TABLE 10.3. FT–ICR Upper Mass Limit Formulas for Various Definitions of Frequency-Scale Linewidth $\Delta\omega$, and Various Choices of Maximum Allowed Mass-Scale Linewidth Δm at $m = m_{max}$, for Zero-Pressure Absorption Mode and Magnitude (Absolute-Value) Line Shapes

		% peak height at which both frequency-scale linewidth $\Delta\omega$ and mass-scale linewidth Δm are defined				
		50%	25%	10%	5%	1%
		Absorption				
Upper mass limit m_{max} in amu, as a function of q, B, and T[a]	$\Delta m = 1$ amu at $m = m_{max}$	$(2.545 \times 10^3 qBT)^{1/2}$	$(1.949 \times 10^3 qBT)^{1/2}$	$(1.691 \times 10^3 qBT)^{1/2}$	$(1.613 \times 10^3 qBT)^{1/2}$	$(1.551 \times 10^3 qBT)^{1/2}$
	$\Delta m = 0.5$ amu at $m = m_{max}$	$(1.272 \times 10^3 qBT)^{1/2}$	$(9.745 \times 10^2 qBT)^{1/2}$	$(8.456 \times 10^2 qBT)^{1/2}$	$(8.063 \times 10^2 qBT)^{1/2}$	$(7.754 \times 10^2 qBT)^{1/2}$
	$\Delta m = 0.1$ amu at $m = m_{max}$	$(2.545 \times 10^2 qBT)^{1/2}$	$(1.949 \times 10^2 qBT)^{1/2}$	$(1.691 \times 10^2 qBT)^{1/2}$	$(1.613 \times 10^2 qBT)^{1/2}$	$(1.551 \times 10^2 qBT)^{1/2}$
Upper mass limit m_{max} in amu, as a function of N and m_{min}[b]	$\Delta m = 1$ amu at $m = m_{max}$	$(0.8287 N m_{min})^{1/2}$	$(0.6348 N m_{min})^{1/2}$	$(0.5507 N m_{min})^{1/2}$	$(0.5251 N m_{min})^{1/2}$	$(0.5050 N m_{min})^{1/2}$
	$\Delta m = 0.5$ amu at $m = m_{max}$	$(0.4143 N m_{min})^{1/2}$	$(0.3174 N m_{min})^{1/2}$	$(0.2753 N m_{min})^{1/2}$	$(0.2625 N m_{min})^{1/2}$	$(0.2525 N m_{min})^{1/2}$
	$\Delta m = 0.1$ amu at $m = m_{max}$	$(0.0829 N m_{min})^{1/2}$	$(0.0635 N m_{min})^{1/2}$	$(0.0551 N m_{min})^{1/2}$	$(0.0525 N m_{min})^{1/2}$	$(0.0505 N m_{min})^{1/2}$
		Magnitude (absolute value)				
Upper mass limit m_{max} in amu, as a function of q, B, and T[a]	$\Delta m = 1$ amu at $m = m_{max}$	$(1.272 \times 10^3 qBT)^{1/2}$	$(9.747 \times 10^2 qBT)^{1/2}$	$(8.456 \times 10^2 qBT)^{1/2}$	$(8.060 \times 10^2 qBT)^{1/2}$	$(7.756 \times 10^2 qBT)^{1/2}$
	$\Delta m = 0.5$ amu at $m = m_{max}$	$(6.362 \times 10^2 qBT)^{1/2}$	$(4.874 \times 10^2 qBT)^{1/2}$	$(4.228 \times 10^2 qBT)^{1/2}$	$(4.030 \times 10^2 qBT)^{1/2}$	$(3.878 \times 10^2 qBT)^{1/2}$
	$\Delta m = 0.1$ amu at $m = m_{max}$	$(1.272 \times 10^2 qBT)^{1/2}$	$(9.747 \times 10^2 qBT)^{1/2}$	$(8.456 \times 10^1 qBT)^{1/2}$	$(8.060 \times 10^1 qBT)^{1/2}$	$(7.756 \times 10^1 qBT)^{1/2}$
Upper mass limit m_{max} in amu, as a function of n N and m_{min}[b]	$\Delta m = 1$ amu at $m = m_{max}$	$(0.4143 N m_{min})^{1/2}$	$(0.3174 N m_{min})^{1/2}$	$(0.2753 N m_{min})^{1/2}$	$(0.2625 N m_{min})^{1/2}$	$(0.2525 N m_{min})^{1/2}$
	$\Delta m = 0.5$ amu at $m = m_{max}$	$(0.2072 N m_{min})^{1/2}$	$(0.1587 N m_{min})^{1/2}$	$(0.1377 N m_{min})^{1/2}$	$(0.1372 N m_{min})^{1/2}$	$(0.1263 N m_{min})^{1/2}$
	$\Delta m = 0.1$ amu at $m = m_{max}$	$(0.0414 N m_{min})^{1/2}$	$((0.0317 N m_{min})^{1/2}$	$(0.0275 N m_{min})^{1/2}$	$(0.0262 N m_{min})^{1/2}$	$(0.0253 N m_{min})^{1/2}$

[a] In these expressions, q is ionic charge in multiples of the electronic charge, B is magnetic field strength in kilogauss, and T is data acquisition time in milliseconds.

[b] In these expressions, N is the number of time domain data points, and m_{min} is the lowest mass ion in the FT–ICR spectrum, in amu.

numerical tabulations for FT–ICR spectral linewidth, spectral resolution, and mass range.

10.6. DISCUSSION

Comparison of the absorption-mode, dispersion-mode, and absolute-value FT–ICR spectra of Figure 10.3 shows that the absorption mode provides the narrowest line shape, and will thus be the display of choice when high mass resolution is desired. As has been pointed out,[12] the absolute-value display provides somewhat better signal-to-noise ratio and eliminates the need for manual phase correction during Fourier transformation of the time domain data set, and is thus the display of choice when signal-to-noise ratio and convenient data reduction are more important than high mass resolution.

Since both absorption-mode and magnitude (absolute-value) spectra are now in use in FT–ICR spectroscopy because each has distinct advantages[12] and since several different definitions of "linewidth" are also in use in mass spectroscopy, we have expressed the various formulas for FT–ICR linewidth, resolution, and upper mass limit for all combinations of spectral mode and the five common linewidth criteria shown in Tables 10.2 and 10.3, based on the zero-pressure FT–ICR line shape formulas of Table 10.1.

For example, if the absorption-mode linewidth is to be measured at 10% of maximum peak height, then the analytical expression for mass-scale linewidth in amu is $5.91 \times 10^{-4}\ m^2/qBT$, in which m is ionic mass in amu, q is ionic charge in multiples of the elementary charge, B is the magnetic field strength in kilogauss, and T is data acquisition time in milliseconds. The practical significance of the formulas of Tables 10.2 and 10.3 is most readily illustrated from the numerical examples in Tables 10.4–10.6, calculated for typical experimental choices of the remaining parameters.

Table 10.4 shows the significance of Fourier data reduction in improving experimental ICR mass resolution. Even though FT–ICR resolution decreases as $(1/m)$ [see equation (10.17)], it is still possible to obtain mass resolution $m/\Delta m_{25\%}$ of better than 10,000 at $m/e = 1000$, using a total data acquisition time (see Table 10.4) of less than 0.5 s. Finally, equation (10.17) shows that FT–ICR resolution increases linearly with applied magnetic field strength.

Equation (10.22) and the associated Table 10.5 show that for the finite data sets encountered in actual FT–ICR experiments, FT–ICR linewidth depends only on specified observed mass, specified minimum desired mass in the same spectrum, and number of words (data points) of computer data storage N. For a sampled time domain FT–ICR signal, FT–ICR

TABLE 10.4. FT–ICR Spectral Resolution, $\omega/\Delta\omega_{25\%} = m/\Delta m_{25\%}$, at Various Ionic Masses as a Function of Data Acquisition Time T, for Zero-Pressure Absorption-Mode [or Magnitude-Mode in Brackets] Line Shape, $\Delta m_{25\%}$ Is Defined as the Mass-Scale Full Linewidth at 25% of Maximal Spectral Peak Height, in amu[a]

	m (amu)						
T (ms)	15	30	60	120	240	500	1000
5	13,000 [6,500]	6.500 [3,250]	3,250 [1,620]	1,620 [812]	812 [406]	390 [195]	195 [98]
10	26,000 [13,000]	13,000 [6,500]	6,500 [3,250]	3,250 [1,620]	1,620 [812]	780 [390]	390 [195]
20	52,000 [26,000]	26,000 [13,000]	13,000 [6,250]	6,500 [3,250]	3,250 [1,620]	1,560 [780]	780 [390]
50	130,000 [65,000]	65,000 [32,500]	32,500 [16,200]	16,200 [8,120]	8,120 [4,060]	3,900 [1,950]	1,950 [980]
100	260,000 [130,000]	130,000 [65,000]	65,000 [32,500]	32,500 [16,200]	16,200 [8,120]	7,800 [3,900]	3,900 [1,950]
200	520,000 [260,000]	260,000 [130,000]	130,000 [62,500]	65,000 [32,500]	32,500 [16,200]	15,600 [7,800]	7,800 [3,900]
500	1,300,000 [650,000]	650,000 [325,000]	325,000 [162,000]	162,000 [81,200]	81,200 [40,600]	39,000 [19,500]	19,500 [9,800]
1000	2,600,000 [1,300,000]	1,300,000 [650,000]	650,000 [325,000]	325,000 [162,000]	162,000 [81,200]	78,000 [39,000]	39,000 [19,500]

[a] Entries in the table are based on the formula (see Table 10.2), $(m/\Delta m_{25\%}) = (1.9495 \times 10^3\ qBT/m)$ for absorption-mode line shape and $(m/\Delta m_{25\%}) = (9.7476 \times 10^2\ qBT/m)$ for magnitude-mode line shape, in which $B = 20$ kG, q = ionic charge = 1 elementary charge, T is data acquisition time in ms, and m and $\Delta m_{25\%}$ both have the same units (in this case, amu).

linewidth *in mass units* is therefore *independent* of magnetic field or sampling rate. Table 10.5 shows that an FT–ICR zero-pressure absorption mode linewidth (measured as full width at half-maximal spectral peak height) of less than 1 amu at m/e values greater than 1000 amu/electronic charge can be realized for an FT–ICR mass spectrum from $m/e = 60$ upward, based on time domain data obtained in a single acquisition period lasting only 33 ms, using a computer data size of 32,768 words. For a different magnetic field than the 19.6 kG of Table 10.5, the data acquisition time (and linewidth *in frequency units*) would be different, but FT–ICR linewidth *in mass units* at the specified mass would be the same.

There is no theoretical upper limit to the potential FT–ICR mass range. However, a useful practical upper limit may be defined by the largest mass-to-charge ratio at which FT–ICR linewidth (measured at some

TABLE 10.5. FT–ICR Zero-Pressure Mass-Scale Full Linewidth (in amu) at Half-Maximum Height, at Various Ionic Masses m as a Function of Number of Time Domain Data Points N, for Absorption-Mode [Magnitude-Mode in Brackets] Line Shape, for Singly Charged Ions. Minimum Ionic Mass Included in the Spectral Range Is m_{min} = 60 amu

	m (in amu)							
N	60	120	240	480	600	1000	Acquisition time, T[a]	FT–ICR zero-pressure linewidth (Hz)[a]
128	0.566 [1.13]	2.26 [4.53]	9.05 [18.1]	36.2 [72.4]	56.6 [113]	157 [314]	128 μs	4,710 [9,430]
256	0.283 [0.566]	1.13 [2.26]	4.53 [9.05]	18.1 [36.2]	28.3 [56.6]	78.6 [157]	256 μs	2,360 [4,710]
512	0.141 [0.283]	0.566 [1.13]	2.26 [4.53]	9.05 [18.1]	14.1 [28.3]	39.3 [78.6]	512 μs	1,180 [2,360]
1,024	0.071 [0.141]	0.283 [0.566]	1.13 [2.26]	4.53 [9.05]	7.07 [14.1]	19.6 [39.3]	1 ms	589 [1,180]
2,048	0.035 [0.071]	0.141 [0.283]	0.566 [1.13]	2.26 [4.53]	3.54 [7.07]	9.82 [19.6]	2 ms	295 [589]
4.096	0.018 [0.035]	0.071 [0.141]	0.283 [0.566]	1.13 [2.26]	1.77 [3.54]	4.91 [9.82]	4 ms	147 [295]
8,192	0.0088 [0.018]	0.035 [0.071]	0.141 [0.283]	0.566 [1.13]	[0.884] [1.77]	2.46 [4.91]	8 ms	74 [147]
16,384	0.0044 [0.0088]	0.018 [0.035]	0.071 [0.141]	0.283 [0.566]	0.442 [0.884]	[1.23] [2.46]	16 ms	37 [74]
32,768	0.0022 [0.0044]	0.0088 [0.018]	0.035 [0.071]	0.141 [0.283]	0.221 [0.442]	0.614 [1.23]	33 ms	18 [37]

[a] Equation (10.29) shows that FT–ICR zero-pressure *mass*-scale linewidth (in amu) is determined only by N and m_{min}. The *particular* values for acquisition time and resultant FT–ICR zero-pressure absorption-mode [magnitude-mode] linewidth in Hz shown in this table are based on *particular* choice of magnetic field of 19.6 kG, so that for minimum ionic mass, m_{min} = 60 amu, ICR frequencies range from 500 to 0 kHz, corresponding to mass-to-charge ratio range, 60 amu/elementary charge to ∞ amu/elementary charge, requiring charge, requiring a digitizing rate of at least 1 MHz.

TABLE 10.6. FT–ICR Upper Mass Limit for Various Choices of Minimum Specified Ionic Mass m_{min} in amu, and Number of Time Domain Data Points N for Singly Charged Ions. Upper Mass Limit in This Table Is Defined (See Table 10.3) as the Largest Ionic Mass in amu at which the Zero-Pressure Absorption-Mode [Magnitude-Mode] FT–ICR Spectral Full Linewidth at Half-Maximum Height Is 1 amu or Less

	m_{min}				
N	28	60	100	200	400
128	54 [39]	80[a]	103[a]	[a]	[a]
256	77 [54]	113 [80]	146 [103]	206[a]	
512	109 [77]	160 [112]	206 [146]	291 [206]	412[a]
1,024	154 [109]	226 [160]	291 [206]	412 [291]	583 [412]
2,048	218 [154]	319 [226]	412 [291]	583 [412]	824 [583]
4,096	308 [218]	451 [319]	583 [412]	824 [583]	1,165 [824]
8,192	436 [308]	638 [451]	824 [583]	1,165 [824]	1,648 [1,165]
16,384	617 [436]	903 [638]	1,165 [824]	1,648 [1,165]	2,330 [1,648]
32,768	872 [617]	1,276 [903]	1,648 [1,165]	2,330 [1,648]	3,296 [2,330]

[a] For these cases, the time domain data set is so small that unit mass resolution is not possible even at $m = m_{min}$.

specified fraction of spectral peak height for specified spectral mode) is less than or equal to 1.0 amu. Table 10.6 shows that the common mass spectrometric decade, $m/e = 60$ to $m/e = 600$, can be accommodated by Fourier data reduction of a single time domain ICR transient signal, with a zero-pressure FT–ICR absorption mode linewidth (measured at half-maximum spectral peak height) of less than 1.0 amu, using a computer having the relatively modest data set of 8192 words.

In equations (10.19)–(10.22) and Tables 10.5 and 10.6, it is assumed that the time domain FT–ICR signal is sampled directly, so that FT–ICR linewidth and resolution are determined by T in equation (10.18), which is in turn determined by F in equation (10.19) or (10.19a). However, a useful means for enhancing FT–ICR resolution is based on the extraction of a particular spectral segment only.[13] In this procedure, the time domain FT–ICR signal is multiplied by a reference signal to produce output signals at the sum and difference frequencies between the two input signals. This output signal is then subjected to a low-pass filter, which extracts just the difference frequency signal. The net effect of this process is to extract a band of ICR frequencies that are now shifted down in frequency (an ICR signal exactly at the reference frequency will be shifted to zero frequency; an ICR 10 kHz higher than the reference frequency will be shifted to 10 kHz; etc.). The advantage of this technique is that the sampling rate F in equations (10.19) and (10.19a) [and thus acquisition time in equation (10.18) and resolution in equation (10.21)] are now determined, *not* by the largest natural

ICR frequency [as in equations (10.19) and (10.19a)], but by the (much lower) largest *difference* frequency in the extracted band. Thus,

$$F = 2(\nu_{max} - \nu_{ref}) = \left(\frac{qB}{\pi m_{min}} - 2\nu_{ref}\right) \quad \text{(mks)} \tag{10.28}$$

$$T = \frac{N}{F} = \frac{\pi N m_{min}}{qB - 2\pi m_{min}\nu_{ref}} \quad \text{(mks)} \tag{10.29}$$

or equivalently,

$$F = 2(\nu_{max} - \nu_{ref}) = 2\left(\frac{3.071 \times 10^6 qB}{m_{min}} - \nu_{ref}\right) \tag{10.28a}$$

$$T = \frac{N}{F} = \frac{1000 N m_{min}}{2(3.071 \times 10^6 qB - m_{min}\nu_{ref})} \tag{10.29a}$$

Equations (10.28a) and (10.29a) give the sampling rate F in s^{-1} and the acquisition time T in ms, for FT–ICR spectral segment extraction, as a function of the ionic charge q in multiples of the elementary charge, the magnetic field strength B in kilogauss, the minimum ionic mass m_{min} in amu, the number of points N in the time domain data set, and the reference frequency ν_{ref} in hertz. Equations (10.28) and (10.29) are the analogous expressions in mks units. [In equations (10.28)–(10.29a), ν_{ref} has obviously been chosen to be smaller than any of the ICR frequencies in the desired band.] Clearly, by choosing ν_{ref} to be arbitrarily close to the ICR frequency band of interest, the acquisition time T (and hence the FT–ICR resolution) may be made arbitrarily high. For example, if the FT–ICR mass spectrum is to include $m_{min} = 28$ amu for singly charged ions at an applied magnetic field strength of 20 kG, then direct sampling of the resultant FT–ICR time domain signal [equation (10.19) or (10.19a)] would require a sampling rate of 2.194×10^6 points/s. Thus for a computer data storage size $N = 8192$ words, the data acquisition time would be limited to $T = N/F = 3.73$ ms, to give an absorption-mode zero-pressure FT–ICR full linewidth at half-maximum height of 1.015×10^3 s^{-1}, and a resolution [equation (10.17a)] of 6.789×10^3 at $m/e = 28$. *However*, by choosing the frequency of a reference signal to be $\nu_{ref} = 1.095881 \times 10^6$ Hz (i.e., 1 kHz smaller than the ICR frequency for an ion of $m/e = 28$), the sampling rate required for the $m/e = 28$ ICR time domain signal is now reduced to 2 kHz, the acquisition time may be increased to $T = 8192/2000 = 4.096$ s, corresponding to an absorption-mode zero-pressure FT–ICR full linewidth at half-maximum height of 0.9255 s^{-1}, and an (enhanced) resolution of 7.446×10^6 at $m/e = 28$, an improvement of more than three orders of magnitude in resolution.

The Fourier transform ion cyclotron resonance experiment is the newest of the applications of Fourier methods that are discussed in this monograph and only a few FT–ICR experiments have as yet been performed. The results of one early FT–ICR experiment are shown in Figures 10.8

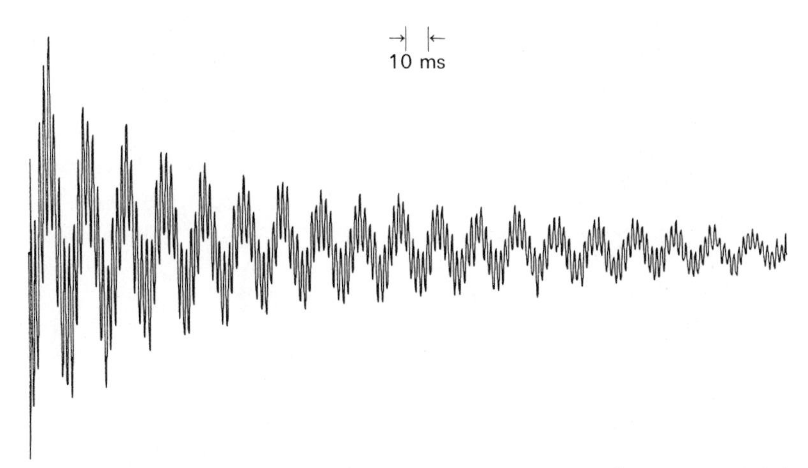

FIGURE 10.8. A single time domain ICR signal for $C_{10}H_{14}{}^{79}Br^+$ and $C_{10}H_{14}{}^{81}Br^+$ fragment ions formed by electron bombardment of parent dibromoadamantane neutrals at a nominal pressure of 3×10^{-7} Torr, using frequency sweep excitation of 23 V (p–p) from 98–271 kHz in 0.14 ms to excite the spectrum. This signal was then mixed and filtered and then digitized at 10 kHz to give the trace shown in the figure. The total acquisition period was 0.1024 s. From reference 13.

and 10.9. Figure 10.8 shows the transient signal that was obtained by the spectral segment extraction technique described earlier in this section. Fourier transformation of the signal of Figure 10.8 yields the frequency domain mass spectrum shown in Figure 10.9.

As mentioned in Section 10.1, an important feature of the ion cyclotron resonance experiment is that the spectral linewidth is easily varied by the experimentalist. This control of ICR linewidth arises because the damping of the excited ion cyclotron motion (Figure 10.2) is a second-order kinetic process: first-order in the ion density and first-order in the density of neutral molecules. Thus by reducing the neutral concentration, i.e., by reducing the sample pressure, the relaxation time of the excited motion can be made arbitrarily long. Indeed, in this chapter, the relaxation time has been assumed infinite. By observing the long-lasting transient signal that can be obtained at low sample pressures for a longer period of time, the frequencies that make up that transient signal may be more accurately determined. This is equivalent to saying that the frequency spectrum is measured with greater high resolution.

Graphical illustration of the increasing resolution obtainable by increasing the observation time of a long-lasting ICR transient signal is provided in Figure 10.10. Figure 10.10 shows three ICR spectra derived from three different choices for the acquisition time of the transient signal of Figure 10.11 and subsequent Fourier transformation.

The measurement of cyclotron frequencies with greater accuracy can be used to extend the ICR upper mass limit [equation (10.25a)] and can be used to obtain very high resolution mass spectra [equation (10.17a)]. Experimental examples that corroborate the predictions of each of equations (10.25a) and (10.17a) have been obtained by the FT–ICR method. Figure 10.12 shows an example of a mass spectrum, obtained by the FT–ICR spectral segment extraction technique, in which the mass range extends to over $m/e = 1000$. Figure 10.13 shows an FT–ICR ultra-high-resolution mass spectrum near $m/e = 28$ of the molecular ions in a ternary mixture of nitrogen, ethylene, and carbon monoxide.

The fact that the relaxation time of an excited ICR sample is controllable and inversely dependent upon the sample pressure is of course not unique to the FT–ICR experiment but applies to conventional ICR spectroscopy as well. Thus the theory developed above should also apply to conventional marginal oscillator-type ICR spectrometers. It can be shown[14] that in the limit of zero damping of the excited ICR motion, the conventional ICR linewidth is given by

$$\Delta\omega_{50\%} = 5.566/T \qquad \text{(conventional ICR)} \tag{10.30}$$

where $\Delta\omega_{50\%}$ is the conventional ICR linewidth in rad/s and T the detection

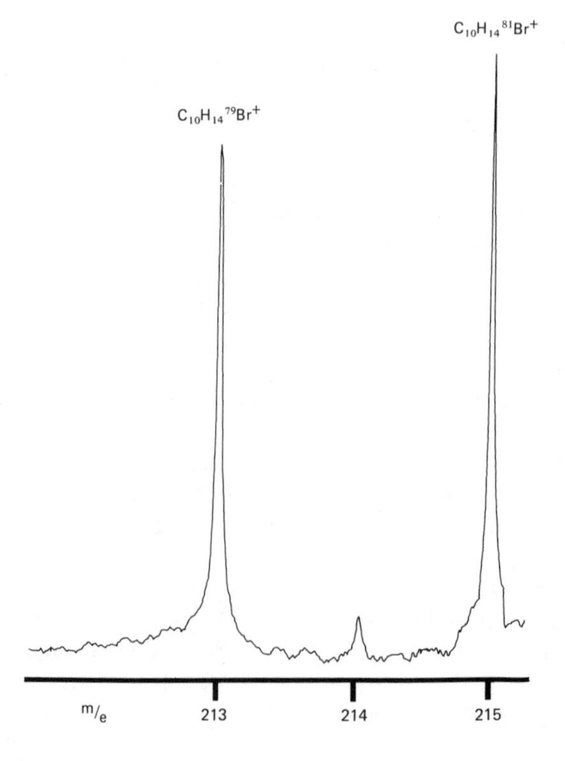

FIGURE 10.9. ICR mass spectrum obtained by Fourier transformation of the digitized time domain signal shown in Figure 10.3. The signal at $m/e = 214$ is due to the ^{13}C species, $^{12}C_9\,{}^{13}CH_{14}\,{}^{79}Br^+$. From reference 13.

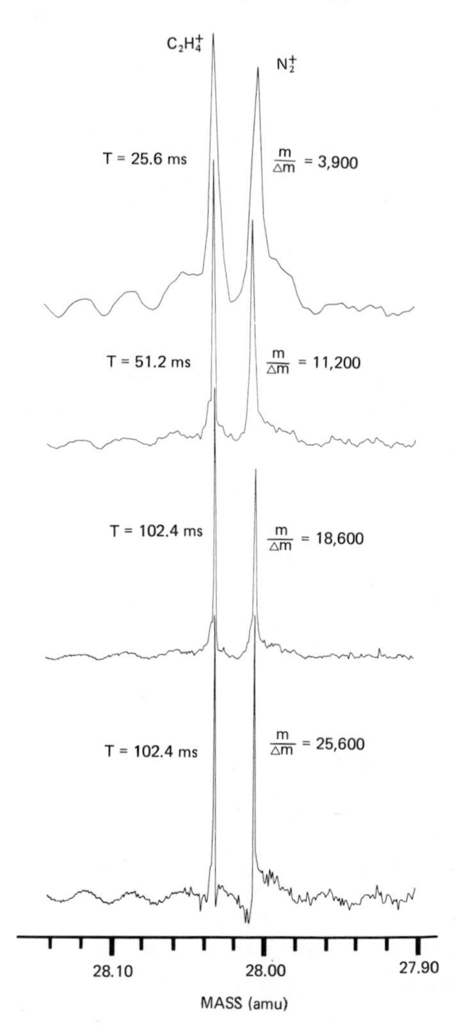

FIGURE 10.10. ICR mass spectra obtained by Fourier transformation of the digitized time domain signal shown in Figure 10.11, for three different choices of acquisition time *T*. The top three spectra are displayed in "absolute value" mode in order to simplify phase corrections and to provide uniform displayed line-shape. The bottom spectrum shows that even better resolution is obtained from "absorption mode" display.

time in seconds. Combination of equations (10.30) and (10.8) yields

$$\frac{m}{\Delta m_{50\%}} = \frac{qBT}{5.569 \times 10^{-4} m} \quad \text{(conventional ICR)} \qquad (10.31)$$

which gives the conventional ICR mass resolution as a function of the ion charge q, the magnetic field B, the ion mass, m, and the detection time T [units for q, B, m, and T as in equation (10.17a)]. Equation (10.31) should be compared with equation (10.17a), which is the analogous equation for the FT–ICR experiment. Similarly, equation (10.31) may be combined with equation (10.24a) to yield equation (10.32), which gives the conven-

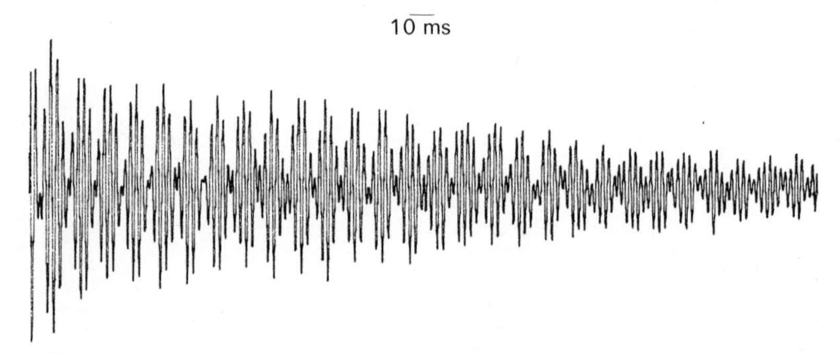

FIGURE 10.11. A single time domain ion cyclotron resonance signal for N_2^+ and $C_2H_4^+$, formed by electron bombardment of a mixture of N_2 and C_2H_4 at a nominal pressure of 10^{-7} Torr, using pulsed excitation of 500 mV (p–p) at 307 kHz for 0.5 ms to excite the ions. The signal was then mixed and filtered and then digitized at 10 kHz to give the trace shown in the figure. The total acquisition period was 0.1024 s.

tional ICR upper mass limit as a function of q, B, and T:

$$m_{\max} = (1.735 \times 10^3 qBT)^{1/2}, \quad \Delta m = 1.0 \quad \text{amu} \qquad \text{(conventional ICR)} \tag{10.32}$$

[units for q, B, and T as in equation (10.25a)]. Equation (10.32) should be compared with equation (10.25a). It is clear from equations (10.31) and (10.32) that conventional ICR spectrometers, like FT–ICR spectrometers, can in principle be used to obtain high-resolution mass spectra [equation (10.31)] and can be used to obtain mass spectra of high-molecular-weight ions [equation (10.32)]. In practice this has not worked out, and as a consequence ion cyclotron resonance spectrometers have in the past been considered low- to medium-resolution mass spectrometers with an upper mass limit of about $m/e = 200$. The reason why it is possible to obtain

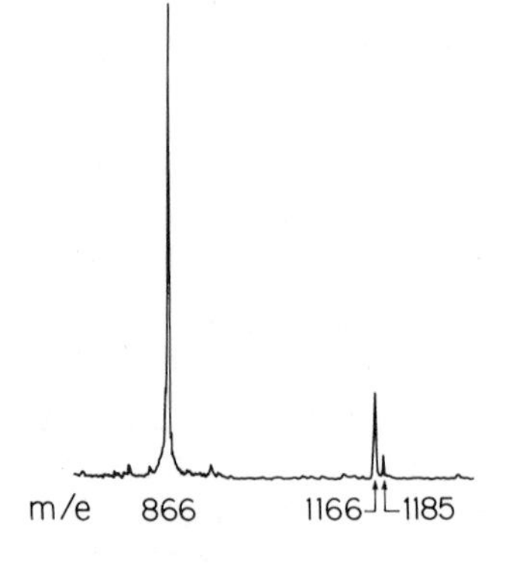

FIGURE 10.12. FT–ICR mass spectrum from m/e = 850 to m/e = 1200 of tris-(perfluoroheptyl) azine, $C_{24}N_3F_{45}$ (mol. wt. = 1185).

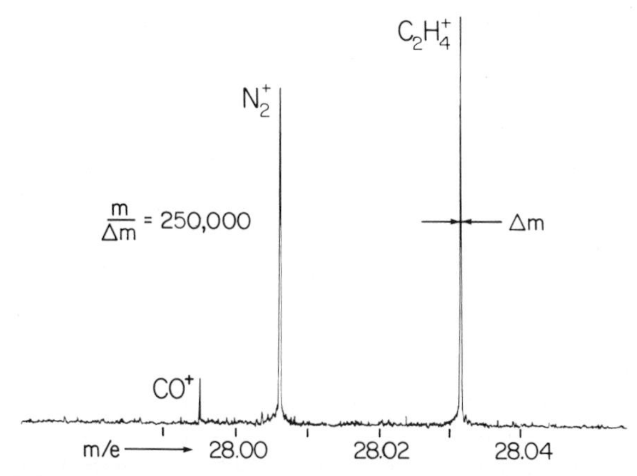

FIGURE 10.13. Ultrahigh resolution FT—ICR mass spectrum of a ternary mixture of nitrogen, ethylene, and carbon monoxide near $m/e = 28$. The mass resolution $m/\Delta m$ is 250,000.

high mass resolution in the Fourier transform ion cyclotron resonance experiment but not in the conventional marginal oscillator ICR experiment is *not* related to the Fourier transform process *per se* but rather to the temporal separation between the cyclotron resonance excitation and detection processes that exists in the FT–ICR experiment. In the conventional ICR experiment, ions are excited by a radio-frequency electric field applied across the ICR cell and execute the spiral motion shown in Figure 10.1. The detection of the excited cyclotron motion is achieved by measuring the power absorption of the ions from the radio-frequency electric field. This detection is readily achieved with a marginal oscillator type of radio-frequency oscillator. Thus in the conventional ICR experiment, ion excitation and ion detection are simultaneous. In the FT–ICR experiment ion detection is achieved by measuring the time domain signal induced in the plates of the ICR cell by the excited ion motion *after* the ion motion has been coherently excited. In the absence of ion–molecule collisions and other processes that interrupt the coherent ion motion, the excited ICR motion will be as shown in Figure 10.2. It is clear from Figure 10.1 that during *the detection period* the excited ions in the conventional ICR experiment will experience the whole magnetic field over the volume of the spiral path shown in Figure 10.1. In contrast, during *the detection period*, the excited ions in the FT–ICR experiment (Figure 10.2) will experience a more restricted region of the magnetic field and therefore a more limited range of magnetic inhomogeneities. For example, in the hypothetical case that the magnetic field inhomogeneity consisted solely of a radial gradient from the center of the ICR cell (i.e., the magnetic field increased or decreased from the center of the ICR cell) the ion in a conventional ICR experiment would experience the entire magnetic inhomogeneity during its detection period, whereas the ion in the FT–ICR experiment would experience a constant magnetic field during its detection period. Another factor that should be considered is that the FT–ICR motion shown in Figure 10.2 may effectively average out the magnetic field inhomogeneities experienced by the ion via a process that is analogous to the technique of "spinning the sample,"[18] which is used in high-resolution NMR spectroscopy to allow the NMR sample to "see" a magnetic field that is more homogeneous than if the sample were not spun. This effect will apply to the FT–ICR motion shown in Figure 10.2 because the ion retraces the same path on each cycle, but will not apply to the conventional ICR motion shown in Figure 10.1 because the average magnetic field seen by the ion changes since the radius of the ion motion is increased during the detection period.

ACKNOWLEDGMENTS

This research was supported by the National Research Council of Canada. The author would like to thank the American Institute of Physics for permission to reproduce material from references 13 and 14.

REFERENCES

1. J. Henis, *in: Ion–Molecule Reactions,* Chap. 9 (J. L. Franklin, ed.), Vols. 1 and 2, Plenum Press, New York, 1972.
2. J. L. Beauchamp, *Ann. Rev. Phys. Chem.* **22**, 527 (1971).
3. C. J. Drewery, G. C. Goode, and K. R. Jennings, *in: MTP International Review of Science, Mass Spectroscopy, Physical Chemistry,* Series One, Vol. 5, p. 183 (A. D. Buckingham and A. Maccoll, eds.), Butterworth, London, 1972.
4. J. I. Brauman and L. K. Blair, *in: Determination of Organic Structures by Physical Methods,* Vol. 5, p. 152 (F. C. Nachod and J. J. Zuckerman, eds.), Academic Press, New York, 1973.
5. G. A. Gray, *Adv. Chem. Phys.* **19**, 141 (1971).
6. T. A. Lehman and M. M. Bursey, *Ion Cyclotron Resonance Spectrometry,* Wiley (Interscience), New York, 1976.
7. J. D. Baldeschwieler, *Science* **159**, 263 (1968).
8. R. T. McIver, Jr., E. B. Ledford, Jr., and J. S. Miller, *Anal. Chem.* **47**, 692 (1975).
9. H. Sommer, H. A. Thomas, and J. A. Hipple, *Phys. Rev.* **82**, 697 (1951).
10. M. B. Comisarow and A. G. Marshall, *Chem. Phys. Lett.* **25**, 282 (1974).
11. M. B. Comisarow and A. G. Marshall, *Chem. Phys. Lett.* **26**, 489 (1974).
12. M. B. Comisarow and A. G. Marshall, *Can. J. Chem.* **52**, 1997 (1974).
13. M. B. Comisarow and A. G. Marshall, *J. Chem. Phys.* **62**, 293 (1975).
14. M. B. Comisarow and A. G. Marshall, *J. Chem. Phys.* **64**, 110 (1976).
15. M. B. Comisarow, *J. Chem. Phys.* **55**, 205 (1971).
16. D. C. Champeney, *Fourier Transforms and Their Physical Applications,* p. 9, Academic Press, New York, 1973.
17. A. B. Carlson, *Communication Systems, An Introduction to Signals and Noise in Electrical Communication,* p. 279, McGraw-Hill, New York, 1968.
18. F. Bloch, *Phys. Rev.* **94**, 596 (1954).

Chapter 11

Fourier Domain Processing of General Data Arrays

John O. Lephardt

11.1. INTRODUCTION

In previous chapters, the Fourier transformation has been discussed from the vantage point of three types of multiplex spectrometers. The reader should be aware, however, that infrared, NMR, and ICR are not the only techniques that utilize Fourier transforms to advantage. X-ray crystallography and holography depend in a fundamental way upon Fourier analysis. Engineers also commonly employ Fourier transforms in the solution of electrical circuit response functions. These nonspectroscopic uses of Fourier transforms have been discussed extensively elsewhere,[1-3] and it would be impossible to discuss the ramifications of Fourier transforms to even these three areas within the confines of this chapter. These other major uses of Fourier transforms are mentioned at the outset only so the reader is aware that Fourier transforms are applicable to many disciplines besides spectroscopy.

In this chapter, attention will be diverted away from any discipline-specific applications of Fourier transforms and toward the general application of Fourier transforms as a data-manipulating tool. Examples of the use of the techniques to be described will be drawn, where possible, from the chemical literature of disciplines that historically have not utilized Fourier techniques. It is hoped that from this discussion the readers will

John O. Lephardt • Philip Morris U.S.A., Research Center, P.O. Box 26583, Richmond, Virginia 23261

gain some insight into the possible utilization of Fourier techniques in their own fields, and also gain sufficient familiarity with the specific manipulations to initiate their own investigations.

11.2. FOURIER TRANSFORMATION AND A GENERAL DATA ARRAY

Prior to discussing the generally applicable features of Fourier transforms, it is important to understand the nature of a general data array as it relates to its Fourier transform. Since we will assume throughout this chapter that Fourier transform connotes a discrete transformation utilizing a fast Fourier transform (FFT) routine, our general data array will have some restrictions. In this chapter, a general data array will consist of j finite complex-valued measurements spaced equally along some relevant coordinate system. For the purpose of transformation, j must be less than

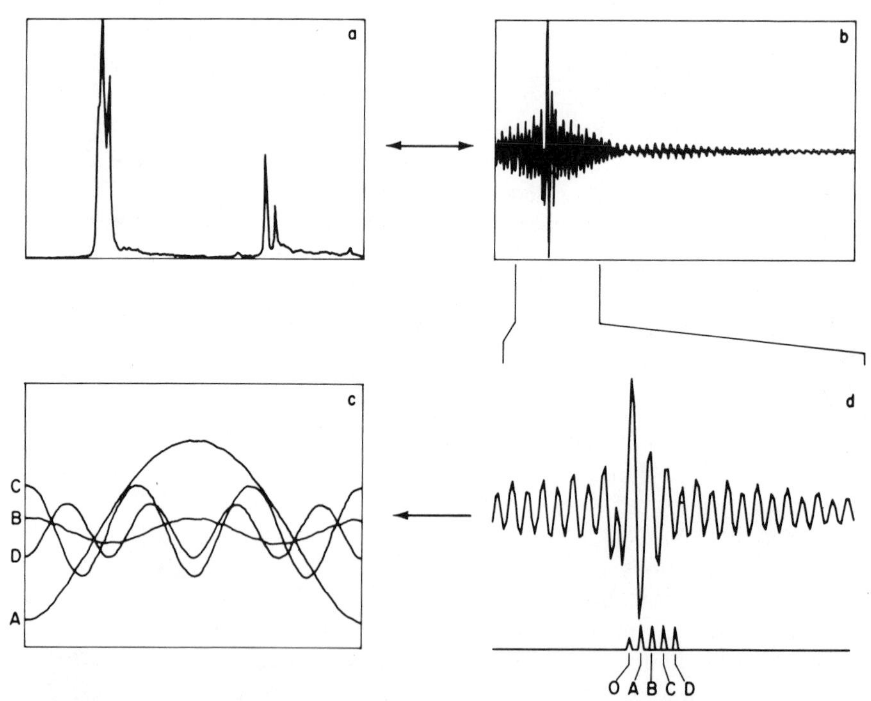

FIGURE 11.1. Spatial frequencies and Fourier transformation. The spectrum indicated in (a) produces upon transformation the real Fourier array shown in (b). Reverse transformation of individual data points from this array (c) results in the spatial frequency components shown in (d). Summation of all the spatial components would result in the regeneration of spectrum (a).

or equal to 2^K for some real value of K. The length of any array will be denoted by $2^K = L$. From this definition of a general data array, it is apparent that almost any real measured function that can be digitized with respect to time or some other parameter is suitable for transformation. This is essentially true.

Once it has been established that our data will fit in a box of length L, we can discontinue any further regard for the specific origin of the data and treat all data similarly—at least as far as the Fourier transformation is concerned. To explain why this is so, the concept of spatial frequencies must be introduced. Spatial frequencies are basically those sine or cosine waves that will exactly fit in a box. In Figure 11.1 a data array and its Fourier transform are depicted. If each of the points in the Fourier space array is reverse transformed individually, it is apparent that with the exception of the zero point, each corresponds to a harmonic of a sine or cosine wave with wavelength L. It is further apparent that the highest spatial frequency will be determined by the number of points in the array. Of paramount importance to the analysis of general data arrays is the independence of these spatial frequencies from the actual data. The spatial frequencies depend only on the box size and number of points for their frequency definition. The specific data define only the magnitude of each spatial frequency required to replicate the original data array upon their addition. The Fourier transform operation simply provides the means to determine the magnitude of each spatial frequency present in a data set. From this spatial frequency view of Fourier transforms, it is reasonably apparent that any manipulations in Fourier space could be described as alterations in either the number or magnitude of spatial frequencies that describe the data. Before pursuing this direction further, however, some additional inspection of the Fourier domain data format and the transformation operation between domains is necessary.

11.3. AMPLITUDE AND PHASE ARRAYS

It has been stated that the Fourier domain data describe the magnitude of various spatial frequencies. Specifically, for each spatial frequency, the Fourier domain contains one real and one imaginary point corresponding respectively to a cosine and a sine wave originating at the first point of the data array. The amplitude of each frequency can be determined from the magnitudes in the sine and cosine arrays using

$$A(\nu) = \{[I_{\cos}(\nu)]^2 + [I_{\sin}(\nu)]^2\}^{1/2} \tag{11.1}$$

In addition to the amplitude distribution of the Fourier spatial frequencies, it is also often useful to determine the phase distribution of the spatial frequencies. The phase of each spatial frequency can be determined

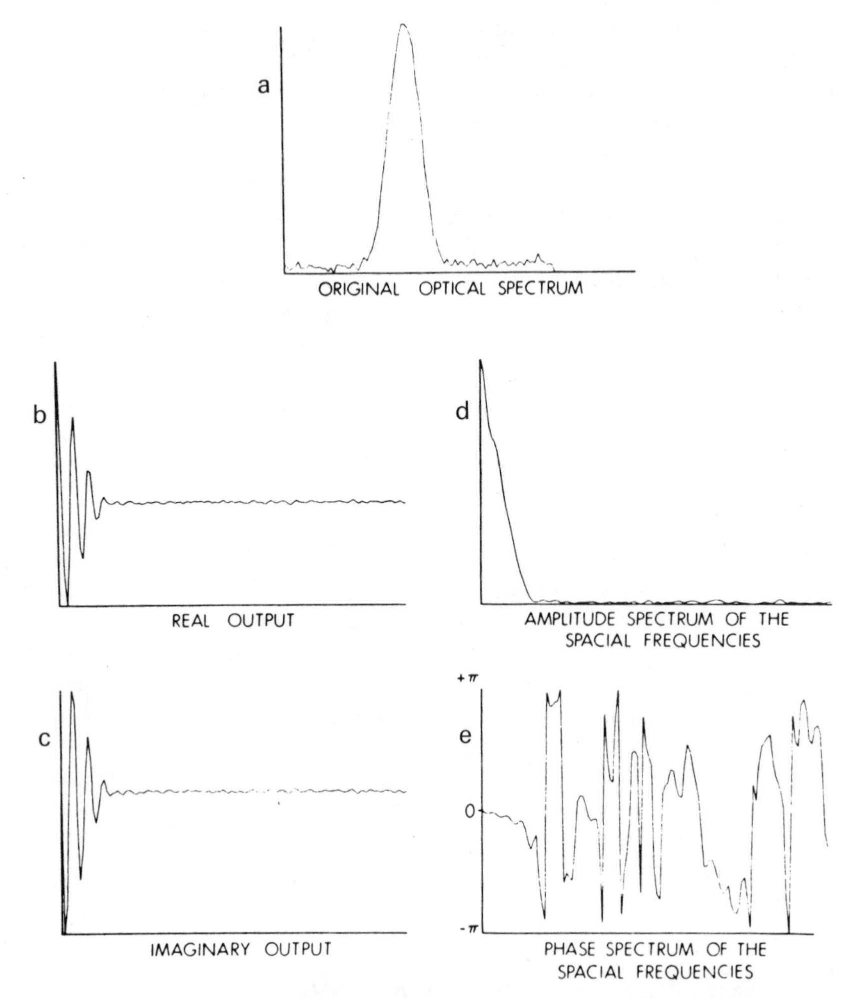

FIGURE 11.2. Data formats in the Fourier domain: (a) original spectrum, (b) real (cosine) Fourier array, (c) imaginary (sine) Fourier array, (d) Fourier amplitude spectrum, and (e) Fourier phase spectrum. (Reproduced from reference 4 by permission of the author and the American Chemical Society; copyright 1972.)

using

$$\phi(v) = \tan^{-1}[I_{\sin}(v)/I_{\cos}(v)] \tag{11.2}$$

Shown in Figure 11.2 are the sine, cosine, amplitude, and phase arrays obtained by Horlick[4] upon transformation of a visible spectral band. At this point, the reader's attention is drawn to the fact that the amplitude spectrum drops to near zero very quickly, indicating that lower spatial frequencies contribute the most substantial portion to the spectral band being analyzed. In addition, as noted by Horlick, the phase spectrum remains relatively flat over the range of spatial frequencies that contribute significantly to the spectral information, while large phase oscillations are apparent at higher frequencies. This behavior is consistent with a noise origin for these higher frequencies in the original data.

11.4. TRANSFORMATION AS A REVERSIBLE OPERATION

Before entering into the discussion of specific data manipulations in the Fourier domain, it is important to cite one additional important and seemingly obvious feature of the Fourier transform operation. The Fourier transform is a reversible operation. This feature, which may appear obvious, was not a necessary condition for any of the transformations performed in previous chapters of this book. Reversibility of the transformation operation is, however, essential if any advantage is to be realized by Fourier domain manipulations. The procedure employed to forward or reverse transform a data array is dependent upon the specific FFT program used. While the forward transformation of real data usually poses no major problems, the reverse transformation of a cosine (real) and sine (imaginary) array is not always easily accomplished. If possible, the cosine component and the negative of the sine component should be used as inputs to the reverse transform.[5] In this instance, if the original data were real, then the regenerated data will be in the real array. When the FFT program used is not designed for complex inputs, the cosine array may be used alone as an input into the reverse transform and the resultant real output array again taken as the regenerated data.[4]

Practical considerations relating to the accuracy of each of these methods would depend on the nature of the original data and this subject will not be discussed further here.

11.5. SPECIFIC MANIPULATIONS OF DATA IN THE FOURIER DOMAIN

As stated previously, all Fourier domain manipulations may be regarded as either changes in the amplitude or changes in the frequency boundaries of the spatial frequencies. Generally, these effects are combined to describe the manipulation as the multiplication of the Fourier domain data by some weighting function. This model is particularly useful in that it enables the effect of the weighting function to be predicted using the convolution theorem.[5] The convolution theorem of Fourier transformation states that the Fourier transform of the product of two functions is equivalent to the convolution of the Fourier transforms of the individual functions.

This theorem was utilized in previous chapters to describe the effect of various apodization functions on the interferograms and FIDs. In general, this theorem predicts that by multiplication with an appropriate weighting function in the Fourier domain, a data array can be convoluted with almost any desired function.

11.5.1. Fourier Domain Data Manipulations without Using Weighting Functions

Before examining the effect of various weighting functions on data arrays, it seems relevant to discuss two types of data manipulation that are not easily describable by the weighting-function approach: zero filling and contrast enhancement.

11.5.1.1. Zero Filling

Zero filling was initially employed to enable arrays of data that contained less than 2^m points to be expanded to 2^m points so that an FFT program could be used for processing. The consequence of zero filling in the original data domain is to decrease the fundamental spatial frequency represented in the Fourier domain. By zero filling in the Fourier domain, in effect one interpolates between the original data points using the transform of the filled box function as an interpolation function. While originally zero filling was understood to mean adding zero to reach the next value of 2^m, the popular use of the term today may also connote the addition of one or three times 2^m excess zeros to the array. Excess zero filling can be used to advantage when the sampled data have a certain relationship to the spatial frequencies obtained on transformation, or when the number of data points obtainable is limited. An example of the use of zero filling to improve photometric accuracy is reproduced[6] in Figure 11.3. In this instance, the spectral linewidth of the CO absorption at 2150 cm^{-1} was comparable to the resolution (resolution $\simeq 1/\text{box length}$) of the measurement, and the sparsity of data points across the individual vibrational–rotational lines resulted in substantial photometric error. By adding sufficient ($3\times$) zeros to the interferogram before transformation thereby quadrupling the effective box length, and applying a suitable weighting function on the nonzero data, additional points were extrapolated across the spectral lines and improved photometric accuracy was obtained. The reader should note that the best result was obtained by a combination of zero filling with an appropriate weighting function. Some improvement is seen, however, for each operation when used alone.

A second use for zero filling can occur when insufficient data are obtained by the measurement technique and a smoother function is required for further processing. An example of this situation is shown in Figure 11.4,

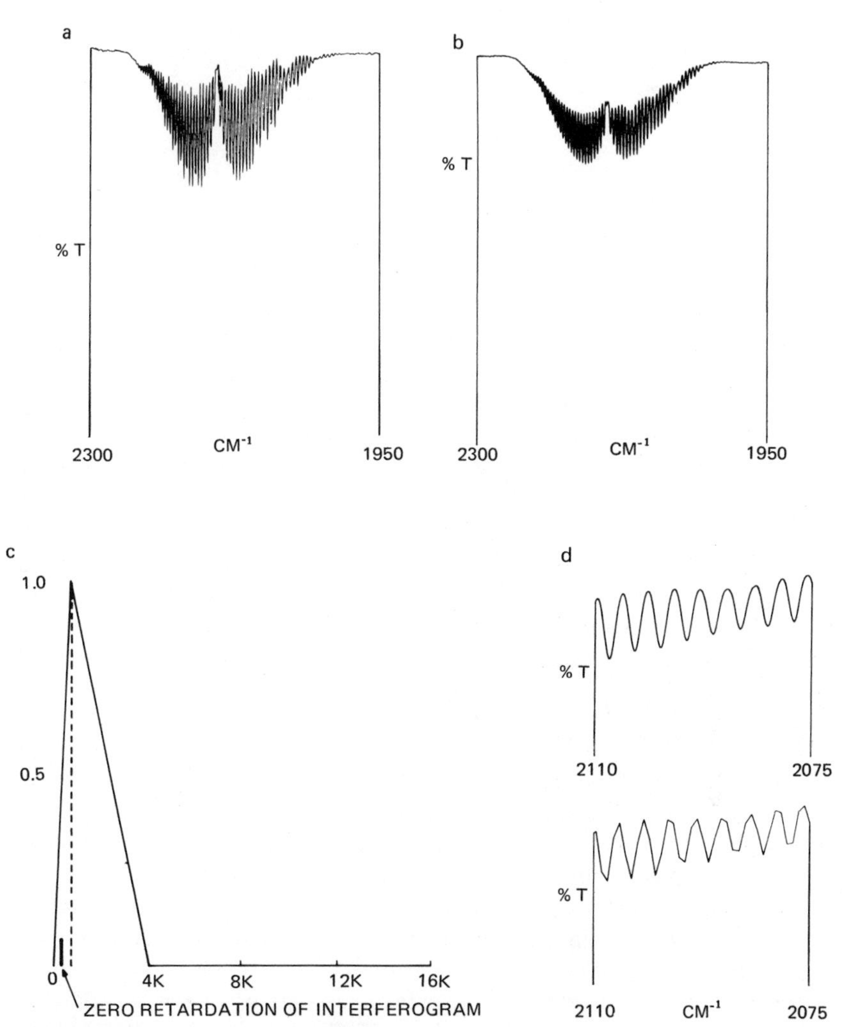

FIGURE 11.3. Use of excess zero filling and Fourier weighting functions to improve photometric precision. (a) Absorption spectrum of CO measured at 2 cm^{-1} nominal resolution, using only the 4K measured data points weighted with a triangular apodization function. (b) The spectrum of the same sample computed from the same 4K data points and 12K zeros (3× zero filling). (c) The apodization function used on the complete zero-filled array for (b). (d) Expanded spectra of (a), plotted with linear interpolation between computed data points (lower trace), and (b), computed using 3× zero filling (upper trace), to show the effect of computing four times as many output points per unit frequency interval.

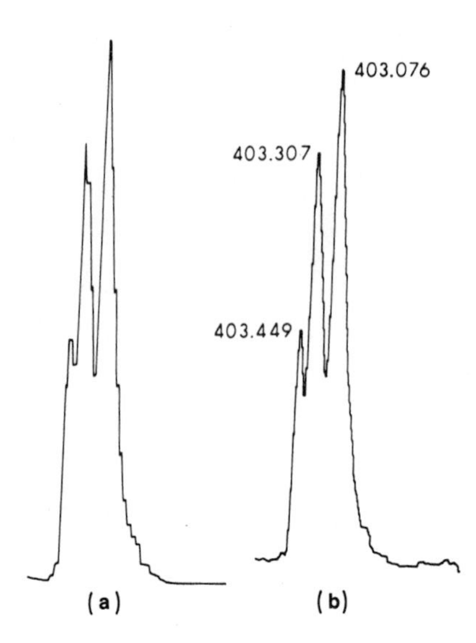

FIGURE 11.4. Use of excess zero filling to improve interpolation of spectra. Spectrum of the Mm 403 nm triplet recorded at 15-μm slit width using a discrete array detector (a) without zero filling and (b) after 3× excess zero filling in the Fourier domain. (Reproduced from reference 7 by permission of the author and the American Chemical Society; copyright 1976.)

where Horlick[7] utilized zero filling to improve the accuracy of measurement of peak frequencies obtained from a photodiode array spectrometer. Considerable improvement in the readability of the maxima is apparent after a threefold zero-filling operation.

11.5.1.2. Contrast Enhancement

A second manipulation operation, which is not readily apparent from the weighting function model, but which can be useful in some cases, is labeled here as contrast enhancement. Contrast enhancement differs from resolution enhancement, which will be discussed later, in the fact that no special weighting functions are needed. This factor is important if the data system being used cannot generate or accommodate a significant variety of weighting functions.

Basically, contrast enhancement takes advantage of the fact that the ultimate contrast is between zero and one. Translating this thought into practical terms, reduction of a signal to the point where it must equal one or zero gives it the highest contrast possible. To illustrate how this fact can be used to advantage, a spectral line containing a shoulder is shown in Figure 11.5a. From the figure, the contrast of the shoulder against the main band is not dramatic. To improve this contrast, we first divide the spectral data until the shoulder is comparable to a digital step and then multiply the data back to the original scale. At this point, the shoulder effectively corresponds to a zero or one (Figure 11.5b) situation, but the entire curve does not look particularly attractive. If we now utilize the 3× zero-filling technique described previously, we obtain the spectrum in Figure 11.5c where the contrast of the shoulder against the main band is more pronounced than in the original curve of Figure 11.5a.

Use of this contrast-enhancement technique does have some restrictions that affect its utility. The spectrum prior to scaling should be reasonably quiet. The noise level should be at least four times smaller than the shoulder to be resolved to ensure that scaling will eliminate the noise components while retaining the signal. Software routines for scaling the data down and back up again are also needed to use this method. While it may seem that the data need not be scaled back up, the interpolation will have no values to draw points from unless the data are scaled back up. Both routines are therefore required. In the case shown, scale down was accomplished by subtracting 0.998 of the spectral magnitude from the total, and scale up was accomplished by performing progressive additions of the scaled-down spectrum to itself or its multiples until a digital factor of at least 32 existed between the digital levels of the scaled-down array. Since a 3× zero filling will result in the addition of three interpolated points between each two original points, a scale-up factor of 32 or more is needed to give the inter-

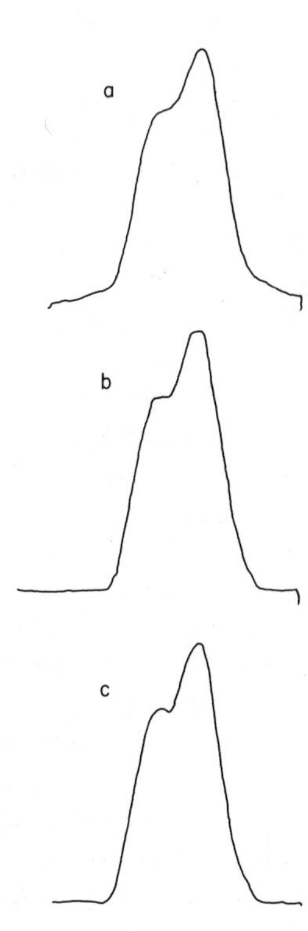

FIGURE 11.5. Effect of contrast enhancement on a spectral doublet. (a) Original spectrum, (b) spectrum scaled to digital limit, and (c) spectrum obtained with 3× zero filling in the Fourier domain.

polated points a digital noise level of less than 10% of the distance between the points.

The relatively small gain in contrast achieved using the contrast-enhancing techniques would suggest it to have limited utility. This is probably a reasonable evaluation, but in the absence of other enhancing capability, the contrast-enhancement technique may serve the experimental requirements.

11.5.2. Fourier Domain Data Manipulations Using Weighting Functions

11.5.2.1. Smoothing

It is rare that scientific measurements are reported where no noise is apparent in the data. In the more common instance, the researcher has made every effort to minimize the noise level during measurement but often such efforts cannot achieve the desired signal-to-noise ratio for the data. The researcher usually then turns to some form of weighted-moving-average smoothing function that hopefully will reduce the noise present in the data without substantially affecting the signal information. Perhaps the most familiar and commonly employed smoothing functions of this type are the convolution functions of Savitsky and Golay.[8] These functions can be very useful, but they do require a convolution program capability in the data processor, and unless some prior information is available concerning the functionality of the data, selection of the best weighting function may be difficult. As with the contrast enhancement technique described previously, however, the weighted-average smoothing functions may not produce the optimum signal-to-noise ratio, but in the absence of other processing capabilities they can be the technique that will solve the problem.

The introduction of the Fourier domain into smoothing considerations provides some interesting new options for the researcher with noise problems. Fourier transformation of a data array containing noise results in a spatial frequency distribution for both the signal and the noise. In many instances, the signal has a different spatial frequency content from the noise, and one method to improve the signal-to-noise ratio is to zero those spatial frequencies arising from noise prior to reverse transformation. While this procedure sounds attractive, often there will be some noise contribution at the same spectral frequencies that contribute most significantly to the signal of interest, and all noise cannot be eliminated by this method. In addition, since deleting information from the Fourier space array can be described as multiplication by some weighting function comprised of ones or zeros depending on the position in the array, upon transformation of the array the original data will be convoluted with the transform of this weighting function. Some familiarity with the weighting function behavior is therefore desirable

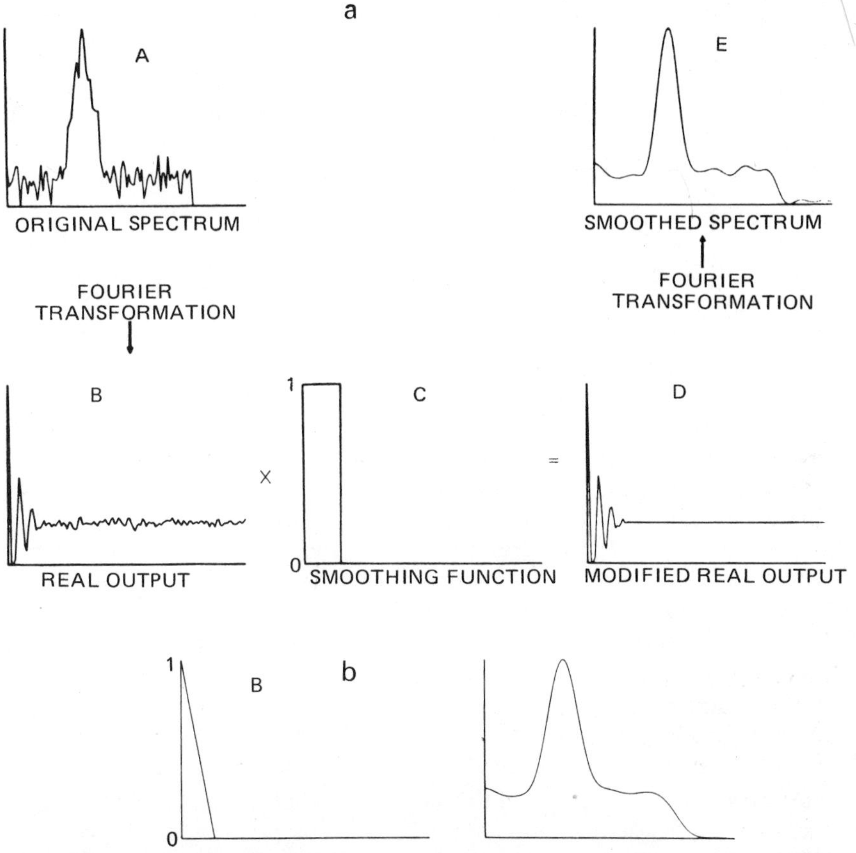

FIGURE 11.6. Use of weighting functions in the Fourier domain to smooth data. (a) Effect of a rectangular truncation function on a visible spectrum, (b) effect of a triangular truncation function on the same spectrum. (Reproduced from reference 4 by permission of the author and the American Chemical Society, copyright 1972.)

to ensure that major artifacts are not introduced into the regenerated data by the weighting function.

An example of this simple approach of data smoothing via truncation is shown in Figure 11.6a. The spectrum displayed in this figure is similar to that shown in Figure 11.2 except for the presence of substantial noise. The reader should note that the major apparent difference between the real Fourier domain data for the two spectra is an increase in the magnitude of the higher spatial frequencies for the noisy data. By the application of the weighting function shown, the higher spatial frequencies are set to zero and only the lower frequencies are retained. Upon reverse transformation, a significant improvement in the signal-to-noise ratio is apparent, while little perturbation of the shape of the band occurs. Some artifacts next to the band are visible, however, that may be attributed to the weighting function selected. As in the earlier chapters, replacement of the rectangular window with a triangular window will reduce the intensity of the side lobes and the result of using a triangular function is shown in Figure 11.6b.

The selection of boundaries for Fourier domain weighting functions is perhaps the most difficult problem associated with their use. At least three approaches have been used. In the first approach,[(9)] the amplitude spectrum is calculated from the cosine and sine arrays, and a criterion (such as the first spatial frequency contributing less than 0.1% to the array) is used to set the truncation point. In the second approach,[(10)] the standard deviation of the upper spatial frequencies around zero is calculated. The array is then analyzed working down from the highest frequency until a significant change (5–40%) in the standard deviation occurs, indicative of the presence of signal information. This point is then selected as the truncation point. In the third approach,[(11)] the maximum noise amplitude (P) in the final 50 points of the Fourier array is determined first. The entire array is then examined working down from the highest frequency until a frequency n is located such that $P(f_n) \geq 5P$. From frequency f_n, one examines higher spatial frequencies until the first instance when the amplitude decreases below P is observed. The point immediately prior to this point is selected as the truncation point. The selection of these or other criteria for truncating a Fourier domain array should depend upon the objective of the truncation operation and the amount of error acceptable in the regenerated data. These factors will vary for different experiments. In Figure 11.7 a circular dichroism measurement[(10)] is shown where the Fourier domain was truncated to produce a smoothed version of the data. Because of the nature of these data, it was possible to zero all but 5 of the 256 Fourier points and still obtain an acceptably accurate curve for the desired data. In fact, the nonzero array was reduced to such an extent that Bush even proposed the possible use of the Fourier coefficients in the place of the data curve as a method of reducing the data to be stored. When data storage space is scarce, eliminating all but the necessary Fourier terms and storing these in place of the original data can allow data to be compressed. In general, this compression would probably be closer to a factor of two than the factor of approximately 50 achieved with the CD data, but even doubling on data storage can be significant for many experiments.

Up to this point, the only weighting functions mentioned in conjunction with smoothing have been the rectangular and triangular functions. Other functions such as exponential or Gaussian functions may also be used to reweight data in the Fourier domain, with the result that the data will be convoluted with the Lorentzian or Gaussian function that corresponds to such functions. Such operations generally distort the band shape of the data, however, and should not be used if such distortions would not be acceptable.

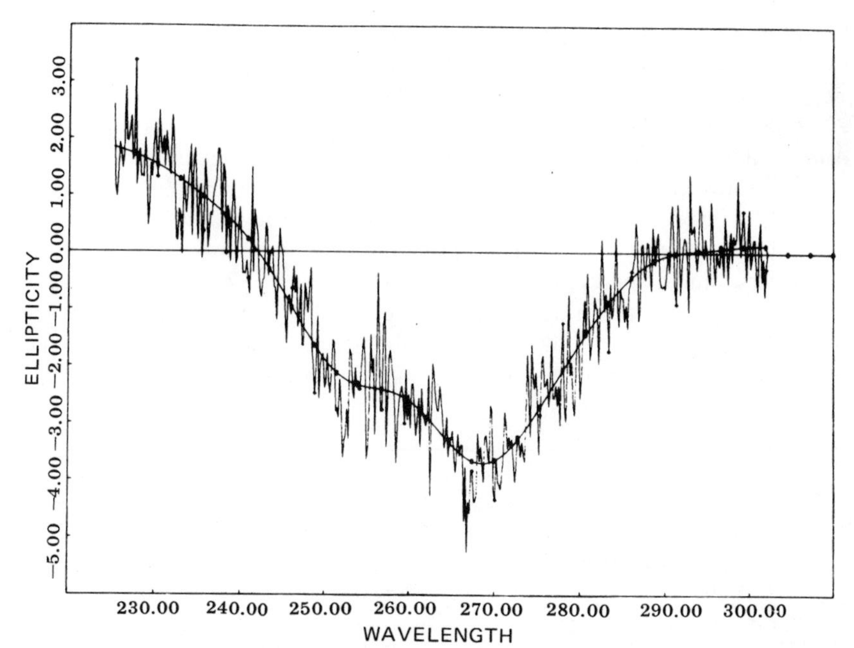

FIGURE 11.7. Use of a weighting function in the Fourier domain to smooth data and reduce data array size. Single-scan CD spectrum of adenosine in 10^{-3} M phosphate buffer. Smoothed spectrum is result of retaining 5 of 256 complex Fourier coefficients. (Reproduced from reference 10 by permission of the author and the American Chemical Society, copyright 1974).

A final weighting function that can be employed to smooth data is the "optimum signal-to-noise filter" of Turin.[12] This filter is basically the spatial frequency spectrum of the measurement instrument line shape function. This filter is derived by calculating the amplitude array for a measured line that has a very high signal-to-noise ratio. In effect, this filter convolutes the measured data with a noise-free version of itself. The band-shape of the resultant data is widened using this function but the signal-to-noise ratio is optimized. If lineshapes are not important and a high signal-to-noise measurement is available, this filter can be quite useful. For other instances, the rectangular and triangular functions serve quite well.

11.5.2.2. Elimination of Low-Frequency Interferences

While the more common experimental need is the elimination of higher-frequency noise from lower-frequency signals, there are some instances where the interference is comprised primarily of lower-spatial-frequency components than the signal of interest. One such situation occurs in Raman spectroscopy. In Raman spectroscopy, the Raman lines of interest may be quite narrow with a broad interference present from an overlapping fluorescence signal. It is generally desirable to minimize or eliminate the fluorescence component during the collection of the spectral data. but this objective is often not obtained. Fourier processing of the data after collection can provide partial separation of the Raman and fluorescence components. An example of the use of Fourier processing of a Raman spectrum to eliminate fluorescence, obtained using Nicolet Instrument Corporation's NIC-1180 data system, is shown in Figure 11.8. In this instance, separation of the fluorescence and Raman components was accomplished by first using a Fourier domain smoothing function of the form $I_i = e(iTC/N)$, where N is the array size, i the index of the current data point, and TC the time constant (in this case, -400). Use of this operation produced a spectrum that was smoothed to the extent of elimination of the Raman components

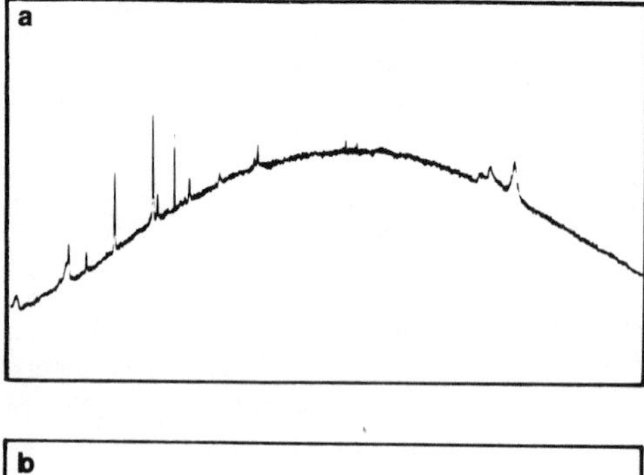

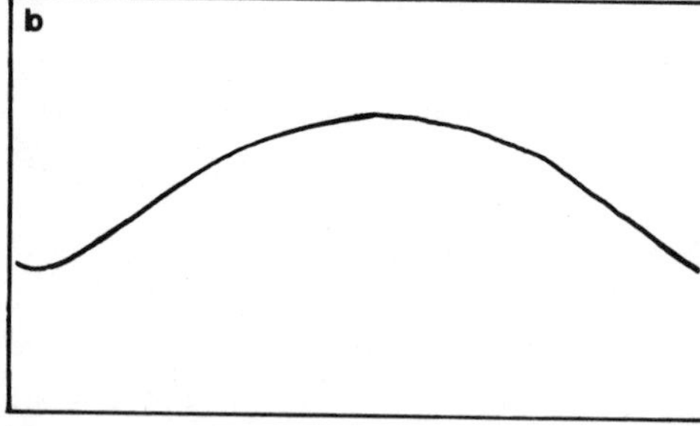

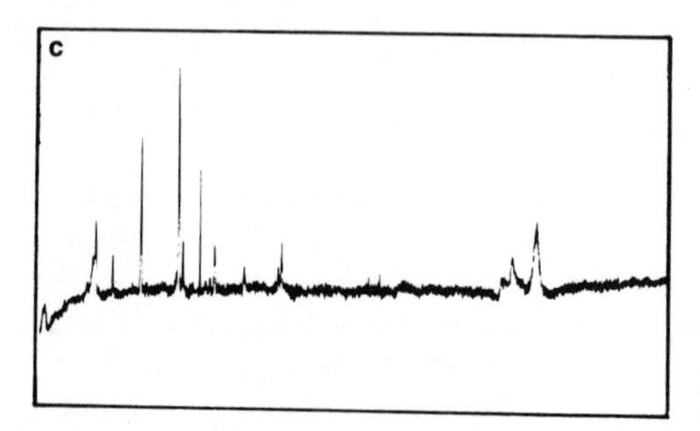

FIGURE 11.8. Use of Fourier domain manipulation to eliminate low-frequency information. Removal of fluorescence from a Raman spectrum of toluene on hydrated zeolite between 200 and 3800 cm^{-1}. (a) Original spectrum, (b) fluorescence spectrum obtained by smoothing the original spectrum using Fourier domain truncation, and (c) Raman spectrum obtained by subtraction of the fluorescence spectrum from the original spectrum. (Reproduced by permission of Nicolet Instrument Corporation.)

leaving only the fluorescence spectrum. This fluorescence spectrum was then subtracted from the original composite spectrum to yield the fluorescence-free Raman spectrum. This same objective could also have been accomplished by a single weighting function, which would perform both of these steps in one operation. Such a weighting function would attenuate the initial points in the Fourier array (i.e., those corresponding to low spatial frequencies) while leaving the later points (higher spatial frequencies) unchanged. The selection of suitable functions to separate different contributors to a composite data set will be discussed further under the category of resolution enhancement and additional data acquisition. Such operations represent a very useful application of the spatial frequency concept, and the use of Fourier processing.

11.5.2.3. Differentiation and Integration

The use of Fourier domain weighting functions for smoothing data may be considered as the art of removing unwanted information from a data array. Commencing with this section, the viewpoint will change to the use of Fourier domain weighting functions to acquire additional information not readily apparent in the original data array. The first such data modification to be considered is differentiation.

The derivative of a data array can in some instances be as valuable as the array itself. Unless a suitably differentiable function can be assigned to the function, however, the derivative of the array may be difficult to acquire with reasonable accuracy. By the use of an interesting feature in Fourier space, however, the nth derivative of a data array can be determined relatively easily. This fundamental property of Fourier-related arrays is stated below and the reader is directed to reference 13 for the details of its derivation.

> If $f(x)$ has the Fourier transform $F(S)$, then $f'(x)$ has the Fourier transform $(i2\pi s)F(s)$.

Extending this property to the general nth derivative, the relationship can be written

$$f(y) \rightleftarrows F(\chi), \qquad d^n f(y)/dy^n \rightleftarrows (i2\pi\chi)^n F(\chi) \tag{11.3}$$

From equation (11.3), it is apparent that by applying a weighting function proposition to $(ix)^n$ to the Fourier domain data, the transform operation will result in the generation of the nth derivative of the original data. The reader should note, however, that the derivative weighting functions all place the major weight on the higher spatial frequencies (i.e., large X). From the previous discussion on noise and smoothing, it is apparent that these functions will amplify the high-frequency noise components unless some data smoothing is done in conjunction with the differentiation. An

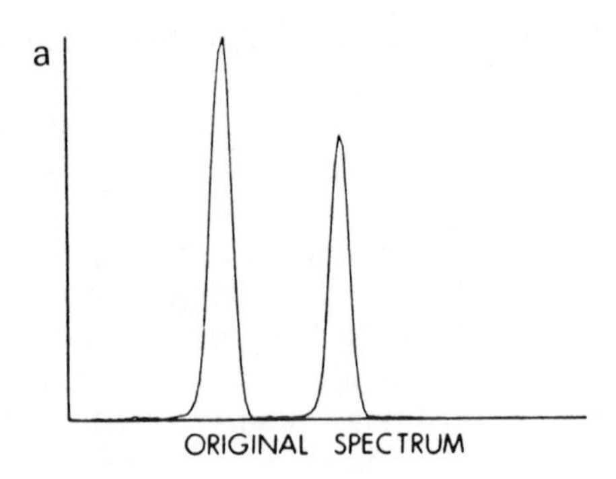

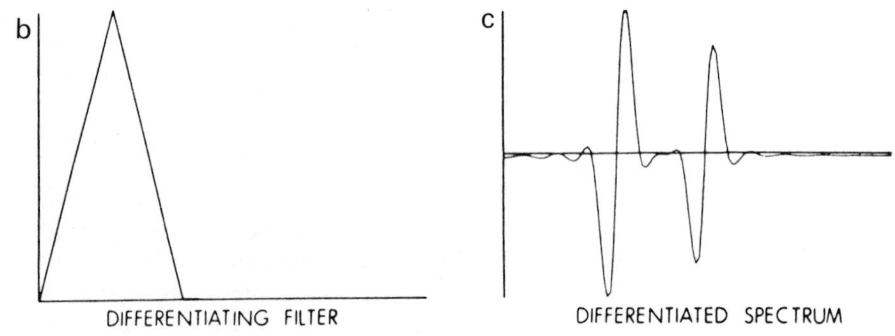

FIGURE 11.9. Differentiation using the Fourier domain. (a) Original spectrum, (b) weighting function applied to the Fourier arrays, and (c) derivative spectrum obtained. (Reproduced from reference 4 by permission of the author and the American Chemical Society, copyright 1972.)

example of a differentiated spectral data array along with the weighting function used to generate the differentiation is shown in Figure 11.9. The reader will note that the weighting function used employs three operations, the (ix) function at low frequencies to produce the derivative, truncation of high frequencies to eliminate noise, and a triangular function to connect these two regions without a sharp transition—thereby reducing the likelihood of sideband artifacts in the final data.

When using the Fourier weighting functions for differentiation it is important not to forget the i in the $(ix)^n$ term. The consequence of this term on odd derivatives is to necessitate switching of the original real and imaginary arrays prior to reinsertion of the weighted arrays for reverse transformation. When only the real array is normally reinserted, either the imaginary output may be taken as the derivative, or the imaginary array may instead be used as the input and the real output taken as the derivative.

Integration of data arrays entails basically the reversal of the logic used for differentiation.[13] If multiplication in the Fourier domain by $(i2\pi s)$ is equivalent to differentiation, then division in the Fourier domain by $(i2\pi s)$ is equivalent to integration. This argument is valid up to inclusion of an integration constant that can be added by the simple addition of a delta function at the zero point of the Fourier domain. The reader will recall that the Fourier domain zero point corresponds to the DC level in the data and addition of a constant to the data may be accomplished by adding to the zero point via a delta function.

11.5.2.4. Resolution Enhancement and Functional Isolation

Using various weighting functions in the Fourier domain, we have now the capacity to improve the appearance of data as they stand, reduce them to their various derivatives, or integrate them. In this section, attention will focus on the Fourier operations that allow us to acquire new or additional information from data by interrelating them to other data. It is in this area that perhaps the largest number of applications await the use of Fourier methods.

The concept of resolution enhancement has been discussed and investigated for many years. It was probably originally considered by the first investigator who thought he might have two contributors to his observation even though his equipment was only capable of resolving one feature. The problem is usually caused by interferences introduced by the measurement technique that translate the fundamental phenomenon into the output data. The measurement technique often does not have the resolution of the data, and there may be information obscured on transfer. For example, a spectral doublet will appear as a single band unless the spectrometer spectral slit-width is sufficiently narrow to discern the gap between the bands. Similarly a large expansion chamber at the detector of a gas chromatograph may mix separated peaks and thereby obscure their resolution. Over the years, there have been a variety of methods employed in attempts to deconvolute the effects of this data distortion, ranging from pseudodeconvolution[14] to use of differentiation techniques.[15] The Fourier domain, however, provides a particularly useful place for data to be deconvoluted since inversion of the Fourier convolution theorem stated earlier results in the Fourier transform of the deconvolution of two functions, which is equivalent to the Fourier transform of the composite function divided by the Fourier transform of the deconvolution function. Therefore, a perturbation function should be deconvoluted from a data array by dividing the Fourier transform of the data array by the Fourier transform of the perturbation function. This technique works up to a point where the perturbation function has virtually eliminated the spatial frequency information. As a result of this

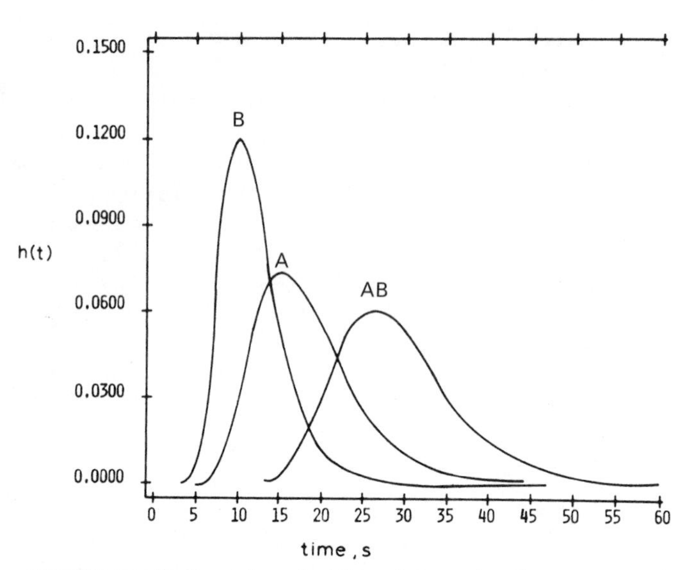

FIGURE 11.10. Extraction of additional data using Fourier processing. Resolution of an experimental argon elution curve *AB* into its column *A* and end *B* components. (Reproduced from reference 16 by permission of the author and the American Chemical Society, copyright 1973.)

nonretrievable loss of some information, it is apparent that Fourier domain operations cannot fully remove the effects of various instrumental convolution functions. Some improvement in resolution using this approach is possible, however, provided the perturbation function can be obtained and transformed in pure form.

To illustrate the practical use of deconvolution via division in the Fourier domain three examples from the chromatography literature will be discussed. In 1973, Dwyer employed the Fourier deconvolution approach in the separation of various parameters contributing to a gas chromatogram[16] (Figure 11.10). Using argon as a nonretained gas, a chromatogram of the column plus end fittings (*AB*) and a chromatogram with the column removed and just the end fittings connected together (*B*) were obtained. By the transformation of both of these chromatograms, followed by ratioing of their Fourier arrays and reverse transformation, the effects of the column alone (*A*) on the chromatogram were obtained. Continuing further in his investigations, Dwyer subsequently employed a gas phase diffusion curve derived from the nonretained argon chromatogram to remove the diffusion contribution to a retained butane peak, and thus obtained the time interaction function of butane with the substrate, exclusive of the diffusion effect (Fig. 11.11). This experiment illustrates dramatically how resolution enhancement can provide additional data.

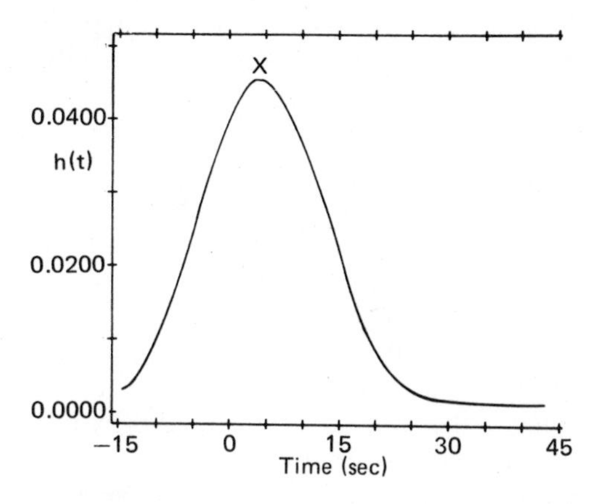

FIGURE 11.11. Extraction of additional data using Fourier domain processing. Derivation of the time distribution of the interaction of butane with cellulose acetate by deconvolution of the diffusion and other components from the gas chromatogram. (Reproduced from reference 16 by permission of the author and the American Chemical Society, copyright 1973.)

In a similar investigation by Maldacker *et al.* in the area of steric exclusion chromatography,[9] Fourier domain deconvolution was accomplished, again using a nonretained material to generate the deconvolution function. The original and enhanced chromatograms for a sample of 2:1 diethylene glycol-pentethylene glycol separated on a column of Sephadex G-25 are reproduced in Figure 11.12. In this case, the deconvolution function was derived from a separate injection of Carbowax 20M in a nonretained material onto the column. A significant increase in the resolution of the two materials is apparent as a result of the resolution enhancement operation.

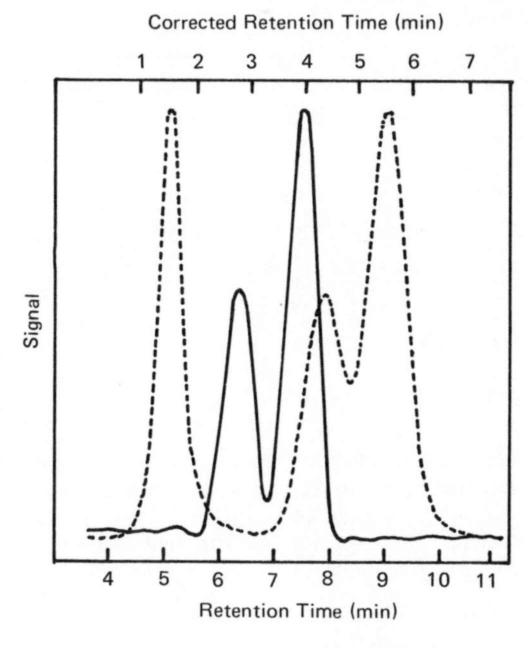

FIGURE 11.12. Resolution enhancement of steric exclusion chromatography data using Fourier processing. Chromatogram of a sample of 2:1 diethylene glycol : pentaethylene glycol on a column of Sephadex G-25. Dashed curve: original chromatogram; solid curve: after resolution enhancement. (Reproduced from reference 9 by permission of the authors and the American Chemical Society, copyright 1974.)

The two previous examples from the literature employed experimentally determined deconvolution curves to enhance the data. In some instances, however, this approach may not be practical or sufficient to achieve the enhancement necessary. In such cases, synthetic enhancement functions can be employed. Kirmse and Westerburg[11] have examined the use of a Gaussian function as an enhancing function for chromatograms, and found that optimum enhancement without generating significant artifacts resulted when the bandwidth of the enhancement function was equal to the chromatogram peak bandwidth. When multiple peaks were present in the chromatogram, the minimum peakwidth in the chromatogram defined the optimum enhancement function bandwidth. Shown in Figure 11.13 are some of the enhancements realized by Kirmse and Westerburg on partially or completely overlapped peaks. In the absence of a suitable experimental deconvolution function, the use of such artificially generated enhancement functions of the same form as the data can enable some enhancement to be realized.

In all instances where resolution enhancement functions are applied to data, they must be accompanied by smoothing functions to ensure that noise is not enhanced with the signal information. In this respect, the resolution enhancement operation is similar to the derivative, and similar consideration to the inclusion of a smoothing factor is required.

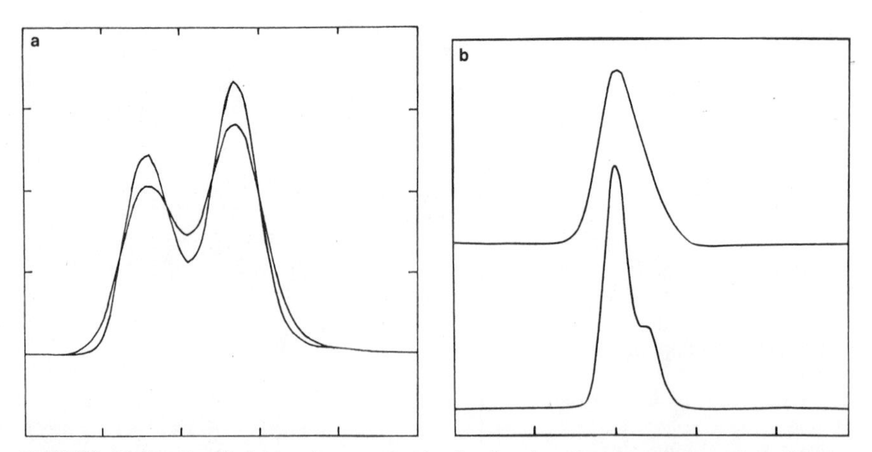

FIGURE 11.13. Resolution enhancement in the Fourier domain using a synthetic enhancement function: Enhancement of (a) partially and (b) completely overlapped Gaussian peaks using a Gaussian enhancement function. (Reproduced from reference 11 by permission of the authors and the American Chemical Society, copyright 1971.)

11.6. SUMMARY

Through the use of various weighting factors and other manipulations in the Fourier domain, it has been shown that almost any data array may be smoothed, enhanced, differentiated, integrated, and even cleaned of interferences from the method of its measurement. The incorporation of these capabilities into physicochemical measurements is at a very early stage, and substantial additional investigations should be expected. The only two uses of the resolution enhancement technique of interest to this author, and as yet not explored completely, are the deconvolution of kinetics information from spectral bandshapes distorted by concentration changes during scanning and the use of deconvolution techniques to determine the interaction functions relating intensity and frequency of molecular vibrations in different environments.

Fourier transformations have provided the chemist with several new instrumentations in recent years, but perhaps the most important feature that Fourier transforms will provide the chemist in the long term is a greater ease and flexibility in the manipulation of data.

REFERENCES

1. H. M. Smith, *Principles of Holography,* Wiley, New York, 1969.
2. H. Lipson and C. A. Taylor, *Fourier transforms and X-Ray Diffraction*, Bell, London, 1958.
3. H. H. Skilling, *Electrical Engineering Circuits,* Wiley, New York, 1957.
4. G. Horlick, *Anal. Chem.* **44**, 943–947 (1972).
5. R. Bracewell, *The Fourier Transform and its Applications*, McGraw-Hill, New York, 1965.
6. P. R. Griffiths, *Appl. Spectroscop.* **29**, 11–14 (1975).
7. G. Horlick and W. K. Yuen, *Anal. Chem.* **48**, 1643–1645 (1976).
8. A. Savitsky and M. J. E. Golay, *Anal. Chem.* **36**, 1627 (1964).
9. T. A. Maldacker, J. E. Davis, and L. B. Rogers, *Anal. Chem.* **46**, 637–642 (1974).
10. C. A. Bush, *Anal. Chem.* **43**, 890–895 (1974).
11. D. W. Kirmse and A. W. Westerberg, *Anal. Chem.* **43**, 1035–1039 (1971).
12. G. L. Turin, *IRE Trans. Inform. Theory* **IT-6**, 311 (1960).
13. D. C. Champeny, *Fourier Transforms and their Physical Applications,* Academic Press, New York, 1973.
14. R. N. Jones, R. Venkatarachavan, and J. W. Hopkins, *Spectrochim. Acta* **23A**, 925 (1967).
15. L. C. Allen, H. M. Guadney, and S. H. Guarum, *J. Chem. Phys.* **40**, 3135 (1964).
16. R. W. Dwyer, *Anal. Chem.* **45**, 1380–1383 (1973).

Chapter 12

Fourier and Hadamard Transforms in Pattern Recognition

Charles L. Wilkins and Peter C. Jurs

12.1. INTRODUCTION

The advent of high-speed, stored-program, general-purpose digital computers as powerful information-handling devices has led to revolutionary developments in many fields. Among these are several fields that deal with important problems that previously appeared to have only extremely difficult, complex, or even unrealizable solutions. A large number of such problems have proven amenable to attack by a set of techniques known as pattern recognition. Pattern recognition has become a fertile area for the development of concepts and techniques now being applied routinely to problems formerly considered to be approachable only by humans. It is for this reason that pattern recognition is often considered to be a subset of the artificial or machine intelligence field.

A basic attribute of humans, as well as other living things, is an ability to recognize patterns. A pattern is just a description of an object or an event. Objects or events to be recognized include spatial patterns such as physical objects, models of physical objects, images of physical objects (photographs), abstractions of physical objects (e.g., maps), fingerprints, and more abstract patterns such as letters or numbers. Examples of temporal objects to be recognized include speech, waveforms, EKGs, and time series.

Recognition of objects is generally really a weighing of the odds that the pattern to be classified can be associated with a particular set or popula-

Charles L. Wilkins • University of Nebraska, Lincoln, Nebraska 68588, and *Peter C. Jurs* • The Pennsylvania State University, University Park, Pennsylvania 16802

tion known from past experience. Thus the problem of pattern recognition is that of placing a pattern in a category. The criteria used are measurements or features extracted from the pattern to be classified. Often a pattern class is only defined by the existence of a number of labeled examples.

Pattern recognition can be defined as the classification of input data into categories by using features or attributes of the data extracted from the total available information. The overall problem breaks down into two halves: (1) extraction of useful invariant properties from the welter of meaningless (or even confusing) background information, and (2) use of the extracted features to classify the input into the proper category.

An astounding variety of practical problems has been attacked with pattern recognition techniques. Several excellent reviews of the pattern recognition literature have appeared that dramatize the breadth of pattern recognition applications.[1-8] A large number of books dealing with the fundamentals of pattern recognition and applications have appeared as well.[9-14]

Some of the motivations underlying the research effort accorded to pattern recognition are as follows:

(1) The substitution of machines for humans in performing routine information-processing tasks. The substitution of machines for humans is attractive because tasks can often be done more accurately, safely, or inexpensively by machines.

(2) The development of effective and efficient interfaces between humans and computers requires the development of pattern recognition capabilities on the part of the computer. Since man's preferred means of communication are printed and spoken characters, the machine must have the capability to decode them.

(3) Pattern recognition is an attractive field for research for its own sake. The problem to be solved is: What is required in general to develop machines that provide performance in perceptual problems similar to that achieved by humans?

The individual, specific applications of pattern recognition are broadly dispersed. Lengthy lists of examples with references to the original literature can be found in the reviews cited.[8] A partial list of applications includes medical diagnosis and treatment, drug interaction studies, multiphase screening and analysis, neurobiological signal processing, sonar detection and classification, image processing, analysis of photomicrographs of tissue cells, clinical data analysis (EKG and EEG), analysis of aerial photographs, weather radar data, fingerprint identification, signature verification, industrial process control, speech recognition, and interpretation of seismic signals for geological exploration.

Within chemistry there have been applications in several areas including the following: spectral analysis (mass spectra, infrared spectra, proton and ^{13}C NMR spectra, gamma-ray spectra); material science, classification of mixtures (petroleum samples and biological samples); modeling of chemical experiments; and structure–activity studies. Citations to the original literature describing these studies can be found in the reviews cited.[15-20]

12.1.1. Basic Pattern Recognition System

A general pattern recognition system must be capable of accepting a sample of data, preprocessing or transforming it, and classifying the pattern into the probable class. A schematic diagram of a pattern recognition system is shown in Figure 12.1. It consists of a number of interrelated parts: a measurement unit, a preprocessing–feature selection unit, and a classification unit. Although these components are highly interdependent in any implementation of a pattern recognition system, it is convenient to separate them for pedagogical reasons. A feedback loop is included in the schematic diagram of the system to indicate that information generated during the development of a classifier can be used to seek the features of the patterns that are most useful in performing the classification of interest.

The measurement unit accepts the raw data of the objects to be classified. The raw data are often in the form of pictures or waveforms. The first operation applied to such data must be to convert each pattern of raw data into an n-dimensional vector in an n-dimensional euclidean space called the pattern space. That is, the pattern must be decomposed into a set of measurements. The string of measurements forming the pattern vector must contain the essence of the raw input data. Thus the actual implementation of the transducer is entirely dependent on the nature of the raw data. When the raw data consist of time series, such as interferograms or EEGs, a sampling procedure in time might be in order. When they are a function of frequency, for example, infrared spectra, a sampling procedure in frequency is appropriate. When the raw data are pictures, the field can be examined for darker and lighter areas, edges, or geometric forms. If the raw data were naturally digital data, such as a low-resolution mass spectrum, the measurement step might be unnecessary. The pattern vectors, or n-dimensional points, developed by the transducer are then passed to the feature extraction or preprocessing unit of the system.

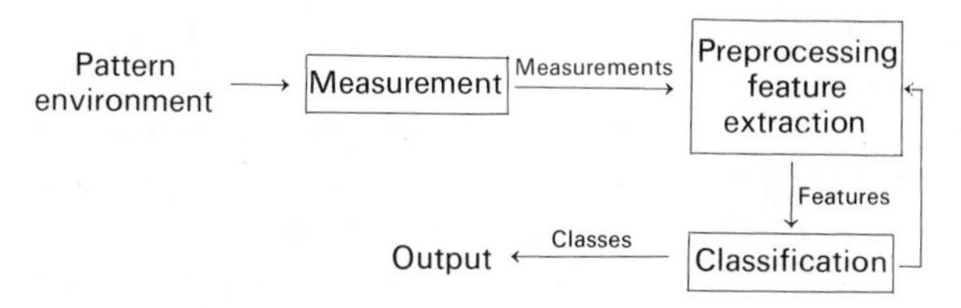

FIGURE 12.1. Basic pattern recognition system.

12.1.2. Preprocessor–Feature Extractor

The preprocessor–feature extractor subunit accepts pattern vectors produced by the transducer and operates on them in order to pursue the following goals:

1. To eliminate or at least reduce the fraction of information contained in the raw data that is irrelevant or even confusing. For example, pictorial data should be independent of translations, scale changes, or rotations, and waveforms should be independent of translations, scale changes, or rotations, and shifts in time or phase.
2. To preserve sufficient information to allow discrimination among the pattern classes, that is, to discover invariances among patterns of common class.
3. To preserve the information in the pattern vector in a form that can be effectively utilized by a linear classifier, if possible.
4. To reduce the total number of measurements per object to a manageable number.

12.1.3. Classifier

The transformed patterns are classified by the third subunit of the pattern recognition system. Classifiers have been developed by utilizing various branches of applied mathematics: statistical decision theory, information theory, geometric theory, and so on.

The task of the classifier can be stated in general as follows. A set of transformed pattern vectors, termed the training set, is used to determine a decision function $f(X)$ such that

$$\begin{aligned} f(X) &> 0 \qquad \text{for } X \text{ members of class 1} \\ &\leq 0 \qquad \text{for } X \text{ members of class 2} \end{aligned}$$

The procedure used to develop $f(X)$ is commonly known as the adaptation, training, or learning phase. The goal is to minimize the probability of error in the classifications.

Implementations of $f(X)$ fall naturally into two categories: parametric methods and nonparametric methods. Parametric training methods begin by estimating the statistical parameters of the samples forming the training set. The estimates are subsequently used for specification of the discriminant functions. The most common parametric discriminant function is Bayes' rule, since it is optimum for a well-defined class of problems.

In order to apply Bayes' rule to a two-class pattern recognition problem, the functional form for the conditional density functions for each class and the parameters for these conditional density functions must be known. The usual assumption is that the patterns are normally distributed about their class means. Then the requirement is that the mean vectors and covariance matrices of the class be known. Additionally, the loss functions that define the severity of misclassifying a pattern must be specified. Then the discriminant function is

$$f(X) = \frac{P_1 L_1 F_1(X)}{P_2 L_2 F_2(X)} - 1 \qquad (12.1)$$

where P_1 is the *a priori* probability of occurrence of a pattern in class 1; $P_2 = 1 - P_1$ is the *a priori* probability of occurrence of a pattern in class 2; L_1 and L_2 are the losses associated with misclassifying a member of class 1 or class 2; and $F_1(X)$ and $F_2(X)$ are the probability density functions of class 1 and class 2.

On the assumption that the mean vectors and covariance matrices are truly representative of their classes, then the Bayes discriminant function is optimum. Therefore the Bayes criterion has been used as a benchmark for comparison of other discriminant functions.

If the pattern of the training set cannot be described by statistical measures, then a nonparametric discriminant function must be employed. During the development of nonparametric discriminant functions, the only data used are the training set patterns themselves. In order for the training method to yield reliable results, the training set size must be large enough to be representative of the data set from which it is drawn. (A large data set is a necessary, but not sufficient, condition to allow the possibility of development of parametric discriminant functions by the estimations of probability functions.)

A widely studied nonparametric binary pattern classifier is the threshold logic unit (TLU). Besides the pattern vectors of the training set, the only adjustable parameters of TLUs are their linear coefficients, which are determined during training. An adaptive TLU is provided with the means to monitor its own performance in relation to a specified index of performance, and it is capable of modifying its own parameters so as to improve its performance. TLUs are usually adaptive only during the design stage. Individual TLUs can be combined in interdependent grids to implement more sophisticated classifiers known as piecewise linear or layered classifiers.

Another popular nonparametric discriminant method is called the K-nearest-neighbor classification. An unknown pattern is classified as a member of the class most often represented among its K nearest neighbors. Nearest is usually defined by a euclidean distance measure, but any metric can be employed.

12.2. BINARY PATTERN CLASSIFIERS

12.2.1. Pattern Vectors

A wide variety of types of data can be represented by points in a euclidean space of suitable dimensionality. For example, properties such as position or momentum can be specified as having three numbers forming a three-dimensional vector, or equivalently as a point in three-dimensional space. The implication of this notation is that three linearly independent axes can be defined that completely describe the given property in three-space when the values are specified. This method of description can readily be extrapolated to more than three dimensions. For example, in mechanics it is common to define the position and momentum of a particle in a six-coordinate space referred to as phase space. A single six-dimensional point then simultaneously defines the position and momentum of a particle. There is a one-to-one correspondence between points in the six-dimensional space and possible values of position and momentum of any particle. Any set of six values can be equivalently considered as a point in the space or as a vector pointing from the origin of the space to that point, hence *pattern vector*.

Many forms of chemical data can be represented as an n-dimensional vector of the following form:

$$X = x_1, x_2, \ldots, x_n$$

The individual components of the pattern vector, x_j, are observable quantities. For example, to represent a low-resolution mass spectrum as a vector, x_j could be set equal to the intensity of the peak in the m/e position j. In the case of tabular data, a dimension can be assigned to each data column. For example, a table of melting points, boiling points, atomic weights, indices of refraction, and densities could be converted into a set of five-dimensional points or pattern vectors, one for each compound in the table.

In the case of graphical data, which are seemingly continuous, the problem of transcribing the data into the vector format is slightly more difficult. As long as data are incremental, such as those that might be represented as a bar graph, the way to assign dimensions is obvious. For example, in a low-resolution mass spectrometer with a nominal resolution of 1 mass unit and a scan from 1 to 200 mass units, a 200-dimensional vector can represent the mass spectrum of any compound. However, many devices produce analog data that are continuous. Continuous-data collection devices, such as scanning spectrometers, however, must have some resolution limit. Usually, the nature of the electronics and optics of these instruments is such that they integrate over a range, thereby defining the resolution limit. In such cases it is usually simplest to determine the resolution r and then divide the spectrum into R/r components or dimensions, where R

is the scan range. Thus an infrared spectrum covering the wavelength range 2.0–15.0 μm might be digitized into 130 elements if $r = 0.1$.

It is important to realize that the hyperspace notation is only one way of representing data. In general, geometric properties cannot be associated with the actual data. Thus while the orthogonal dimensions of a vector are independent, it is certainly not the case that fragments of a mass spectrum represented in this fashion are always independent.

Because of the orthogonal nature of the vector components, various operations may be performed on individual dimensions in a reversible fashion. Therefore, an operator that treats each dimension separately may be applied without destroying the original data. This is true as long as an inverse operation exists that will reproduce the original vector. For example, various normalization procedures such as taking the square root or logarithm of each component, while changing the dynamic range of the data, are transformations for which the original vector can be generated by the inverse operation. This feature plays an important part in certain discriminant training operations.

12.2.2. Similarity and Clustering

One of the primary purposes of pattern recognition is the correct classification of data into categories. Any collection of data representing some fundamental entity, process, or collection therefore is referred to as a pattern. For example, the mass spectrum of a chemical compound is considered a pattern resulting from the complex chemical and physical process that produced it and, furthermore, that pattern is characterized by certain fundamental properties of the compound and the measurement process. Pattern recognition techniques are then applicable to such problems as classifying mass spectra into chemical categories. An example of such a chemical category is oxygen-containing compounds vs. compounds not containing oxygen.

Representing a set of mass spectra as points in a hyperspace yields a set of pattern points containing all the information inherent in the original spectra. The problem then becomes one of dividing the set of points into subsets as defined by the classes to be recognized, that is, how to transform pattern space into classification space.

In order to deal with points in a high-dimensional space, similarity must be defined. The intuitive notion that points that are close to each other represent objects that are similar is appealing. To put this idea on a firm footing necessitates that the pattern space be a metric space. The space must satisfy several conditions. For the three n-dimensional points named $\mathbf{X}_1$, $\mathbf{X}_2$, and $\mathbf{X}_3$, where $d(\mathbf{X}_1, \mathbf{X}_2)$ indicates the distance between points one and two the following conditions must be satisfied:

$$d(\mathbf{X}_1, \mathbf{X}_2) = d(\mathbf{X}_2, \mathbf{X}_1) \tag{12.2}$$

$$d(\mathbf{X}_1, \mathbf{X}_2) > d(\mathbf{X}_1, \mathbf{X}_3) + d(\mathbf{X}_2, \mathbf{X}_3) \tag{12.3}$$

$$d(\mathbf{X}_1, \mathbf{X}_2) > 0 \tag{12.4}$$

$$d(\mathbf{X}_1, \mathbf{X}_2) = 0 \quad \text{if } \mathbf{X}_1 = \mathbf{X}_2 \tag{12.5}$$

These conditions are met if the distance between the points labeled i and j is defined as the normal euclidean distance:

$$d_{ij} = \left[\sum_{k=1}^{n} (x_{ik} - x_{jk})^2 \right]^{1/2} \tag{12.6}$$

For real situations points that represent similar objects may not only be relatively close to one another but also form clusters. A cluster of points is characterized by isolation and coherence; a cluster has a relatively high local density of points and a lower density outside the cluster. As an example of clustering, think of mass spectra of organic compounds. One might expect the points representing the mass spectra of ketones to fall in one relatively limited region of space and the spectra of non-ketones to be elsewhere. The ketone spectra points could form a cluster. This expectation is indeed met with real spectra of real organic molecules. The general properties of clusters will often be related to the types of pattern recognition techniques that are effective for a particular problem. Properties include (a) the central tendency, (b) the dispersion, (c) the shape—sheets, spheres, etc., (d) the connectivity, and (e) the sizes of the gaps between clusters.

12.2.3. *K*-Nearest-Neighbor Classification

Given a set of points in the n-dimensional pattern space and a formalized definition of similarity such as equation (12.6), then unknown points can be classified using the K-nearest-neighbor method. To classify the point X_i, the distance between it and each of the points in the data set is computed. The class of the unknown is taken to be the same as the nearest point ($K = 1$) or the class most often represented among the three nearest points ($K = 3$), etc. This procedure is logically equivalent to closest-matching schemes often employed in information retrieval systems.

The K-nearest-neighbor classification procedure suffers from the fact that all pattern vectors must be stored and that many computations must be performed to make a classification. However, it has been shown that the error rate of the first nearest neighbor method is at most twice that of the optimal Bayes' error in which all the underlying probabilities are known.[13] Its conceptual and computational simplicity are also appealing. Later investigations with nearest-neighbor classification have shown that a carefully chosen subset of the data sample can be used, thereby reducing storage requirements and computational burden.[14]

12.2.4. Decision Surfaces

One method of accomplishing the classification of pattern points is to find groups that belong in the same class and locate decision surfaces between them. In simple two-space this amounts to drawing lines (not necessarily straight ones) between points of different classes, for example, cows can be separated from horses by building a fence.

In hyperspace it might be expected that pattern points corresponding to compounds with similar characteristics would cluster. For example, the pattern points representing a group of mass spectra of alcohols would be expected to cluster in one limited region of the hyperspace, and pattern points representing the mass spectra of a group of alkenes would be expected to cluster elsewhere. This expectation is often met by sets of points representing chemical data such as mass spectra. When clusters occur, it is often possible to locate decision surfaces which pass between the clusters. The simplest such decision surface is a hyperplane with the same dimensionality as the hyperspace being employed.

While this hyperplane need not be linear or "flat" with respect to the dimensionality of the space, a convenient mathematical simplification occurs when the decision surface is linear and passes through the origin. Under these conditions the hyperplane can be represented by a normal vector from the origin. Or, more formally, every vector from the origin defines a plane that is the locus of points perpendicular to the vector.

Since it is very convenient to have the decision surface pass through the origin of the pattern space, it is worthwhile to ensure that this will always be possible with no loss of separability. An extra, orthogonal dimension is added to the pattern space, and all the pattern vectors are augmented with an $(n + 1)$th component. The $(n + 1)$th component of the pattern vectors can be assigned any value, but it is usually given the value of unity.

In addition to being able to represent a linear decision surface by specifying its normal vector, another important feature arises for this simple situation. The dot product of the normal vector **W** and a pattern vector **X** defines on which side of the hyperplane a given pattern point lies:

$$\mathbf{W} \cdot \mathbf{X} = |\mathbf{W}|\,|\mathbf{X}| \cos \theta \tag{12.7}$$

where θ is the angle between the two vectors:

$$\begin{aligned} \cos \theta &> 0 \quad \text{for } -90^\circ < \theta < 90^\circ \\ &< 0 \quad \text{for } 90^\circ < \theta < 270^\circ \end{aligned}$$

Since the normal vector is perpendicular to the plane, all patterns having dot products that are positive lie on the same side of the plane as the normal vector, and all those with negative dot products lie on the opposite side.

(Points with zero dot products lie in the plane, which constitutes another definition of the location of the plane.)

While decision surfaces need not be linear, their simplicity when linear is appealing. Additionally, it can be shown that more complex decision surfaces can be implemented by linear decision surfaces preceded by appropriate preprocessing.

12.2.5. TLUs as Binary Pattern Classifiers

The TLU is so named because it is analogous to a logical circuit element that exists in one of two states depending on whether an input is above or below a certain level (threshold).

When a binary decision is to be made, that is, when patterns are to be placed in one of two classifications, a TLU is one frequently successful method. In general some function must be used that generates one of two results based on the input.

While linear TLUs are frequently used and make good examples, TLUs of any functionality are possible. The only requirement is that a clear discrimination can be made between the two classifications of interest.

A simple bimetallic thermostat as used in a typical household heating system is a good example of a TLU. The input, room temperature, is transformed by the bending of bimetal and, as long as it is above a certain threshold, which can be defined in degrees, the thermostat is at zero voltage and does not activate the relay to increase the furnace output. When the temperature drops below the desired threshold, a voltage is generated that can be used to activate the furnace.

By using hyperspace notation, a TLU for pattern vectors can be described as any algorithm that generates two different states from the vector as an input. While not required, zero is often used as a threshold for mathematical convenience. The linear discriminant function described earlier can be used as a TLU with great convenience. As shown above, a normal vector **W** may describe the discriminating hyperplane; the dot product of it and a pattern vector is positive for pattern vectors on the same side of the plane as the normal vector, and negative for those on the opposite side. Hence, with a threshold of zero, the linear discriminant function defines a classification space with points lying in two groups as defined by the hyperplane normal to the discriminating vector.

Note that no real restriction has been placed on the classification process by having only two results, because several TLUs may be used in conjunction to give any desired degree of discrimination. (Later examples show this process.) Coin-operated vending machines use the sizes of coins to make a series of binary decisions, for example.

Another aspect of interest in the use of linear discriminant functions as TLUs is the concept of the weight vector. A second and equivalent way to define the dot product of two vectors is

$$\mathbf{W} \cdot \mathbf{X} = |\mathbf{W}|\,|\mathbf{X}| \cos \theta = \omega_1 x_1 + \omega_2 x_2 + \cdots + \omega_n x_n + \omega_{n+1} \tag{12.8}$$

that is, each of the components of W weights each of the terms of X. When used as a TLU, this weighting yields discrimination by causing the scalar product to fall above or below the threshold.

12.2.5.1. Training of TLUs Using Error Correction Feedback

As seen above, a TLU can be used to dichotomize a set of data represented as points or vectors in hyperspace. The problem then becomes one of finding a successful dichotomizer for a given set of classifications. This is what was meant by the earlier statement concerning the transformation of pattern space into classification space.

In many cases it is possible to classify data successfully into two classes with a linear discriminant function. However, finding such a function for multidimensional data poses an interesting problem. It is not possible simply to plot a high-dimensional graph and look at the points. Furthermore, it is frequently impractical to calculate all possible decision surfaces in order to find a successful one. However, a heuristic method has proven quite successful in many cases.

This heuristic approach is based on selecting a starting classification surface or discriminant function, either arbitrarily or by some approximation scheme, and then "training" the classifier by modifying it as it accumulates experience in making its decisions. This training is done with a set of patterns for which the classifications are known (the training set). The patterns are presented one at a time to the classifier being trained and, when incorrect classifications are made, the decision surface is altered in order to correct the error.

There are several schemes that are called error correction feedback. One method, which can be shown to converge for any linearly separable set of data, corrects the decision plane by reflecting it about the misclassified point. As described earlier, the dot product of the weight vector and a pattern vector gives a scalar whose sign indicates on which side of the decision surface the pattern point lies:

$$\mathbf{W} \cdot \mathbf{X} = s \tag{12.9}$$

(An arbitrary decision must be made as to which subset of the data is to be called the positive class and which the negative class.) When pattern i of the training set is misclassified, then

$$\mathbf{W} \cdot \mathbf{X}_i = s \tag{12.10}$$

in which s has the incorrect sign for classifying X_i. The object is to calculate an improved weight vector $\mathbf{W}'$, such that

$$\mathbf{W}' \cdot \mathbf{X}_i = s' \tag{12.11}$$

where the sign of the scalar result s' is opposite what it was previously. The new weight vector is calculated from the old one by adding an appropriate multiple of X_i to it:

$$\mathbf{W}' = \mathbf{W} + c\mathbf{X}_i \tag{12.12}$$

Combining equations (12.11) and (12.12) gives

$$s' = \mathbf{W}' \cdot \mathbf{X}_i = (\mathbf{W} + c\mathbf{X}_i)\,\mathbf{X}_i \tag{12.13}$$

which can be algebraically rearranged to give

$$c = \frac{s' - s}{\mathbf{X}_i \cdot \mathbf{X}_i} \tag{12.14}$$

It remains only to choose a value for s' to complete the derivation. An effective method is to let $s' = -s$. This moves the decision surface, so that after the feedback correction the point X_i is the same distance on the correct side of the decision surface as it was previously on the incorrect side. If $s' = -s$ is put into equation (12.14), then

$$c = -2s/\mathbf{X}_i \cdot \mathbf{X}_i \tag{12.15}$$

and $\mathbf{W}'$ can be calculated directly by using equations (12.12) and (12.15).

The training procedure involves iterating over all the pattern points in the training set and correcting the weight vector whenever an error is committed, until the discriminant function converges on one that successfully classifies all the points. The process is completely analogous to asking repeatedly a series of questions until the device being trained responds correctly to all of them. It is this type of approach that had led to the use of such terms as "learning" and "learning machine," that is, the definition of learning used in this sense is the improvement of performance with experience.

As mentioned above, this error correction procedure can be shown to find a solution if one exists. Therefore the weight vector can be initialized arbitrarily, although it is obviously better practice to use whatever information is available to estimate a starting weight vector.

A word of caution is in order with respect to the question of the training set size N versus the number of components per pattern d. While the exact ratio necessary for obtaining valid results is in question, it is generally agreed that the ratio N/d should be as large as possible. An added problem is that the value of d is not so important as the number of descriptors in the patterns necessary for separation of the two classes. This number is not usually known *a priori*. It is now generally accepted that, if N/d is greater than approximately 3, the results obtained will not be in question.

12.2.5.2. Properties of TLUs

Four properties are discussed with respect to TLU and other pattern classification devices. These are recognition, convergence rate, reliability, and prediction.

Recognition is defined as the ability of a discriminant function to correctly classify those patterns with which it was developed, that is, its training set. How well can it answer questions that were used to develop it? This is analogous to asking on examinations virtually the same questions that were covered in class; however, it may have great utility, by reducing the classification procedure to a simple mathematical operation rather than requiring the retention and retrieval of a large library of data. This may be much more economical in many applications. In this book 100% recognition is analogous to perfect or error-free training of a classifier.

Convergence rate refers to the speed with which a training algorithm converges toward 100% recognition. This is of interest with regard to the economics of developing useful pattern classifiers. Rapid convergence is useful to those who have limited computer budgets.

While slow convergence does not mean a problem is insolvable, it often means that it cannot be done economically. Therefore in many cases it may be necessary to trade or compromise other advantages of classifiers for improved convergence rate.

Reliability refers to the ability of a classifier to classify correctly data that were used in its development but that are classified after undergoing some distortion. All processes of information transfer have some noise level. For example, while a classifier might have been developed with mass spectra from a certain standard set of chemical compounds, repeated runs of the same compound do not produce exactly the same mass spectrum. The degree of reliability indicates the classifier's ability to handle such distorted or noisy data. Furthermore, it is related to the redundancy of the decision procedure itself. Certainly, human pattern recognition systems have incredibly high reliability for certain classifications. Only a small subset of a large body of possible data is necessary for a human being to recognize pussycats optically. Such fundamental properties as size, color, and weight have little to do with the decision process. Rather, general features of shape, type of movement, and so on, are more important. It is certainly important for the application of pattern recognition to chemical problems that noise levels may be tolerated. Furthermore, this is one of the inherent advantages that pattern recognition techniques may have over more conventional direct comparisons of data with library compilations.

Prediction is probably the most exciting aspect of pattern recognition applications to chemistry. Prediction refers to the ability of the classifier to classify correctly patterns that were not members of the training set.

This amounts to the "unknown" test. If a pattern classifier can be shown to predict unknown data successfully, then it is implied that some of the fundamental relations between data and their classifications have been extracted in the development of the discriminant function. Prediction may not only answer questions on the unknown data, but the way it is accomplished may suggest fundamental relations and lead to further understanding of cause-and-effect relations in chemistry.

12.2.6. Preprocessing and Transformations

The success of pattern recognition techniques can frequently be enhanced and/or simplified by prior treatment of the data before the classification stage. In terms of the n-dimensional euclidean space, called the pattern space, being used, the goals of preprocessing or transformation are (1) to separate clusters of related points further apart to make classification easier, (2) to reduce the dimensionality of the pattern space to make classification more economical.

These two goals are often contradictory. However, if separation between classes is improved, classification becomes easier. Furthermore, reducing the dimensionality of data makes classification more economical, since the time required for calculation of the discriminant function is frequently proportional to the dimensionality of the data. Also, reduction of dimensionality can reduce the size of the data set required to avoid underdetermined situations.

Several terms frequently used in conjunction with the pretreatment of data are preprocessing, transgeneration, or transformation; feature selection; and feature extraction. Preprocessing and transgeneration refer to altering the individual components of pattern vectors. While various functionalities may be applied, this constitutes what we call a dimensionally independent transformation. Feature selection refers to selecting those features of the raw data believed to be more important. Feature extraction refers to combining individual primitive features into higher-level ones. This involves transformations that do not preserve dimensional independence, such as for the formation of cross terms, Fourier transformations, and processes such as factor analysis.

The simplest preprocessing involves normalization and other methods for scaling the components of the pattern vector. One such normalization involves setting the sum of the components of each pattern vector (or their squares) equal to an arbitrary, convenient constant. Another procedure involves allowing the weight of a particular descriptor (dimension) to be inversely proportional to the variance of that descriptor in the pattern space. A more sophisticated procedure involves utilizing a covariance matrix to set up a matrix equation that is solved for eigenvectors and eigenvalues. This yields a set of orthogonal dimensions that define a new pattern space, which is a rotation of the old one, in which classification might be more easily done. This procedure, known as principal components analysis or Karhunen–Loeve analysis, can also be used to decrease the dimensionality of pattern vectors. Only the new, transformed dimensions that have large eigenvalues associated with them are saved, while the remainder are discarded. These are just a few of the linear transformations that have been utilized in pattern recognition systems.

Several more complex transformations have also been employed. For example, pattern vectors can be subjected to a Fourier transformation, and then the power spectrum can be developed. Iteratively defined transformations have also been developed. This involves minimizing an error criterion iteratively, for example, the difference between the distances between all pairs of points in the original pattern space and in the new, lower-dimensional pattern space. Pattern vectors can also be represented by polynomial expansions.

Templates or prototype patterns have been used for comparison with the patterns to be classified to identify important features. Interactive techniques, sometimes involving computer graphics and mapping routines, have been used. Calculations of statistical parameters such as moments and histograms directly from patterns have been made. As is evident from these few paragraphs, the number of feature extraction–preprocessing algorithms is enormous. And it is growing rapidly, since the methods chosen for application to any problem are highly dependent on the problem. Depending on how well it is done, the preprocessing step can lead to success or failure of a pattern recognition study. It is a widely held opinion that the feature extraction–preprocessing stage is the area in which the major new advances will come in the study of pattern recognition systems.

A review of the wide variety of feature extraction and feature selection methods reported can be found in the review by Kanal.[8]

12.3. FOURIER AND HADAMARD TRANSFORMS IN PATTERN RECOGNITION

In a definitive paper on generalized spectral analysis, Andrews and Caspari[21] have clearly discussed the relationships between a variety of transforms including the Fourier, Hadamard, Haar, generalized Walsh, and generalized Haar transforms. In particular, they described the general spectral analysis potential of employing various transforms. In the present context, pattern recognition use of only two of these, the Fourier and the Hadamard, will be discussed in detail. They can be shown to represent the two extreme variations of the several categories of related transforms.

One of the earliest pattern recognition applications of *translation invariant* transforms such as the Fourier or Hadamard transforms was the use of another type, the autocorrelation transform, for character recognition work.[22] In chemical pattern recognition, Fourier and related transforms have been applied in several different ways. First, Fourier transforms have been examined as possible tools for feature reduction in mass spectrometry[23,24] and NMR spectrometry.[25] Second, frequency domain ^{13}C NMR spectra have been processed via Fourier transformation in order to permit evaluation of the possibility of direct pattern recognition analysis of time domain experimental data.[24] Third, it has been observed that the convenient relation between the autocorrelation function $A(f)$ and the Fourier transform of the square of a time series $G(t)$, as in equation 12.16 provides an alternate means of carrying out autocorrelation transforms for feature selection purposes[26]:

$$A(f) = \int G(t)^2 e^{i2ft} dt \tag{12.16}$$

A generalized Walsh transform has been used to construct nonlinear discriminant functions for mass spectral interpretation.[27] Hadamard transforms have also been explored as a preprocessing method for both mass[28] and ^{13}C NMR[29,30] spectral data, and combined results using discriminants derived from Fourier and Hadamard transform processed ^{13}C NMR spectra have been employed in a multiple discriminant function analysis of NMR data.[31]

12.3.1. Feature Reduction

Jurs[23] investigated the merits of using Fourier-transformed mass spectra for development of binary pattern classifiers by the linear learning-machine method. The impetus behind that study and subsequent similar studies of the use of Hadamard-transformed mass spectra[28] was that fact that both discrete Hadamard and Fourier transforms of that type of spectral data result in functions in which each point arises from a linear combination of *all* points in the original spectrum. Thus, it was expected that the desirable, but often mutually exclusive, goals of reducing the dimensionality of the problem and simultaneously increasing ease of discriminant function separation might be better achieved via use of the transformed representations. Discarding some of the descriptors from Fourier or Hadamard transform patterns is equivalent to losing *some* information about *all* of the original spectrum, rather than losing *all* information about *some* of the original spectrum, which results if descriptors are discarded from the untransformed patterns.

Both transforms were also applied to ^{13}C NMR data[25,29] and their use for feature elimination examined in that application for exactly the same reasons. In NMR, particularly, the elimination of all information about some of a spectrum is likely to cause difficulty in successful pattern recognition. The reason for this is obvious. If mass spectra at unit resolution with approximately 250 resolution elements and 50–60 peaks per spectrum are compared with ^{13}C NMR spectra at 1 ppm resolution ($\sim$200 resolution elements) and with 3–15 peaks per typical spectrum, it is clear that the number of peaks per resolution element in mass spectra is much higher than in ^{13}C NMR. Thus, it is more possible to discard mass spectral descriptors while preserving sufficient information to make categorizations. With NMR data, significant information loss can more readily occur in the feature selection process. In fact, use of selected descriptors chosen from untransformed spectra can result in discarding *all* information about spectra with small numbers of peaks. Of course, this is also true with mass spectra. However, as explained above, this is a less likely occurrence with those data. This reasoning might further suggest that there would be more to gain by pattern recognition using transformed ^{13}C NMR data than with use of transformed mass spectral data. However, the experimental data, shown in Table 12.1, do not yield this result.

The available data suggest that only modest improvements in feature reduction are to be expected as a result of using Fourier transformed mass spectra rather than intensity-coded spectra. On the other hand, the one study available on the use of Hadamard transform processing for mass

TABLE 12.1. Effects of Feature Elimination on Prediction Accuracy for Mass Spectral and ^{13}C NMR Data

				Number of features/% correct		
Data	Reference	Question	Method	Intensity	Fourier	Hadamard
Mass spectra	23	$2(nC) \geq nH$	LLM[a]	111/93.9	75/94.5	—
	28	6, 7, or 8 C	KNN[b]	128/87.9	—	128/94
				64/80.2	—	64/93
^{13}C NMR	25	Aldehyde and ketone	LLM[a]	90/96	142/96	108/96
		Aliphatic alcohol		126/79	193/87	138/92
		Carboxylic acid		92/95	135/87	107/91
		Phenyl		106/80	202/81	104/85

[a] The linear learning-machine method used to develop discriminant functions. Feature elimination was performed by discarding those descriptors whose signs changed in the final weight vector when the starting weight vector was first initialized with +1 then with −1 values for all elements.

[b] The K-nearest-neighbor method ($K = 3$) was used for classification. Feature elimination was carried out by deleting the lowest intensities. Deletion of high and low sequences was also examined and yielded essentially identical results.

spectra of hydrocarbons shows that use of this form of the data is far less susceptible than intensity data to degradation of prediction performance as numbers of descriptors are reduced. In none of these cases did prediction performance improve as a result of feature elimination.

Turning to the ^{13}C NMR results, where direct comparisons of use of intensity, Fourier-transformed, and Hadamard-transformed representations of the same data are possible, the picture is less clear. From the standpoint of feature reduction, in each of the four functional group problems, the Hadamard-transformed data yield more reliable predictors with fewer features than the Fourier-transformed data. Contrary to expectation, there is not clear marked improvement for either use of Fourier- or Hadamard-transformed spectra over the intensity-coded spectra. In fact, in two of the cases, intensity-coded spectral data produced discriminant functions that were equal or superior to either of those derived from the Fourier or Hadamard representations.

Considering the sparsity of data, it would be premature to conclude that use of these transforms is not efficacious for either mass or NMR spectra feature selection. However, the results of Table 12.1 do seem to indicate that significant advantage is not to be obtained in these applications.

12.3.2. Pattern Recognition Analysis of NMR Data

In an early paper on pattern recognition analysis of proton NMR spectra, it was suggested that the free-induction decay waveform available in the pulsed NMR experiment could be directly analyzed to produce molecular structure information.[26] However, because of the need to recognize patterns due to coupling, the direct use of FID measurements was contraindicated. Fortunately, proton noise-decoupled ^{13}C NMR spectra are generally devoid of patterns arising from this source and their FID data therefore might be directly analyzed. Pattern recognition analysis by the linear learning machine method requires a data base to be used as the training set and, if results are to be evaluated, additional data to serve as test (or prediction) sets. Brunner and co-workers used an inverse fast Fourier transform to generate the necessary data for evaluation of the usefulness of chemical pattern recognition methods in this way.[25] Two different "simulated free-induction decay" data sets, each containing 500 transformed spectra, were produced. The first of these was derived using only the presence or absence of peaks in each of 200 1-ppm resolution elements as source data. The second set utilized intensity-encoded information that were normalized so that each spectrum's summed intensities were a constant. Hadamard transformation of the intensity-encoded spectra and of the simulated FID data, based on intensity-encoded spectra, was also carried out to produce two additional data bases for comparative pattern recognition analysis study. Predictors were tested on sets of experimental free-induction decay data of known compounds. Tables 12.2 through 12.5 summarize the results of that study.

12.3.2.1. Simulated Free-Induction Decay Analysis

Table 12.2 summarizes the results of training, (using 400 spectra) listing both the outcome of using the full real parts of the *simulated FID* function (both positive and negative real-intensity values) and the truncated simulated FID data (using only the positive real-intensity values of the functions). Note that this approach includes the redundant "negative time" information generated by the mathematical procedure. Weight vectors were developed for both the FIDs from the peak–no-peak (PNP) coded data and the normalized absolute intensity (NAI) data sets. The latter of the two data sets yielded significantly better results. In none of the cases was it possible, within the 8000 feedback constraint imposed, to train to 100% using the PNP derived data. Predictions on the remaining 100 spectra using these incompletely trained weight vectors were uniformly poor, although they did improve somewhat when negative values were eliminated from the initial set. It appears necessary to retain intensity information if useful interpretation accuracy is to be attained.

When the training was based on the NAI data, complete training was difficult, but possible, in several cases using the full simulated FID. Nevertheless, elimination of the negative values significantly improved both rate of training and the percentage of correct predictions for almost all of the structural questions. Whereas, before, complete training could not be obtained for three of the six discriminant functions, now all were able to achieve perfect recognition for the training set. Performance on the 100-member test set was now approximately 90% correct.

To test whether the linear discriminant functions generated to answer structural questions using *simulated* FID data would be applicable to *real* FID data, natural abundance ^{13}C FID's for five representative compounds were analyzed. Six additional compounds (chosen at random from the spectra collection) served for comparison purposes to test the functional group predictors. Both direct interpretation of the experimental FID data and interpretation of the same data via a time to frequency conversion (Fourier transform), normalization, and inverse transform followed by application of the weight vectors was carried out, using vectors derived for positive simulated FID data from the normalized intensity starting set. Table 12.3 summarizes these results.

For compounds drawn from the set of spectra from which the weight vectors were developed the results are reasonably good (compounds 6–11). In each of these six cases, one important structural feature was correctly identified directly from the simulated FID pattern. In one case (benzoic acid) a functional group was erroneously predicted to be absent. Results

TABLE 12.2. Error Correction Feedback Training for Functional Group Identification from Simulated ^{13}C NMR Free-Induction Decay Data

			Initial results[b,c]		After iterative feature elimination[d]		
Functional group[a]	Preprocessing method	Initial data	Number of feedbacks	% Correct unknown set	Features retained	Number of feedbacks	% Correct unknown set
Aldehyde and ketone	None[e]	PNP[g]	—[i]	39	—	—	—
	None	NAI[h]	3032	87.5	120	—	87.5
	PV[f]	PNP	—	66	—	—	—
	PV	NAI	271	94.5	142	153	96
Aliphatic alcohol	None	PNP	—	57	—	—	—
	None	NAI	—	75.5	—	—	—
	PV	PNP	—	63	—	—	—
	PV	NAI	1167	85	193	1164	87
Carboxylic acid	None	PNP	—	33	—	—	—
	None	NAI	514	92	116	1219	91
	PV	PNP	—	77	—	—	—
	PV	NAI	419	87.5	135	336	88.5

Alkyl bromide	None	PNP	—	70	—	—	—
	None	NAI	617	92	99	1592	94.5
	PV	PNP	—	70	—	—	—
	PV	NAI	183	95.5	117	128	97.5
Alkyl chloride	None	PNP	—	40.5	—	—	—
	None	NAI	290	100	97	270	96.5
	PV	PNP	—	78	—	—	—
	PV	NAI	140	96.5	138	111	96.5
Phenyl	None	PNP	—	54	—	—	—
	None	NAI	—	75.5	—	—	—
	PV	PNP	—	53	—	—	—
	PV	NAI	2601	81	202	2158	81

[a] Total set of spectra is 500; numbers of spectra containing each of these functional groups are 29, 78, 31, 18, 14, and 130, respectively.
[b] Using 400 spectra in the training set; the remaining spectra comprise the unknown set.
[c] Results are the average of those obtained using a weight vector initialized to -1 and another initialized to $+1$.
[d] After successively eliminating features whose final weight vector elements have different algebraic signs for $+/-$ initialization.
[e] Both positive and negative real values used; 256 features initially.
[f] Only positive real values used, negative values set equal zero; 256 features initially.
[g] Peak–no peak coded data set used.
[h] Normalized absolute intensity data set used.
[i] Failed to train to 100% within 8000 feedbacks or nonconvergence detected. No feature elimination attempted when this occurred on original training.

TABLE 12.3. Functional Group Predictions Using FID and Hadamard-Transformed FID Weight Vectors[a]

Unknown[b]	Actual FID[c]	Simulated FID[d]	Hadamard (FID)[e]	Hadamard (PNP)[f]
1. *p*-Methylacetophenone	ϕ	ϕ	OH	—
2. Cyclohexane	[g]	—	—	C=O
3. *i*-Butyl ketone	—	—	OH	C=O
4. 2-Chloropropionic acid	—	—	—	—
5. *t*-Butanol	—	–OH	—	OH
6. *t*-Butanol		–OH	—	OH
7. 1, 2-Dibromopropane		–Br	—	Br
8. Butyraldehyde		C=O	OH	C=O
9. 4-Methyl-2-pentanol		–OH	—	ϕ
10. Benzoic acid		$-\phi$	—	ϕ
11. Toluene		$-\phi$	OH	ϕ

[a] Procedure followed was to apply each of the appropriate six functional group weight vectors to the unknown data. Listed are those groups predicted to be *present* for each unknown using forms of the data indicated at the heads of the columns.
[b] Data for compounds 1–5 were collected with the XL-100-15 spectrometer system.
[c] Using only the first 256 points of the compound's actual FID data and the PV (NAI) FID weight vectors.
[d] Generated from the frequency domain representation as described in the experimental section and using the PV (NAI) FID weight vectors.
[e] Generated by applying the Hadamard transform to the 256 real parts of the simulated FID of NAI data. The PV Hadamard weight vectors were used.
[f] Generated by applying the Hadamard transform to peak–no peak spectra augmented with 65 zeros to yield a 256-point transform. HSP vectors using positive values only were applied (see reference 29).
[g] None of the six functional groups predicted to be present in the unknown.

were less impressive with the actual FID data, in that only one compound (cyclohexane) yielded error-free prediction. The others were erroneously found to have fewer functional groups than they actually had. Treatment of the actual FID data, in a manner analogous to the other data, to generate simulated FID spectra, which were then analyzed, did improve the prediction accuracy slightly. This suggests that discriminant functions based on a data set of actual FID spectra would be required, if the desired performance (of a level comparable to that for compounds drawn from the original catalog) is to be attained. Thus, although the feasibility of direct analysis of FID data was established by this study, the lack of a large library of experimental FID data may limit such application. Table 12.4 supports the feasibility of using FID data directly, if discriminant functions are ultimately developed, since it shows that prediction performance is about the same with this form of data as with the other representations investigated.

12.3.2.2. Hadamard-Transformed Data Analysis

Table 12.5 lists corresponding facts for Hadamard-transformed data. It is seen, as for the FID data, that training is significantly more difficult

TABLE 12.4. Comparison of Error Correction Feedback Training for Functional Group Identification Using Frequency, Hadamard-Transformed Frequency, and Simulated Free-Induction Decay ^{13}C NMR Spectra

Functional group[a]	Intensity coding	Frequency features/% correct[b]	Hadamard-transformed[c] frequency features/% correct	Simulated FID[c] features/% correct	Hadamard-transformed[c] simulated FID features/% correct
Aldehyde and ketone	PNP	17/100	78/99	256/66	—
	NAI	90/96	108/96	142/96	111/88.5
Aliphatic alcohol	PNP	100/91	119/88	256/53	—
	NAI	126/79	138/92	193/87	176/84.5
Carboxylic acid	PNP	88/95	97/92	256/77	—
	NAI	92/95	107/91	135/87	117/93
Alkyl bromide	PNP	96/96	89/95	256/78	—
	NAI	105/96	121/99	117/98	135/95
Alkyl chloride	PNP	92/96	106/98	256/63	—
	NAI	114/97	100/99	138/97	114/98
Phenyl	PNP	86/80	114/76	256/49	—
	NAI	109/80	104/85	202/81	256/78

[a] See Table 12.2 for description of the data set and explanation of notation.
[b] Minimum number of features obtained by feature elimination. % Correct refers to performance on 100 spectra not used in training.
[c] Setting all negative real intensity values equal to zero.

TABLE 12.5. Error Correction Feedback Training for Functional Group Identification from Hadamard Transform ^{13}C NMR Simulated Free-Induction Decay Data

		Initial results[b,c]		After iterative feature elimination[d]		
Functional group[a]	Preprocessing method	Number of feedbacks	% Correct unknown set	Features retained	Number of feedbacks	% Correct unknown set
Aldehyde and ketone	None[e]	1358	86.5	149	1368	88
	PV[f]	456	94	111	286	88.5
Aliphatic alcohol	None	—[g]	76	—	—	—
	PV	2170	82.5	176	1812	84.5
Carboxylic acid	None	334	91.5	129	291	91
	PV	376	90.5	117	253	93
Alkyl chloride	None	303	94	145	286	94
	PV	151	98.5	114	120	98
Alkyl bromide	None	195	98	118	125	91.5
	PV	314	94.5	135	267	95
Phenyl	None	—	79	—	—	—
	PV	5586[g]	78	—	—	—

[a] Total set of spectra is 500; numbers of spectra containing each of these functional groups are 29, 78, 31, 18, 14 and 130, respectively.
[b] Using 400 spectra in the training set; the remaining spectra surprise the unknown set.
[c] Results are the average of those obtained using a weight vector initialized to -1 and another initialized to $+1$.
[d] After successively eliminating features whose final weight vector elements have different signs for $+/-$ initialization.
[e] Both positive and negative Hadamard-transformed values used; 256 features initially.
[f] Only positive Hadamard-transformed values used, negative values set equal zero; 256 features initially.
[g] Failed to train to 100% within 8000 feedbacks, or converged so slowly that computer time limit exceeded. No feature elimination attempted when this occurred on original training.

when negative intensity values are retained in the Hadamard data set. Furthermore, in all cases except prediction of the presence of the phenyl group, use of only the positive elements of the transformed intensity data improved the prediction results for the test set. Even in the case of phenyl, although the results did not improve, they were not seriously degraded. When the Hadamard results are compared with those found for direct analysis of simulated FID data, it is seen that number of features retained and performance on the test set are not much different (i.e., approximately 90% correct predictions) for most of the structural questions examined.

12.3.2.3. Autocorrelation Transforms

In the first reported study of pattern recognition analysis of NMR spectra, it was noted that a serious hindrance to success in this endeavor was the translational variance of spectral patterns arising from coupling of protons.[26] Use of part of the positive portion of the autocorrelation transforms of simulated NMR spectra to develop weight vectors by the linear learning-machine method permitted removal of this variance. Consequently, perfect recognition of training set members became possible. Prediction performance was not extensively examined in this study. It was observed that the relationship between the autocorrelation transform and the Fourier transform of the square of the free-induction decay, equation (12.16) provides a ready route to such a transformed spectrum. Although this relation was not used in that or the subsequent study where the K-nearest-neighbor classification procedure was applied,[32] it is worth noting that use of the Fourier transform in this way can provide autocorrelation functions for pattern recognition purposes. In this latter study, autocorrelated spectra that were not linearly separable were interpreted using both the linear learning-machine and K-nearest-neighbor procedures. Significantly better results were obtained with the cluster analysis method.

12.4. CONCLUSIONS

The research thus far reported on applications of the Fourier and Hadamard transforms to pattern recognition analysis in chemistry seems to indicate that these kinds of transforms may have some utility in feature elimination applications, but that the results obtained may be very highly dependent upon the characteristics of the particular data to which they are applied. It has been demonstrated that direct analysis of time domain data by pattern recognition analysis is feasible. Accordingly, due to the predominance of frequency domain data in the literature, for NMR as well as other spectral methods, the inverse Fourier transform may be employed to produce the data bases necessary for development of pattern recognition based methods. Because of the limited number of investigations of use of such transforms for pattern recognition purposes, it is to be expected that more information on their utility will be forthcoming. For the present, it is safe to say that, although the potential of such preprocessing methods has been revealed, it has not yet been realized.

REFERENCES

1. M. Minsky, *Proc. IRE* **49**, 8 (1961).
2. R. J. Solomonoff, *Proc. IEEE* **54**, 1687 (1966).
3. C. A. Rosen, *Science* **156**, 38 (1967).
4. G. Nagy, *Proc. IEEE* **56**, 836 (1968).
5. M. D. Levine, *Proc. IEEE* **57**, 1391 (1969).
6. Special Issue on Technology and Health Services, *Proc. IEEE* **57**(11), 1969.
7. Special Issue on Digital Pattern Recognition, *Proc. IEEE* **60**, Oct. 1972.

8. L. Kanal, *IEEE, Trans.* **IT-20,** 697 (1974).
9. N. J. Nilsson, *Learning Machines,* McGraw–Hill, New York, 1965.
10. E. A. Patrick, *Fundamentals of Pattern Recognition,* Prentice-Hall, Englewood, Cliffs, New Jersey, 1972.
11. K. Fukunaga, *Introduction to Statistical Pattern Recognition,* Academic Press, New York, 1972.
12. W. S. Meisel, *Computer-Oriented Approaches to Pattern Recognition,* Academic Press, New York, 1972.
13. H. C. Andrews, *Introduction to Mathematical Techniques in Pattern Recognition,* Wiley–Interscience, New York, 1972.
14. J. T. Ton and R. C. Gonzalez, *Pattern Recognition Principles,* Addison–Wesley, Reading, Massachusetts, 1974.
15. B. R. Kowalski and C. F. Bender, *J, Am. Chem. Soc.* **94**, 5632 (1972); **95**, 686 (1973).
16. T. L. Isenhour, B. R. Kowalski, and P. C. Jurs, *Crit. Rev. Anal. Chem.* **4**, 1 (1974).
17. B. R. Kowalski, Pattern recognition in chemical research, *in: Computers in Chemical and Biochemical Research,* Vol. 2 (C. E. Klopfenstein and C. L. Wilkins, eds.), Academic Press, New York, 1974.
18. P. C. Jurs and T. L. Isenhour, *Chemical Applications of Pattern Recognition,* Wiley–Interscience, New York, 1975.
19. B. R. Kowalski, *Anal. Chem.* **47**, 1152A (1975).
20. P. C. Jurs, *Proc. Workshop Chem. Appl. Pattern Recognition,* Washington, D. C., May 1975.
21. H. C. Andrews and K. L. Caspari, *IEEE Trans. Comput.* **C-19**, 16 (1970).
22. L. P. Horwitz and G. L. Shelten, Jr., *Proc. IRE* **49**, 175 (1961).
23. P. C. Jurs, *Anal. Chem.* **43**, 1812 (1971).
24. L. E. Wangen, N. M. Frew, T. L. Isenhour, and P. C. Jurs, *Appl. Spectroscop.* **25**, 203 (1971).
25. T. R. Brunner, C. L. Wilkins, R. C. Williams, and P. J. McCombie, *Anal. Chem.* **47**, 662 (1975).
26. B. R. Kowalski and C. A. Reilly, *J. Phys. Chem.* **75**, 1402 (1971).
27. J. B. Justice, Jr., D. N. Anderson, T. L. Isenhour, and J. C. Marshall, *Anal. Chem.* **45**, 2087 (1972).
28. B. R. Kowalski and C. F. Bender, *Anal. Chem.* **45**, 2334 (1973).
29. T. R. Brunner, R. C. Williams, C. L. Wilkins, and P. J. McCombie, *Anal. Chem.* **46**, 1798 (1974).
30. R. C. Williams, R. M. Swanson, and C. L. Wilkins, *Anal. Chem.* **46**, 1803 (1974).
31. C. L. Wilkins and T. L. Isenhour, *Anal. Chem.* **47**, 1849 (1975).
32. B. R. Kowalski and C. F. Bender, *Anal. Chem.* **44**, 1405 (1972).

Chapter 13

Spectral Representations for Quantized Chemical Signals

Russell D. Larsen

13.1. INTRODUCTION

A cardinal precept of management theory is that innovation comes from the outside. What, of relevance, can a spectroscopist learn from related disciplines? In this chapter we attempt to show how some well-known ideas in statistics, mathematics, and electrical engineering contribute to a better understanding of what a spectroscopist is attempting to measure, how certain traditional data-handling techniques are special cases of more general mathematical methods, and how modern signal-processing techniques relate to specific chemical signals. Most of the concepts in this chapter will be illustrated by application to FT NMR, in which the primary measured quantity is the free-induction decay signal (FID). The illustrations involve ^{13}C FIDs but the techniques are applicable to other nuclei as well. In addition, most of what follows applies to FT–IR and the transform techniques discussed in the earlier chapters of this book.

The Name of the Game: Spectrum Estimation

Although it is not absolutely essential that one consider molecular physics to be stochastic or to consider that physical variables are random variables, it is rigorously correct to do so. Fruitful consequences result if we take this as our starting point for chemical signal analysis. The case

Russell D. Larsen • Department of Chemistry, University of Nevada—Reno, Reno, Nevada 89507. *Present address*: Department of Chemistry, University of Michigan, Ann Arbor, Michigan 48109

may be stated as follows. A dynamic chemical system, such as spin systems in a magnetic field, constitutes a stochastic process, the time behavior of which characterizes the process. Simple textbook examples of stochastic processes, such as Markov processes or weakly stationary processes, rarely occur, however, in actual physical situations. This is unfortunate, because a stationary stochastic process $\{X(t)\}$ is, loosely speaking, one that is completely specified by its mean and autocovariance function, i.e.,

$$\begin{aligned} E\{X(t)\} &= m, & -\infty < t < \infty \\ C(\tau) &= E\{X(t)\,X(t+\tau)\}, & -\infty < t < \infty \end{aligned} \tag{13.1}$$

(see the list of notation at the end of the chapter for explanations of symbols). Spectrum analysis depends on the ability to be able to convert time-based data into an estimate of the spectrum of the stochastic process.[1,2] One may convert the autocovariance function to the spectrum in virtually any situation including the common signal-plus-noise nonstationary processes. Generally, one considers that some pseudorandom process generates a member of an ensemble of possible realizations (this ensemble is the stochastic process). Each realization, a time series $X(t, \theta)$, is generated in accordance with the probability distribution of the random variable θ, $\theta \in \Theta$. It is common for a given, fixed θ to suppress the θ dependence and refer to the random variable $X(t)$ as a time series. While each $X(t)$ has a spectrum the goal is to *estimate the spectrum of the stochastic process*—the spectrum of the ensemble of all time series $\{X(t)\}$. The name of the game then is to obtain an *optimal* (best) *statistical estimate* of the spectrum of the stochastic process. These prefatory remarks are extremely important to the whole philosophy of chemical signal analysis. That which one does to a signal, experimentally and postexperimentally, must be done in accordance with the goal of optimizing the spectrum estimate. Digital filtering, smoothing, transformation, etc., all affect the spectrum estimate—generally in a positive way. In fact, one may argue that the most violence is done to the data *if no* postexperimental data analysis is undertaken, i.e., one may improve the spectrum estimate in many ways.

13.2. ^{13}C FID SIGNALS AND THEIR SPECTRA

Most of what we have to say will be illustrated by taking the ^{13}C FID as an illustrative, representative signal. The ^{13}C FID is remarkably rich as far as signals go and nicely illustrates certain aspects for which signal-processing techniques are applicable and useful.

Because of the complexity of an FID a chemist would not think of extracting information from it directly. One may say that the "pattern space" of the FID is too complex. The first step of pattern recognition, in fact, is a reduction of the dimensionality and complexity of the pattern by suitable transformation to a "feature space" in which the dominant features are more clearly recognizable. The Fourier spectrum of the FID achieves this feature extraction. The final step of a pattern recognition scheme is a classification based on these reduced features. In this chapter we are concerned only with the feature extraction operation.

The FID signal $s(t)$, say, is bounded, i.e., $|s(t)| < B$, and band-limited. Band-limited (BL) means that $s(t) \in L^2$ or that the Fourier transform of $s(t)$ vanishes above some upper frequency, or band limit, W Hz. W is called the single-sided bandwidth. If $s(t)$ is band-limited over (W_1, W_2) and sampled over a finite time interval T sec, it is known to be completely specified by its ordinates, which are spaced at uniform intervals of time, $1/2W$ seconds apart. The bandwidth W is given by $W = N/T$ Hz, where $N = N_2 - N_1$, and N_1 and N_2 are lower and upper limits to the fundamental frequency defining the bandwidth $(N_1\Omega, N_2\Omega)$ rad/s, where $\Omega = 2\pi/T$ rad/s [cf. equation (13.12)]. Because an FID signal has a band-limited Fourier spectrum, the signal $s(t)$ is an entire function in the time domain. That which will be said subsequently concerning zero-based descriptions of the FID rests upon such BL signals being entire functions. The signal $s(t)$ possesses $2WT$ degrees of freedom in time T. In conventional amplitude-sampling techniques, these $2WT$ coordinates are either Nyquist samples or Fourier coefficients. However, these $2WT$ coordinates may also be specified by the $2WT$ zeros of $s(t)$ in period T, a viewpoint that will be considered in subsequent sections.

We should note that the FID $s(t)$ is not a stationary signal, although it does have a zero mean. The FID contains sinusoidal, exponential, and noise components. It will be convenient to consider that the noise in $s(t)$ is Gaussian. The functional form of the FID for a single isolated spin i is given by[3]

$$\begin{aligned} M_y(t) = M_y(+0)\cos[(\Omega_i - \omega_1)t]\, e^{-t/T_2} \\ + M_x(+0)\sin[(\Omega_i - \omega_1)t]\, e^{-t/T_2} \end{aligned} \qquad (13.2)$$

One should consider that $M_y(t)$ is a random variable (time series) generated by a stochastic process. Interest concerns the process $\{M_y(t)\}$ and its spectrum estimate $F\{M_y(t)\}$.

13.3. ORTHOGONAL EXPANSIONS AND SPECTRAL REPRESENTATIONS

Given a signal s and a complete set of linearly independent functions $\{\Phi_n\}_{n=1}^{\infty}$, the spectrum of s with respect to $\{\Phi_n\}$ is the sequence of numbers

$\{\alpha_n\}_{n=1}^{\infty}$ where the α_n are coefficients of the expansion

$$s = \sum_n \alpha_n \Phi_n \tag{13.3}$$

In the case that the Φ_n are the usual trigonometrical functions $\{e^{jk\Omega t}\}$, the $\{\alpha_n\}$ are the coefficients of the Fourier spectrum, which is computed in the usual way. Recall that in practice the signal is discrete and consequently the transform actually calculated is the discrete Fourier transform, which is most often computed with a fast Fourier transform algorithm.

Although the spectra with respect to two different sequences $\{\Phi_n\}$ and $\{\psi_n\}$ of a given signal contain equivalent information and completely determine the signal, the facility with which one can utilize and interpret this information can vary radically. The apparent reasons that the trigonometric system is most commonly used are (a) its remarkable properties, which are well documented, (b) the corresponding spectrum has "physical" meaning for certain signals, (c) the rapidity with which the spectrum can be computed due to the advent of the fast Fourier transform, and (d) the long history of its successful use in the physical and engineering sciences. However, it appears that, depending on the nature of the signal, spectra with respect to other systems may be more useful. For example, very extensive and useful applications have been found for the Walsh and related systems (see References 4 and 5, and the references therein).

When analyzing a signal in the presence of noise, one considers the family of such signals to be a stochastic process. That is, for fixed t, $s(t)$ is treated as a random variable and hence the coefficients in equation (13.3), the α_n, are random variables. In this case one desires to compute certain statistical properties of the signal or its spectrum. To obtain reasonable estimates of these quantities certain *a priori* assumptions must be made concerning the process and a large number of samples taken. The number of samples can be considerably reduced through the use of spectral or data filters when estimating properties of s or its spectra, respectively. In the case of classical Fourier expansions see Brillinger[2] and references contained therein.

Depending on the nature of the signal, the use of certain complete systems, other than the traditional trigonometric one, can simplify the analytical and computational nature of the problems involved in estimating the signal or its spectrum.

For example, let $\{\Phi_n\}$ be an orthogonal system of Walsh functions. Walsh functions are complete, orthonormal sequences, which take on only the values $+1$ and -1.[6] Walsh-like functions are the rows of Hadamard matrices of orders $n = 1, 2, 4, 8, 12, \ldots$ (0 mod 4). Rigorous methods for generating Walsh functions, which are special cases of Walsh-like functions, have been given by Larsen and Madych.[7] These methods employ new matrix products, called permuted Kronecker products. Walsh functions are the rows of Hadamard matrices of order $n = 2^p$, $p = 0, 1, 2, \ldots$.

Given a complete system of Walsh functions one may calculate the spectrum of an FID with respect to this basis. Such a spectrum is a Walsh sequency spectrum. Examples of both the Fourier and Walsh spectra are given in Figures 13.1 and 13.2. The Walsh transform (WT), like the Fourier transform, is a unitary transform and is information preserving. As a consequence, no bandwidth reduction results from either a Fourier or a Walsh transformation of the signal. However the WT may also be performed by a fast computational algorithm, the FWT. Since the Walsh basis is considerably more simple in structure than the Fourier basis, Walsh spectra, via the FWT, can be computed much more rapidly than Fourier spectra. Neither the Fourier nor the Walsh transforms are optimal in the sense of totally decorrelating the transform coefficients.

While the Walsh spectrum is completely equivalent in information content to the Fourier spectrum, the Walsh representation is not suitable for FID signals [cf. equation (13.2)]. The resulting WT of an FID has more structure (lines) in the spectrum because additional Walsh coefficients

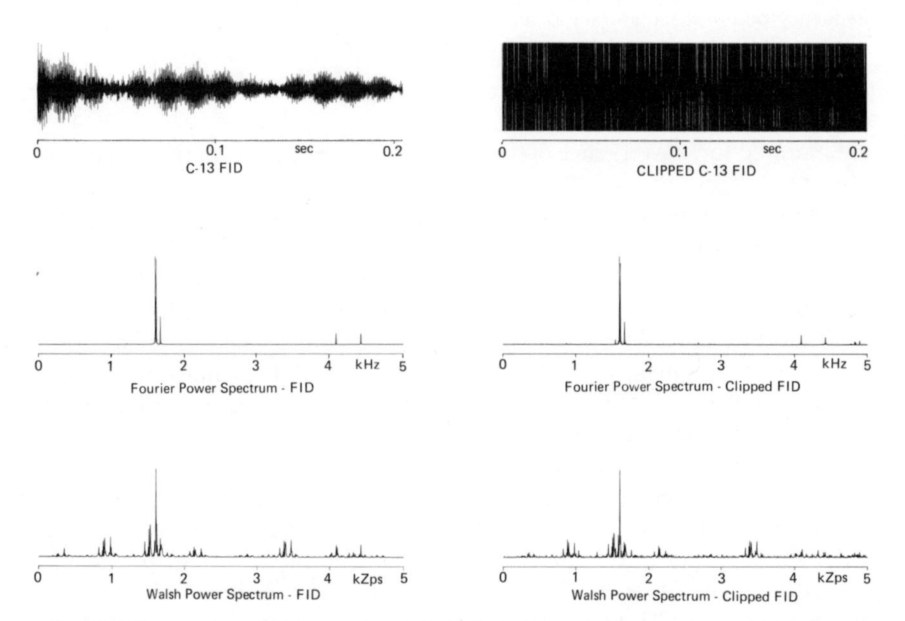

FIGURE 13.1. Free-induction decay signal, clipped free-induction decay signal, Fourier frequency power spectrum of FID and CFID, and Walsh sequency power spectrum of FID and CFID for ethylbenzene (average of 2048 scans).

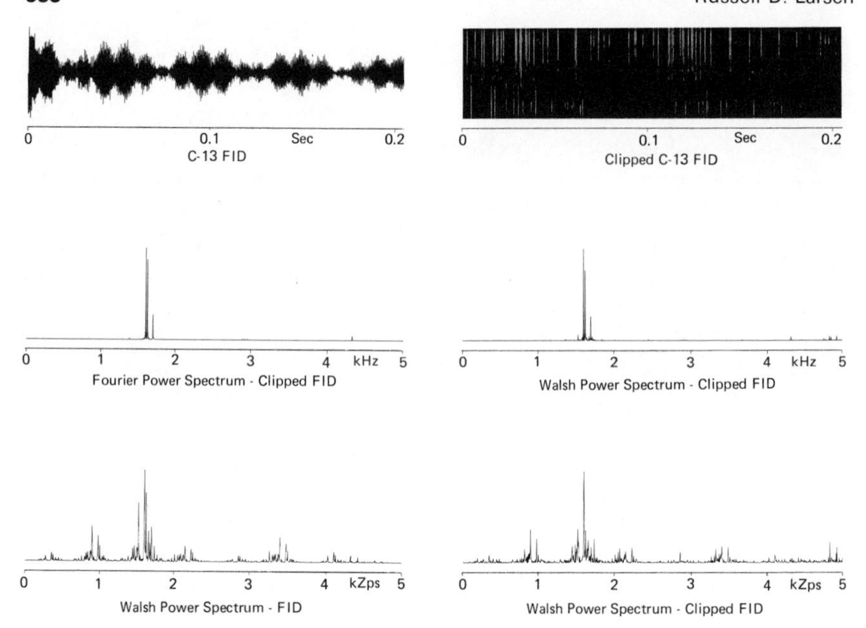

FIGURE 13.2. Free-induction decay signal, clipped free-induction decay signal, Fourier frequency power spectrum of FID and CFID, and Walsh sequency power spectrum of FID and CFID for toluene (average of 2048 scans).

are required for convergence of the orthonormal expansion of the sinusoidal $s(t)$.

One may, however, calculate NMR spectra with respect to other orthogonal systems that are related to the Walsh system. For example, first consider piecewise constant systems such as the various Walsh systems, the Haar system (elements with values $\pm 2^{k/2}$ and 0, $k = 0, 1, 2, \ldots$), and other Walsh-like systems. These systems all have constant segments over sections (pieces) of the domain of definition—over a period T; as illustrations, see the first column of Figures 13.3–13.5. It is possible to expand an FID signal in terms of any of these systems; these expansions have differing convergence properties. These piecewise-constant systems may be used as the starting point to form higher-order systems, i.e., piecewise-linear, piecewise-quadratic, and other piecewise polynomial systems.[8] These higher-order systems are formed by integration and orthogonalization of the parent piecewise-constant system. Piecewise-linear systems are shown in the second column of Figures 13.3–13.5 and piecewise-quadratic systems are illustrated in the third column of Figures 13.3–13.5. It is possible to show that certain expansions in terms of these higher-order systems may be performed more rapidly than even the Walsh and Haar expansions.[8]

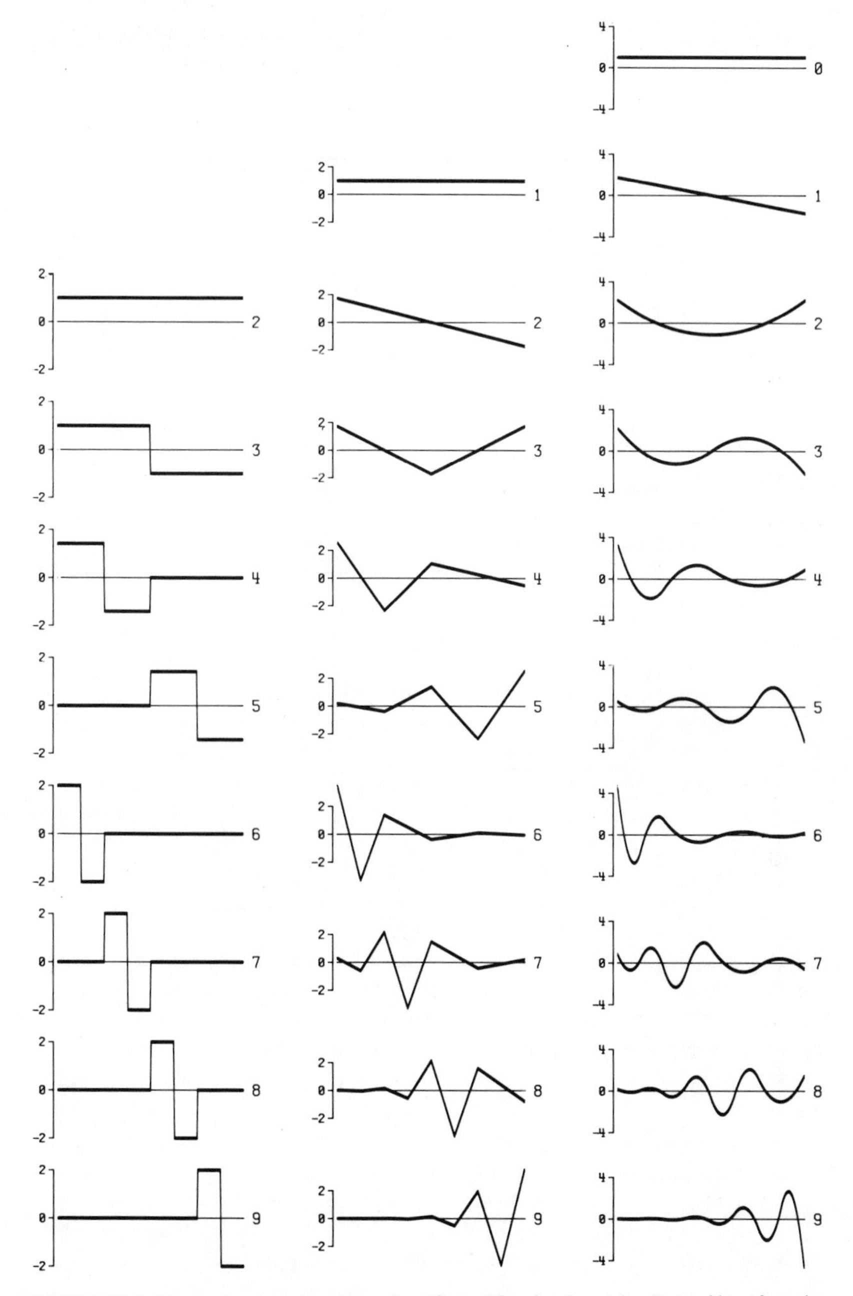

FIGURE 13.3. Piecewise constant Haar functions (Haar), piecewise linear Haar functions (Haar_I), and piecewise quadratic Haar functions (Haar_II). The Haar_I system is obtained from the Haar system by integration and orthogonalization.

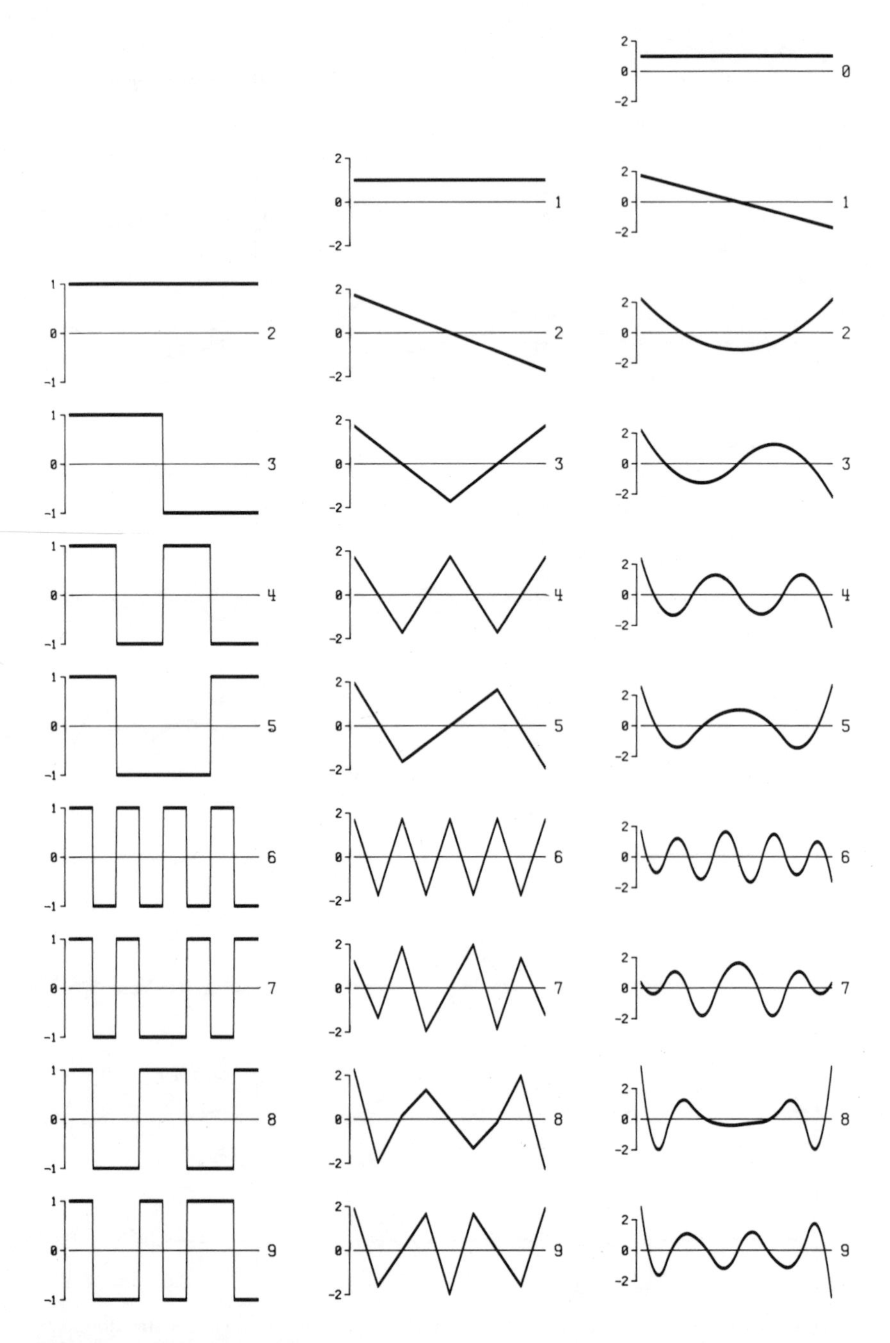

FIGURE 13.4. Piecewise-constant, linear, and quadratic Paley-ordered Walsh systems: Paley, Paley_{I}, and Paley_{II}.

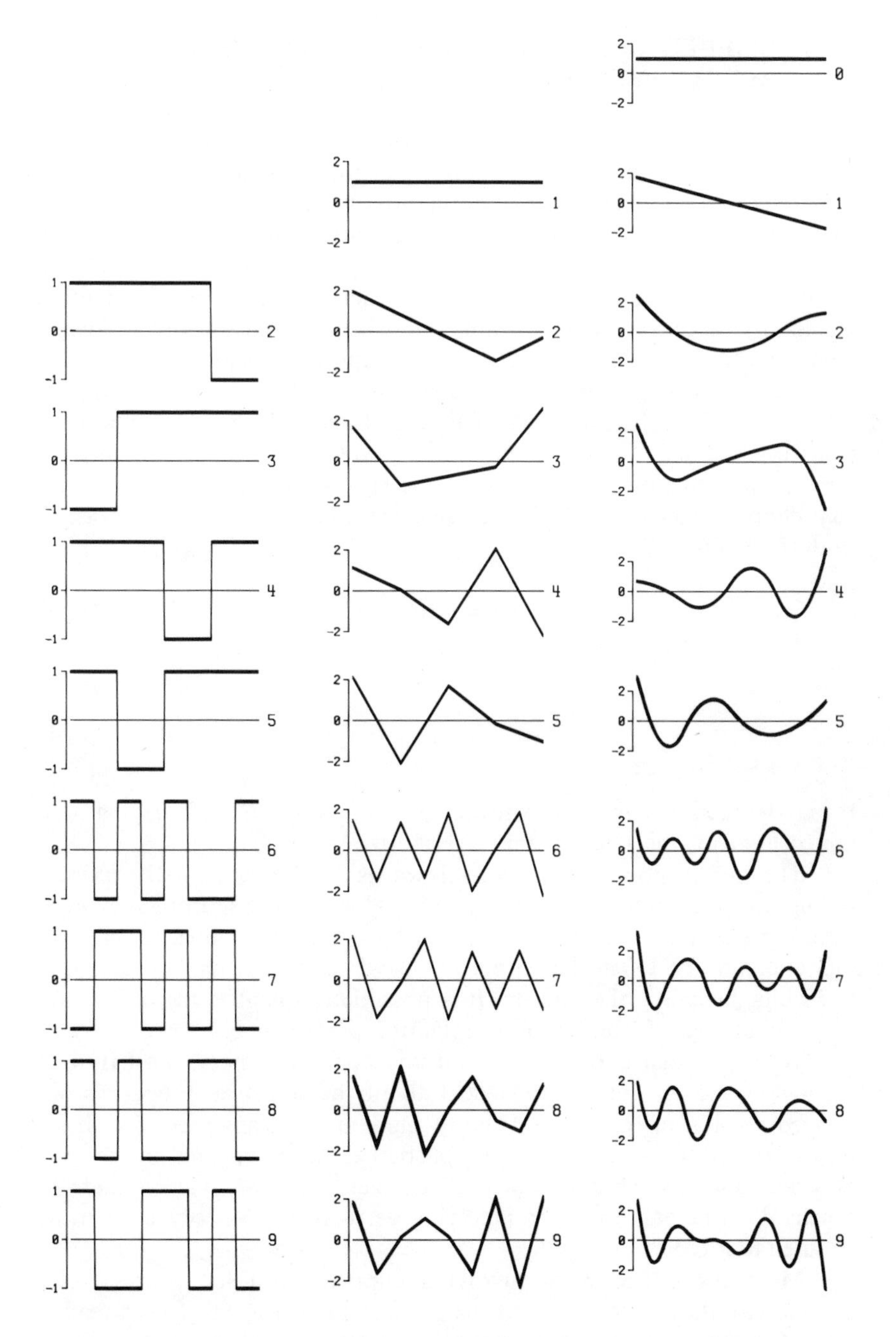

FIGURE 13.5. A Walsh-like system B, and higher-order systems derived therefrom: B_{I} and B_{II}.

13.4. CLIPPED SIGNALS AND THEIR SPECTRAL REPRESENTATIONS

Clipping is a useful type of nonlinear transformation that has been applied to many stochastic signals in engineering applications. Signal clipping (hard clipping or hard limiting) has been primarily used to reduce the number of bits necessary to code data into convenient form for data processing and transmission. In this technique, which is also called 1-bit quantization, only one bit is retained for each sample point of a digitized signal $s(t)$. Because clipping retains only the zero crossings of a signal, a signal analysis becomes an analysis based upon zero statistics rather than amplitude statistics. Zero-crossing methods have found extensive applications in speech analysis and automatic speech recognition,[9] in communications applications,[10] in EEG analysis,[11] and in other areas. These methods have been used rather extensively in the speech area because it is known that clipped speech is highly intelligible.[12] A clipped signal $c(t)$ is defined in the following way:

$$c(t) = s(t)/|(s(t)| \tag{13.4a}$$

or

$$c(t) = \operatorname{sgn} s(t) = \pm 1 \tag{13.4b}$$

It may be seen that the only knowledge of the amplitude of $s(t)$ that is retained in $c(t)$ is the sign (or polarity) of $s(t)$.

The idea of clipping a signal in this way is a consequence of the necessity to digitize a continuous signal by an analog-to-digital process. A given sample value may then be represented by a certain number of bits in a digital computer. Often the output of an A/D converter has a precision of 8–10 bits, including the sign bit. It is possible to quantize the output to a very coarse level of 1 bit by simple retention of the sign bit.

When a signal is basically a sinusoid containing noise (an FID is of this class) one can infer a great deal about the statistical properties of a clipped version of that signal. Recent mathematical analyses of such signals in terms of their zero-crossing properties have appeared in the work of Bond and Cahn,[13] Hinich,[14] Voelcker,[15] Bar-David,[16] and Requicha.[17] The earliest work involving zero-crossing theory is often attributed to Rice.[18]

Most theoretical work involving clipped stochastic processes has started with the assumption that the primary (input) signal $s(t)$ is a realization of a stationary, Gaussian process having zero mean. Under this assumption, the average number of zero crossings per second is given by

Rice's[18] classic result

$$E\left\{\frac{Z}{T}\right\} = \frac{1}{\pi}\left[-\frac{\rho''(0)}{\rho(0)}\right]^{1/2} = 2\left[\frac{\int \omega^2 S(\omega)\, d\omega}{\int S(\omega)\, d\omega}\right]^{1/2} \tag{13.5}$$

where

$$\rho(\tau) = E\{s(t)\, s(t + \tau)\} \tag{13.6}$$

$$S(\omega) = \int \rho(t) \exp(-j\omega t)\, dt \tag{13.7}$$

are the autocovariance of $s(t)$ and the power spectrum, respectively. Note that the average zero-crossing rate is proportional to the square root of the second moment of the spectrum. Denote the clipped and sampled process by $\{c_n\}$. If the sampled signal $\{s_n\}$ is stationary and mean zero, $\{c_n\}$ will also be stationary and mean zero. The discrete covariance of $\{c_n\}$ is given by

$$\rho_m^* = E\{c_n\, c_{n+m}\} \tag{13.8}$$

and may be estimated by the sample covariance r_m^*, where

$$r_m^* = \frac{1}{N} \sum_n c_n\, c_{n+m} \tag{13.9}$$

Hinich[14] has shown that a proper estimator for $S(\omega)$ is given by

$$\hat{S}(\omega) = 1 + 2 \sum_n k\left(\frac{n}{l}\right) \sin \frac{\pi}{2} r_n^* \cos n\omega \tag{13.10}$$

where $k(n/l)$ is a lag window that ensures that the spectrum estimate is consistent. As expected, the asymptotic standard deviation of an estimator based on $\{c_n\}$ is larger than that based upon $\{s_n\}$. Van Vleck,[19] quoted in Hinich,[14] has shown that the spectrum of a clipped narrow-band process contains power in higher harmonic bands centered about multiples of the narrow-band center frequency. Good[20] has argued that, for a Gaussian process band-limited over $(0, W)$, the zeros of the clipped waveform provide only $1/3^{1/2}$ of the information. Masry[21] has shown that, for members of a certain class of Gaussian processes that are band-limited, a process is uniquely determined by its zero crossings. However, Bar-David[16] has argued that Masry's theorems do not guarantee that sample functions of Gaussian processes can be recovered from their clipped versions; however, processes composed of a sum of a given sinusoid and of a bounded, band-limited process are uniquely represented by their clipped versions.[22]

In the light of these estimates of information loss by the clipping operation it is surprising, indeed, that one may obtain accurate NMR spectra if FIDs are clipped and then Fourier transformed. It is necessary, therefore, to explore in somewhat more detail the information content of the zeros of a band-limited signal $s(t)$. The definitive work in this area is that of Voelcker,[15] Voelcker and Requicha,[23] and Requicha.[17]

13.5. RANDOM REAL-ZERO SIGNALS

Voelcker and Requicha, in a series of papers,[15,17,23] have developed a formalism for representing signals that are realizations of a certain stochastic process that is specified in terms of the zero crossings of the process (its zero statistics). They define a class of signals, called random real-zero (RRZ) signals, by specifying the real zeros or zero crossings of the signals that belong to this class. The real zeros (RZs) of these signals are constructed such that they approximate the zeros of a sinusoid with randomness introduced through random perturbations in the nominal location of a zero of the sinusoid. The sample functions in this class are periodic and band-limited to $\pm B$ Hz, where $B = 1/2T$.

Using this model a signal $s(t)$ has the following form:

$$s(t) = \prod_{k=0}^{M-1} 2 \sin\left[\frac{\pi}{MT}(t - kT - \tau_k - \phi)\right] \tag{13.11}$$

where M is the number of RZs in a period of MT seconds. The period is divided into M epochs of T seconds duration each. Each epoch has a RZ that is displaced from the center of the epoch by an amount τ_k, where the τ_k are random variables. The parameter ϕ is a random time displacement. The importance of the RRZ model is that it is perhaps the simplest model for a BL signal for which the autocorrelation and spectrum can be computed exactly from a knowledge of the zeros of the signal.

13.6. ZERO-BASED PRODUCT REPRESENTATIONS FOR BAND-LIMITED SIGNALS

Bond and Cahn,[13] Voelcker,[15] and Requicha[17] have all investigated zero-based descriptions of BL signals. This section is a summary of Requicha's beautiful dissertation, *Contributions to a Zero-Based Theory of Band-Limited Signals.* Requicha's "contributions" are considerable.

A periodic signal $s(t)$ may be represented by a Fourier expansion:

$$s(t) = \sum_{k=N_1}^{N_2} C_k \exp(jk\Omega t) \tag{13.12}$$

On mapping $s(t)$ to the complex plane a signal can be written in terms of the roots of a Fourier polynomial as

$$s(z) = C_{N_2} w^{N_1} \prod_{k=1}^{n} (w - w_k) \tag{13.13}$$

where

$$w = \exp(j\Omega z), \qquad w_k = \exp(j\Omega z_k) \tag{13.14}$$

Then $s(z_k) = 0$ and the $\{z_k\}$ are the roots of $s(z)$. Periodic, real BL signals are such that

$$C_k = C^*_{-k}, \qquad N_2 = -N_1 = WT \tag{13.15}$$

so that

$$s(t) = C \cdot s_{\mathrm{RZ}}(t)\, s_{\mathrm{CZ}}(t) \tag{13.16}$$

where

$$\begin{aligned} s_{\mathrm{RZ}}(t) &= \prod_{i=1}^{2n_r} 2 \sin\left[\frac{\Omega}{2}(t - \tau_i)\right] \\ \tau_k &= \mathrm{Re}(z_k) \end{aligned} \tag{13.17}$$

and

$$\begin{aligned} s_{\mathrm{CZ}}(t) &= \prod_{k=1}^{n_c} 2\{\cosh \Omega\sigma_k - \cos[\Omega(t - \tau_k)]\} \\ \sigma_k &= \mathrm{Im}(z_k) \end{aligned} \tag{13.18}$$

There are $2WT$ zeros: $2n_r + 2n_c = n$. This derivation was first given in Voelcker[15] and is also discussed by Sekey.[24] One should consider that these $2WT$ zeros are completely equivalent in information content to the $2WT$ dimensions provided by Nyquist samples or Fourier coefficients.

Requicha has shown the connection between BL signals and a certain class of entire functions of exponential type (EFET +), which are completely characterized by their zero distributions. A BL finite-energy signal

$$s(t) = \int_a^b \exp(j\omega t)\, S(\omega)\, d\omega \tag{13.19}$$

is an EFET +, which may be expressed in terms of its zeros as

$$s(z) = C \exp(jkz)\, \mathscr{P} \prod (1 - z/z_n) \tag{13.20}$$

where $\mathscr{P}$ denotes the principal value of the infinite product, i.e.,

$$\mathscr{P} \prod (1 - z/z_n) = \lim_{R\to\infty} \prod_{|z_n| \le R} (1 - z/z_n) \tag{13.21}$$

Moreover, for a given set of zeros $\{z_n\}$, it is possible to construct a finite-energy signal that is BL to (a, b) rad/s; thus $s(z)$ is uniquely determined by its zeros and the parameters a and b. The zero density of BL signals coincides with the total bandwidth W Hz of a complex signal, i.e.,

$$W = |a - b|/2\pi \tag{13.22}$$

The sample functions of a wide-sense stationary BL process with ergodic mean and autocorrelation are also EFET+'s. For real signals, $a = -b$ and $k = 0$ in equation (13.20), so the zeros completely determine the signal. W is then a one-sided bandwidth, $W = b/2\pi$ Hz, and the zero density is $2W$, the Nyquist frequency. As with periodic signals,

$$s(t) = s_{\mathrm{RZ}}(t)\, s_{\mathrm{CZ}}(t) \tag{13.23}$$

where $s_{\mathrm{RZ}}(t)$ is a RZ signal, and s_{CZ} is a CZ signal. Separately, these components are not necessarily BL.

The EFET formalism requires an infinite number of zeros in order to have a lossless description of $s(t)$. Practically, however, in a finite time T only a finite number of zeros $\{z_n\}_{n=1}^{N}$ may be observed. Requicha has considered the problem of completing the finite zero set to give an infinite set, which will characterize a BL signal. The problem is analogous to that in conventional sampling theory of L^2 BL signals, i.e., an infinite set of Nyquist samples is required and a finite number of samples in $(-T/2, T/2)$ is not unique. Also shown by Requicha,[17] with several examples, is the effect of inserting, deleting, and replacing zeros. For example, by deleting CZ pairs from a real periodic BL signal and replacing them by single RZs the zero pattern becomes aperiodic for the s_{RZ} component so that s_{RZ} is not BL. Completing procedures may be formulated in terms of a periodic repetition of a given zero set, zero deletion and replacement, or interpolation of a zero set to a sinewave. Completing procedures may be considered to correspond to the application of an infinite time window $w(t)$ to a signal given by N zeros, $p_N(t)$. For example, if

$$p_N(t) = \prod_{n=1}^{N} (1 - t/z_n) \tag{13.24}$$

and

$$w(t) = C \exp(jkt)\, \mathscr{P} \prod_{n=N+1}^{\infty} (1 - t/z_n) \tag{13.25}$$

then

$$s(t) = p_N(t)\, w(t) \tag{13.26}$$

The bandwidth of the resulting signal depends on the completing procedure used, i.e., if the completing procedure involves deleting zeros from a sinc function window the bandwidth depends on the zero density of the sinc function: $W = 1/2\alpha$, where α is the distance between sinc zeros.

Requicha has also investigated zero-based A/D conversion. This procedure, for signals with CZs, requires that a sinewave carrier be added to the signal in order to convert it to a RZ signal. This RZ signal is then clipped to detect the RZs and then the zeros are quantized by dividing each of M time epochs T into q uniform quanta so that there is one zero per epoch. The resulting quantized zero signal requires $M \log_2 q$ bits for a time interval of MT sec. It was shown that the S/N ratio in such a zero-based A/D procedure is comparable to that for conventional Nyquist amplitude sampling and quantization.

13.7. SPECTRA OF CLIPPED FID SIGNALS

Most of the above discussion has been concerned with the zeros of a signal, such as an FID, and the information in those zeros. When an FID is clipped or 1-bit quantized according to the operation given in equation (13.4) a clipped FID (CFID) results. The CFID that results, however, contains in the zero crossings of the FID the essential frequency information required by a spectroscopist.[25]

That this is the case may be seen in Figures 13.1 and 13.2, which show the FID and the CFID, the Fourier power spectrum of the FID, and the Fourier power spectrum of the CFID for ethylbenzene and for toluene. One should note that the Fourier power spectrum of the CFID is virtually identical to the Fourier power spectrum of the unclipped FID, confirming, in essence, that the essential frequency information is embedded in the zero crossings of the FID.

Additional insight into clipped signals and the CFID may be obtained by computing the Walsh sequency spectrum of the CFID (Walsh transforms were briefly discussed in a previous section). Because the CFID $c(t)$ is essentially a Hadamard vector (values of ± 1 only) one might expect its Walsh spectrum to consist of fewer lines (converge faster) than the Fourier spectrum. This is not the case as may be seen in Figures 13.1 and 13.2.

The explanation for the behavior of the Walsh spectrum is as follows. Although the Walsh spectrum contains information that is equivalent to that of the Fourier spectrum, the Walsh spectrum contains additional lines because, despite the clipping operation, the CFID retains the sinusoidal components of the FID in the zero crossings. Thus, additional Walsh coefficients are required for convergence of the orthonormal expansions of both $s(t)$ and $c(t)$. It should be noted that the major Fourier peaks appear at the same position in the Walsh spectrum.

The spectra of the clipped free-induction decay signals contain a few additional lines of low amplitude. These lines are due to intermodulation distortion. (The origin of these lines is discussed in the Appendix.) The presence of these lines should not impede identification of a chemical species: these lines may be attenuated or completely removed in several ways.

13.8. SUMMARY, IMPLICATIONS, AND OPEN QUESTIONS

In this chapter we have considered the consequences of a zero-based spectral analysis of chemical signals, with special reference to the FT–NMR technique. For this case, one measures a clipped free-induction decay signal (CFID) in which the frequency information in the signal is wholly contained in the zero crossings (ZCs) of the signal; i.e., the amplitudes of the CFID are quantized to values of ± 1. The Fourier (power) spectrum of a CFID is virtually identical to that of the FID, except for a few low-amplitude intermodulation peaks. These peaks should not interfere with visual characterization of a chemical compound based upon its CFID/Fourier spectrum. The presence of these additional peaks (terms in a Fourier expansion) is due to the distortion introduced by the highly nonlinear operation of signal clipping. The clipped signal consists of only real zeros (RZs). The work of Voelcker and Requicha,[15,17] who have developed a zero-based theory of band-limited signals, reveals that functions with BL Fourier spectra are entire functions in the time domain. Such functions—aperiodic signals—are defined uniquely in terms of an infinite number of zeros. A CFID has a finite number of RZs; however, it is possible to analytically complete a finite-zero distribution to obtain a strictly BL FID signal. A finite-zero CFID is, however, a reasonable approximation to a BL signal, just as a finite set of Nyquist samples within an interval $(-T/2, T/2)$ is a reasonable, although not unique, set of samples in conventional amplitude-based techniques.

It is also possible to quantize in time the zeros of the FID instead of quantizing the amplitudes as we have done for the examples shown in Figures 13.1 and 13.2. Presumably, however, the S/N ratio for such zero quantization is comparable to that for clipping (or amplitude quantization).[17]

The implications of being able to do routine CFID spectral analyses are interesting. It should be clear that, if it is possible to process a clipped signal having values of ± 1, there will be a substantial saving of computer time and storage. This obtains because it is not necessary to store the amplitudes of the signal for further processing; i.e., one need only store, as bits, $+1$ or -1 (which are equivalent to 0 and 1). A feeling for the storage saved by such a procedure may be obtained by considering, as an illustration, a 2K-point FID (2048 data points). Normally, one would store 2K data points as 2K words that include, of course, a sign bit. With a 60-bit word-length computer (a common CDC word configuration), for example, it is possible to pack the signs of these 2K words, 60 per word, so that only 2K/60 words of storage are required—a savings of 60:1. It is quite simple to store the signs of 2K data points with a binary READ statement. Actually, one can also pack the original 2K data points to achieve a savings of about 2:1, so that storing a CFID vs. an FID provides a data compression ratio of closer to 30:1 than 60:1.

The implication of this storage saving is also interesting to consider. It is possible to store bitwise in an NMR data bank the experimentally obtained CFID. One could then access a given record of interest and then FFT that Hadamard vector (the CFID) to obtain the desired spectrum. A data bank based on this concept would be considerably more compact than conventional storage methods. The presumption here is that, for each and every chemical compound, there is a unique Hadamard vector (CFID); for an 8K-point CFID there exists the possibility of 2^{8K} unique Hadamard vectors! Moreover, the disadvantage of storing an NMR spectrum directly is that spectra are very redundant in their information content. Note, for example, that the ethylbenzene spectrum contains only 5–6 lines over a 5-kHz range.

The technique of storing a CFID is a type of data compression—reduction of the volume of data that are necessary to give a required amount of information. Zero-based signal analysis also provides other advantages. One of these is bandwidth reduction. Although the specific advantage to FT NMR is not clear, a clipped signal containing RZs may be interpolated to give a real-zero signal s_{RZ}:

$$s_{RZ} = I_{RZ} \operatorname{sgn} s(t) \tag{13.27}$$

The resulting spectrum of s_{RZ}, $S_{RZ}(\omega)$, has a smaller bandwidth than either $\mathscr{F}[\operatorname{sgn} s(t)]$ or $\mathscr{F}[s(t)]$. It has been shown that this transformation, called real-zero interpolation (RZI), can yield substantial bandwidth savings.[23,24] Spectrum distortion, if any, can be controlled by CZ insertion.

It should be mentioned, in conclusion, that zero-crossing techniques are applicable to other chemical signals. One can quantize virtually any FID signal including those for ^{31}P, ^{1}H, etc. In fact, we expect even better results with higher-frequency signals than the ^{13}C FID because higher frequencies imply higher density of zero crossings. This leads one to believe that these techniques would work very well with infrared interferograms because of the high-frequency content of these chemical signals.

ACKNOWLEDGMENTS

The author would like to thank Edward F. Crawford for agreeing to do a large portion of the work discussed herein; this work constitutes a substantial portion of Crawford's dissertation. The author would also like to acknowledge helpful discussions with Professor H. B. Voelcker and thank him for sending Dr. Requicha's dissertation to us. Finally, acknowledgment is due Dr. Bernard Saltzberg, who first introduced us to zero-crossing methods.

NOTATION

Symbols

B	upper limit to absolute value of $s(t)$
$c(t)$	clipped signal
c_n	sample value of clipped signal
$C(\tau)$	autocorrelation function
C_k, C^*_{-k}	complex Fourier coefficients
I_{RZ}	real zero interpolation function
j	imaginary number
$k(n/l)$	lag window
l	parameter defining lag window
L^2	space of square integrable functions
m	mean value
M	number of RZs in period of MT seconds
M_y	y component of magnetization after pulse at time t
$\{M_y\}$	ensemble of FID signals
$M_y(t_0)$	y component of magnetization at end of RF pulse
n	parameter defining lag window, equation (13.10)
n	order of Hadamard matrix, $n = 2^p$ for Walsh matrices
n_R, n_Z	number of real zeros and complex zero pairs
N	bandwidth index, $N_2 - N_1$
N_1	lower frequency index
N_2	upper frequency index
$p_N(t)$	finite-length signal defined by N zeros
q	number of uniform quanta
r_m	sample covariance function
$s(t)$	signal
$\lvert s(t)\rvert$	magnitude of $s(t)$
s_n	sample value of signal
s_{RZ}	real-zero signal
sgn	signum operation
$S(\omega)$	frequency spectrum of $s(t)$
$\hat{S}(\omega)$	estimator for $S(\omega)$
t	time variable
T	period of signal in seconds
T_2	transverse relaxation time
$u(t)$	input to nonlinear system
$v(t)$	output of nonlinear system
$w(t)$	time window defined in equations (13.25) and (13.26)
W	bandwidth (equal to $N\Omega$ rad/s or N/T Hz)
W_1, W_2	lower and upper frequency limits (equal to $N_1\Omega$ and $N_2\Omega$, in rad/s)
$\{X(t)\}$	stochastic process
$X(t, \theta)$	time series
z	complex variable; $z = \tau + j\sigma$
$\{z_k\}$	roots of $s(z)$
Z	number of zero crossings

Greek Symbols

α	distance between sinc zeros
α	$\Omega_i - \omega_1$, frequency of FID signal
$\{\alpha_n\}$	set of coefficients calculated with respect to $\{\Phi_n\}$
$\in$	contained in, or belongs to
θ	random variable taking values in Θ
Θ	family of random variables
$\rho(\tau)$	autocorrelation function for signal $s(t)$
ρ^*_m	discrete covariance function of clipped signal
τ	time displacement, lag
τ_k	displacement of real zero from center of epoch
ϕ	random time displacement, in equation (13.11)
$\{\Phi_n\}$	system of linearly independent functions
$\{\psi_n\}$	alternate orthogonal system of functions
ω	angular frequency
ω_1	frequency of RF pulse
Ω	radian frequency ($=2\pi/T$)
Ω_i	resonance frequency of spin i

Abbreviations

A/D analog to digital conversion
BL band limited
CFID clipped free-induction decay signal
CZ complex zero
EFET+ entire function of exponential type
FFT fast Fourier transform
FID free-induction decay signal
FWT fast Walsh transform
RRZ random real-zero signal
RZ real zero
RZI real zero interpolation
S/N signal-to-noise ratio
WT Walsh transform
ZC zero crossing

Special Symbols

$\mathscr{F}$ Fourier transform operator
$\mathscr{P}$ principal value operator

APPENDIX. INTERMODULATION DISTORTION IN THE CFID

One should note the presence of a few extra, low-amplitude lines in the clipped spectra of ethylbenzene and toluene in Figures 13.1 and 13.2. These additional lines are due to intermodulation distortion. It is interesting to consider the origin of these additional lines.

Intermodulation distortion in the Fourier spectrum of the CFID arises from the highly nonlinear operation of clipping the FID to the CFID [equation (13.4)].

To see how extra lines arise in the Fourier power spectrum by a nonlinear operation, consider the following model. Let $v(t)$ be the output of a nonlinear model given by a power series:

$$v(t) = au(t) + bu^2 + cu^3 + \cdots \tag{13.28}$$

where $u(t)$ is an input signal. For a single isolated spin, $u(t)$ may be taken to be $M_y(t)$ as given in equation (13.2). Consider that

$$u(t) = A\cos(\alpha t)\, e^{-\tau} + B\sin(\alpha t)\, e^{-\tau} \tag{13.29}$$

which is the basic functional form of $M_y(t)$. If we compute the quadratic term bu^2 in equation (13.28) we find that

$$bu^2 = e^{-2\tau}[b(A^2 + B^2) + b(A^2 - B^2)\cos 2\alpha t + bAB\sin 2\alpha t] \tag{13.30}$$

Thus, terms arise that are the DC term and second harmonics (terms in 2α). Note that the multiplier is $e^{-2\tau}$. The cubic term in equation (13.28) has the form

$$\begin{aligned} cu^3 = e^{-3\tau}[(\tfrac{1}{4}cA^3 - \tfrac{3}{4}cAB^2)\cos 3\alpha t + (\tfrac{3}{4}cA^3 + \tfrac{3}{4}cAB^2)\cos \alpha t \\ + (\tfrac{3}{4}cB^3 + \tfrac{3}{4}cA^2B)\sin \alpha t + (\tfrac{3}{4}cA^2B - \tfrac{1}{4}cB^3)\sin 3\alpha t] \end{aligned} \tag{13.31}$$

Note the presence of terms in α and 3α with a multiplier $e^{-3\tau}$.

When there are two or more spin systems present so-called intermodulation terms (e.g., $\alpha_1 + \alpha_2$ and $\alpha_1 - \alpha_2$) arise. To see this, consider two spin systems Ω_1 and Ω_2, so that $\alpha_1 = \Omega_1 - \omega_1$ and $\alpha_2 = \Omega_2 - \omega_1$. Then $M_y(t)$ has the form

$$M_y(t) = A\cos(\alpha_1 t)\,e^{-\tau} + B\sin(\alpha_1 t)\,e^{-\tau} + C\cos(\alpha_2 t)\,e^{-\tau} + D\sin(\alpha_2 t)\,e^{-\tau} \tag{13.32}$$

If one computes just the quadratic term bu^2 the following terms arise:

$$\begin{aligned} bu^2 = be^{-2\tau}\{&\tfrac{1}{2}(A^2 + B^2 + C^2 + D^2) + \tfrac{1}{2}(A^2 - B^2)\cos 2\alpha_1 t \\ &+ \tfrac{1}{2}(C^2 - D^2)\cos 2\alpha_2 t + AB\sin 2\alpha_1 t + CD\sin 2\alpha_2 t \\ &+ (AC + BD)\cos(\alpha_1 - \alpha_2)t + (AC - BD) \\ &\times \cos(\alpha_1 + \alpha_2)t + (AD + BC) \\ &\times [\sin(\alpha_1 + \alpha_2)t + \sin(\alpha_1 - \alpha_2)t]\} \end{aligned} \tag{13.33}$$

Thus, terms arise in $2\alpha_1$, $2\alpha_2$, $\alpha_1 + \alpha_2$, and $\alpha_1 - \alpha_2$, in addition to the multiplier $e^{-2\tau}$. The last terms, corresponding to sum and difference frequencies, are intermodulation terms.

The origin of these intermodulation terms in the spectrum of a clipped signal is somewhat different than above. Actually, $u(t) = M_y(t)$, the FID, is clipped and the RZs are observed. In this case the RZs alone do not provide sufficient dimensions to completely specify the signal, although they come close to so doing. If the number of zero crossings actually obtained, n_R, equals the upper band limit $N_2 = N_1 + N$, where N_1 is the lower frequency limit and N is the bandwidth, then $s(t)$ is completely specified by n_R.(23) In the FT NMR case we have the propitious situation that n_R is slightly less than N_2 but as, $n_R > N$, the RZs provide sufficient dimensions, in principle, to completely specify the signal. Unfortunately, since intermodulation distortion appears throughout the CFID spectrum it cannot be removed by standard digital-filtering techniques. It is possible, however, to reinsert complex zeros into the RZ waveform $s_{RZ}(t)$ via Hilbert transformation.(15) This technique has been used to reduce large-amplitude variations in a RZ-interpolated signal, which results when there are wide gaps between zeros.(25)

REFERENCES

1. L. H. Koopmans, *The Spectral Analysis of Time Series*, Academic Press, New York, 1974.
2. D. R. Brillinger, *Time Series Data Analysis and Theory*, Holt, Rinehart and Winston, New York, 1975.
3. R. R. Ernst and W. A. Anderson, Application of Fourier transform spectroscopy to magnetic resonance, *Rev. Sci. Instrum.* **37**, 93 (1966).

4. H. F. Harmuth, *Transmission of Information by Orthogonal Functions*, 2nd ed., Springer-Verlag, New York, 1972.
5. N. Ahmed and K. R. Rao, *Orthogonal Transforms for Digital Signal Processing*, Springer-Verlag, New York, 1975.
6. J. L. Walsh, A closed set of orthogonal functions, *Am. J. Math.* **45**, 5 (1923).
7. R. D. Larsen and W. R. Madych, Walsh-like expansions and Hadamard matrices, *IEEE Trans. Acoust. Speech, Sig. Processing* **24**, 71 (1976).
8. W. R. Madych, R. D. Larsen, and E. F. Crawford, Piecewise polynomial expansions, *IEEE Trans. Acoust., Speech, Sig. Processing* **25**, 579 (1977).
9. R. J. Niederjohn, A mathematical formulation and comparison of zero-crossing analysis techniques which have been applied to automatic speech recognition, *IEEE Trans. Acoust., Speech, Sig. Processing* **23**, 373 (1975).
10. P. C. Jain and N. M. Blachman, Detection of a PSK signal transmitted through a hard-limited channel, *IEEE Trans. Inform. Theory* **19**, 623 (1973).
11. B. Saltzberg, R. J. Edwards, R. G. Heath, and N. R. Burch, Synoptic analysis of EEG signals, *in: Data Acquisition and Processing in Biology and Medicine* (K. Enslein, ed.), Vol. 5, p. 267, Pergamon, Oxford, 1968.
12. J. C. R. Licklider and I. Pollack, Effects of differentiation, integration, and infinite peak clipping upon the intelligibility of speech, *J. Acoust. Soc. Am.* **20**, 42 (1948).
13. F. E. Bond and C. R. Cahn, On sampling the zeros of bandwidth-limited signals, *IRE Trans. Inform. Theory* **4**, 110 (1958).
14. M. Hinich, Estimation of spectra after hard clipping of gaussian processes, *Technometrics* **9**, 391 (1967).
15. H. B. Voelcker, Toward a unified theory of modulation. Part I: Phase-envelope relationships, *Proc. IEEE* **54**, 340 (1966); Part II: Zero manipulation, **54**, 735 (1966).
16. I. Bar-David, Sample functions of a gaussian process cannot be recovered from their zero crossings, *IEEE Trans. Inform. Theory* **21**, 86 (1975).
17. A. A. G. Requicha, *Contributions to a Zero-Based Theory of Band-Limited Signals*, Ph.D. Dissertation, University of Rochester, 1970.
18. S. O. Rice, Mathematical analysis of random noise, *in: Selected Papers on Noise and Stochastic Processes* (N. Wax, ed.), p. 133, Dover, New York, 1954.
19. J. Van Vleck, The spectrum of clipped noise, Radio Research Laboratory Report No. 51, Harvard Univ., 1943.
20. I. J. Good, The loss of information due to clipping a waveform, *Inform. Control* **10**, 220 (1967).
21. E. Masry, The recovery of distorted bandlimited stochastic processes, *IEEE Trans. Inform. Theory* **19**, 398 (1973).
22. I. Bar-David, An implicit sampling theorem for bounded bandlimited functions, *Inform. Control* **24**, 36 (1974).
23. H. B. Voelcker and A. A. G. Requicha, Clipping and signal determinism: Two algorithms requiring validation, *IEEE Trans. Comm.* **21**, 738 (1973).
24. A. Sekey, A computer simulation study of real-zero interpolation, *IEEE Trans. Audio Electroacoust.* **18**, 43 (1970).
25. R. D. Larsen and E. F. Crawford, Clipped free induction decay signal analysis, *Anal. Chem.* **49**, 508 (1977).

Chapter 14

Applications of the FFT in Electrochemistry

Peter R. Griffiths

14.1. INTRODUCTION

Most modern electroanalytical techniques are basically electrochemical relaxation measurements (ERM), in which one observes a time-varying response from some type of electrochemical cell to an applied perturbation such as current, potential, or charge. The observed relationship between the response and the perturbation is known as the *transfer function* and provides analytical data, kinetic information, and/or mechanistic information on a variety of processes that can occur in the electrode, in the electrolyte bulk, or in the interfacial regions. Perhaps the most common experiment is one in which a cell *voltage* perturbation is created and the cell *current* response is measured. Typical techniques that fall into this category are DC, AC, and pulse polarography, linear sweep and triangular wave voltammetry, and potential step chronoamperometry.

Essentially all conventional electrochemical assay procedures are based on the observation of the response of the cell, either at a single frequency (fundamental and second harmonic AC polarography), a single point in time (pulse polarography, DC polarographic measurements at end of drop life), or some form of integral response (DC polarographic average currents and chronocoulometry). The detailed character of the variation of analytical response with potential profile is usually ignored, an assay being based on the response at a single potential (usually the peak potential) or in a region

Peter R. Griffiths • Department of Chemistry, Ohio University, Athens, Ohio 45701

where the response is independent of potential. A detailed examination of an electrode response is not (and perhaps should not be) made in a routine assay procedure, mainly because a careful examination would be too time consuming with conventional equipment.

Recently, however, it has been shown that the application of the FFT to electrochemical procedures, in particular through the use of an on-line minicomputer for data acquisition and processing, has made it feasible to obtain kinetic–mechanistic information in a time-frame that is small relative to the time devoted to an electroanalytical assay using conventional equipment and procedures. In fact it has been stated[1] that, using as a frame of reference the electrochemical analyst's normal experience, one can view the phenomenological kinetic–mechanistic information provided by FFT strategies as being available in real time. Thus it is now possible to deduce the kinetic–mechanistic status of an electrode reaction during the course of an assay run.

In this chapter we shall study several applications of the FFT in electrochemistry, both as a means of acquiring this kinetic–mechanistic information and also for data manipulation (in particular smoothing) as discussed by Lephardt in Chapter 11.

14.2. FARADAIC ADMITTANCE MEASUREMENTS—BASIC PRINCIPLES

One of the more intensively studied properties of an electrochemical cell through the application of the FFT is the variation of faradaic admittance (or impedance) $A_f(\omega)$, with frequency ω. The faradaic admittance is defined through the expression

$$A_f(\omega) = I_f(\omega)/E(\omega) \tag{14.1}$$

where $I_f(\omega)$ is the faradaic current and $E(\omega)$ the potential across the electrical double-layer. From an operational electrical-measurement viewpoint, the measurement of $A_f(\omega)$ is an example of the acquisition of a frequency domain response of a conducting system.

The typical electrochemical cell used in faradaic admittance studies presents three distinct classes of charge transport processes that control the admittance:

(a) charge transport through the solution bulk, which is mainly an ohmic resistance,

(b) virtual charge transport at the electrode–solution interface due to the capacitance of the electrical double-layer (double-layer admittance),

(c) the actual charge transport at the electrode–solution interface associated with electrolysis (the faradaic admittance).

To obtain A_f, the contributions of the two other "nonfaradaic" processes must be subtracted from the overall cell admittance, either by analog compensation schemes[2-5] or by digital data processing. Once the information is obtained, the interpretation of faradaic admittance data is often complicated because its characteristics can manifest effects of numerous rate processes such as diffusion, heterogeneous charge transfer, homogeneous chemical reactions, adsorption, crystallization, etc.[2,6,7] Thus a large, highly precise data set is required for any kinetic–mechanistic or thermodynamic study.

This data set must be composed of two independent frequency domain observables (typically the amplitude and phase angle, or in-phase and quadrature components) over as wide a range of frequency and DC potential as possible. In many cases the data set should be obtained over a range of solution conditions (such as pH, ligand concentration, and electroactive species concentration). Finally it should be noted that in several cases (such as the dropping-mercury electrode, or DME) the admittance is inherently time dependent because of the temporal variation of electrode area, so that the admittance must be measured at precisely defined points in time after the initiation of the experiment.

With these considerations in mind, the nature of the classical approach to faradaic admittance measurements through the use of an impedance bridge should first be recognized. With a DME, for example, one must balance the bridge at a precise point in time after the genesis of each drop under conditions where balancing requires adjustment of a resistance and a capacitance. This operation must be repeated over a range of DC potentials and frequencies to produce a useful data set.

Evolution of electronic instrumentation eventually led to automatic recording instruments, the most common being the AC polarograph, which initially provided a recording of the total admittance vs. DC potential. More recent instruments yield the in-phase and quadrature components, and can also effect the measurement at a particular point in time through the use of sample-and-hold amplifiers. Even with this automated hardware, the operator is still confronted with considerable redundance and tedium during the course of data collection, except in analytical applications that rely on an AC polarogram measured at a single frequency. It is apparent that the area of faradaic admittance measurements is ripe for computerization, and the most significant advances have come through the application of the FFT algorithm.

Because the various frequency domain quantities in equation (14.1) contain both amplitude and phase characteristics (i.e., they are representable as complex numbers), it is convenient to use an alternative form for $A_f(\omega)$:

$$A_f(\omega) = \frac{I_f(\omega)\, E^*(\omega)}{E(\omega)\, E^*(\omega)} \tag{14.2}$$

where $E^*(\omega)$ is the complex conjugate of $E(\omega)$. This formulation confines all phase information to the numerator.

$I_f(\omega)\, E^*(\omega)$ is the cross-power spectrum of the potential and faradaic current signals; this *complex number array* has at each frequency a magnitude equal to the product of the magnitude of $I_f(\omega)$ and $E(\omega)$ and an angle equal to the phase angle Φ between the two signals. $E(\omega)\, E^*(\omega)$ is an autopower spectrum; this *real number array* has a magnitude equal to the square of the magnitude of $E(\omega)$ at each frequency.

Cross- and autopower spectra may, of course, be obtained in FT–IR and FT–NMR spectrometry. However, unlike these types of spectrometry, where the spectral frequencies are heterodyned in some fashion into the audiofrequency range, in ERMs direct acquisition of the individual waveform time domain sequence is readily achieved, greatly simplifying analog circuitry requirements.

Under conditions of most interest, the faradaic admittance is manifested as a nonlinear circuit element. However, provided that the perturbing potential is small enough, the nonlinear properties are minimized to a sufficient extent that, by conventional standards of accuracy and precision, the cell admittance may be approximated as linear. Knowledge of the small-amplitude limit of the cell admittance suffices for most purposes to which the ERM is directed, and it is in this context where most of the important applications of the FFT in electrochemistry predominate. It has also been found that instead of characterizing an admittance spectrum by single-frequency sinusoidal measurements, multiple-frequency applied waveforms can be applied to characterize the admittance spectrum at all frequencies of interest simultaneously at a particular DC potential in just a few seconds. One can then "scan" the admittance spectrum as a function of DC potential to obtain the kind of data normally required for kinetic–mechanistic studies with the same ease as running a classical DC polarogram.

An important property of equation (14.2) is its general applicability to any type of waveform. Thus, in principle, the same computed $A(\omega)$ will result from signals that are aperiodic transients, almost periodic, periodic, or stochastic,[8] provided that conditions are conducive to accurate measurement of the time domain waveforms, and that $E(\omega)$ has sufficient magnitude at the frequencies of interest.[9] In this respect, Smith[10] has pointed out that this property of generality has three significant implications:

(a) Complex perturbation waveforms comprised of numerous frequency components can be utilized conveniently. Analysis of the response of the system to such waveforms via equation (14.2) then enables nearly instantaneous characterization of the frequency response profile in a single step.

(b) Regardless of the shape of the perturbation function, equation (14.2) provides a common format for data presentation and data theory comparison. The applicable empirical rate law framework (theory of AC polarography and faradaic measurements) is the most extensively developed in ERM.

(c) The requirements on accuracy of the waveform with which the electrochemical system is perturbed are greatly relaxed by the Fourier transform approach. Electrochemists have devoted much effort to the development of electronic control devices (e.g., potentiostats) that apply accurate, high-speed perturbations, so that the form and magnitude of the perturbation waveform can be assumed in data processing.[2,3,6,7] Even so, results of non-FFT electrochemical experiments are severely bandpass limited; by requiring only that the perturbing waveforms have significant content at frequencies of interest, higher-frequency measurements are more readily accomplished, even if the potentiostat is not accurately controlling at the high frequencies.[10]

In classical methods for deriving information on rate processes, one has to assume that a specific waveform (such as a square impulse, step function, or constant amplitude sine wave) is actually involved, whereas in practice this may only be an approximation. On the other hand, the Fourier transform methods, in which the *actual* applied waveform is measured, effectively correct for potentiostat nonideality as well as waveform generator imperfections and noise sources at the perturbation source. It is true that for certain reasons the actual applied waveform must at least crudely approximate the desired one. Nevertheless the useful bandwidth of the analog parts of electrochemical instrumentation is greatly enhanced as a result of this relaxation of control network requirements, as has been demonstrated by Doblhofer and Pilla.[11,12]

In situations where it is desired to measure some response property resulting from cell response *nonlinearity*, the FFT can also play an important role. In principle such measurements can be achieved using multiple-frequency waveforms to obtain nonlinearity spectra in a single measurement step.[13] However, to date most electrochemists have adopted the simpler strategy in which a pure harmonic potential is applied to the cell and the harmonic distortion arising from faradaic nonlinearity is evaluated by investigating the signal at all integral multiples (especially the second) of the applied frequency.

14.3. INSTRUMENTATION

A block diagram of the system designed by Creason *et al.*[8] for the measurement of faradaic admittance under AC polarographic conditions using Fourier transform techniques is shown in Figure 14.1. The heart of the analog circuitry is a *dual potentiostat* that controls the potential applied to two cells, the *reference cell*, which contains only solvent and supporting electrolyte, and the sample cell, which contains these constituents together

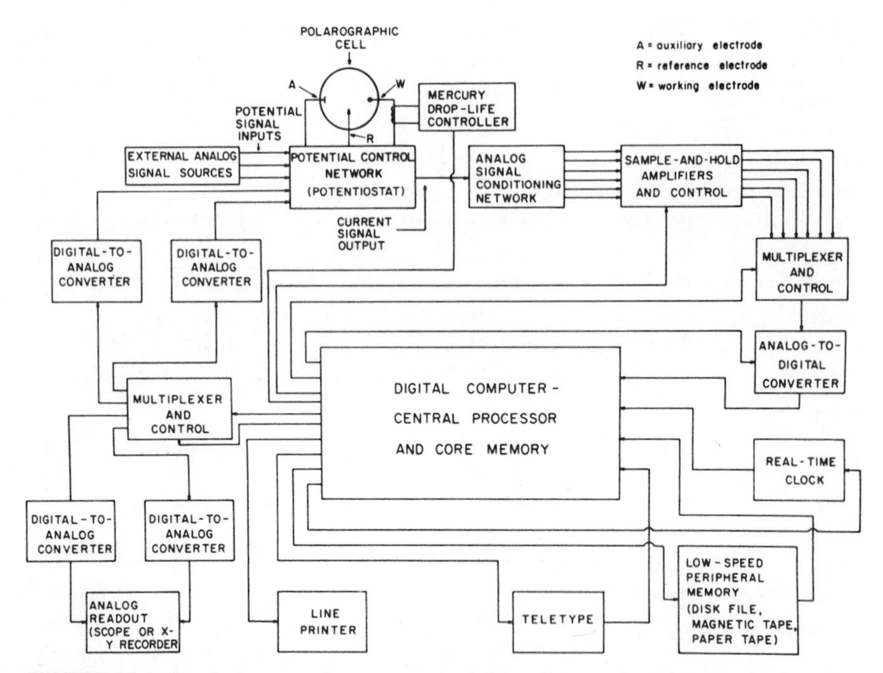

FIGURE 14.1. Block diagram of a computerized AC polarograph with an external analog signal source and an analog signal conditioning network. (Reproduced from reference 2 by permission of CRC Press and the author; copyright 1971.)

with the electroactive species. This dual potentiostat, along with an analog subtractor in the current signal-conditioning network, usually enables the double-layer charging current to be eliminated. This nonfaradaic current does not normally provide direct information on the rate processes associated with electrolysis. The dual potentiostat also provides positive feedback compensation of ohmic potential drop in electrochemical cells,[2,3,5] so that the effective potential applied across the faradaic admittance equals the externally applied potentials derived from the DC and AC voltage sources.

This compensation for nonfaradaic effects can also be achieved through software, since the system is computerized; a schematic of instrumentation with less analog circuitry is shown in Figure 14.2. Computers and digital electronics also allow many other operations to be performed, including the following:

(a) Generation of the applied waveform through digital-to-analog converters.

(b) Synchronization of growth and fall of mercury drops in the sample and reference cells through the use of a programmable clock when DMEs are employed; synchronization of the data acquisition sequence with

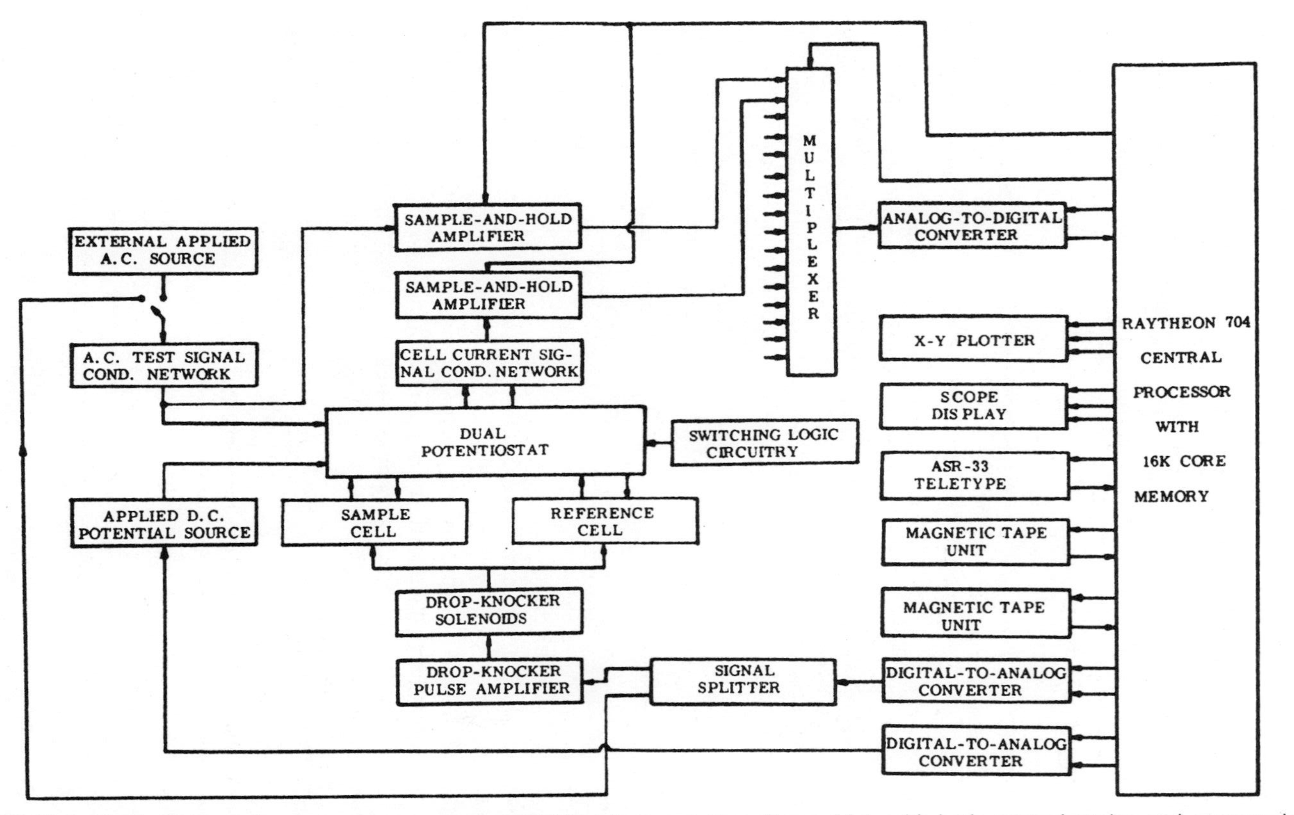

FIGURE 14.2. Block diagram of a more recent computerized AC polarograph than Figure 14.1, with both external analog and computerized generation of test signal waveforms. (Reproduced from reference 8 by permission of the Elsevier Publishing Company and the author; copyright 1973.)

TABLE 14.1. Classification of Perturbation Waveforms Used in Admittance Studies

Waveform type	Examples	Special properties
Almost periodic signals	Waveform typically obtained when outputs of an arbitrarily selected set of sinusoidal oscillators are added	Set of discrete, noncoherently related sinusoidal components; waveform is not periodic, even though its components are
Complex periodic signals	Square wave, triangular wave, full-wave rectified sine wave, pseudorandom white noise	Waveforms described by a conventional Fourier series
Aperiodic transients	Step function, ramp function, triangular impulse, square impulse	Signals with a continuous smoothly varying phase and amplitude spectrum
Stochastic signals	White noise, pink noise, blue noise, flicker noise	Signals with continuous spectra; amplitude (autopower) spectra approach smooth distribution in the long time limit; phase spectrum is randomized

TABLE 14.2. Several Types of Perturbation Waveforms Used in Admittance Studies

Waveform type	Frequency spectrum characteristics	Relationship of signal properties on successive measurement passes
Bandwidth-limited white noise (BLWN) (see Figure 14.3)	Continuous spectrum with random phases (flat distribution function) and amplitudes (Gaussian distribution function)	Randomized phase, amplitude and time domain relationships on successive passes
Combed bandwidth-limited white noise (combed BLWN)	Complex periodic signal with discrete components at harmonics of a certain frequency; randomized amplitudes (Gaussian distribution function) and phases (flat distribution	Same as bandwidth-limited white noise

TABLE 14.2 *(continued)*

Waveform type	Frequency spectrum characteristics	Relationship of signal properties on successive measurement passes
Rectangular pulse	Continuous, smoothly varying phase and amplitude spectrum; amplitude spectrum characterized by decreasing magnitudes as frequencies increase and by periodic zero-amplitude nodes	All signal characteristics identical on successive passes
Almost periodic	Set of discrete frequency components of identical amplitudes at several unrelated frequencies	Amplitude spectrum invariant, time domain waveform changes on successive passes
Pseudorandom white noise (PRWN)	As combed BLWN but with all components of identical amplitudes; frequency components identical to measured frequencies	All signal characteristics identical on successive passes
Phase-varying pseudorandom white noise (phase-varying PRWN)	Same as PRWN	Amplitude spectrum invariant, phase spectrum and time domain waveform change on successive measurement passes (Random-number phase array rotated between each pass)
Old harmonic pseudorandom white noise (odd-harmonic PRWN) (see Figure 14.4)	Complex periodic signal with discrete components of identical amplitudes at frequencies of $(2N - 1)f$ (i.e., includes only odd harmonics of the fundamental)	Same as with PRWN
Phase-varying, odd-harmonic pseudorandom white noise	Same as odd-harmonic PRWN	Same as phase-varying PRWN
M-component odd-harmonic array	Complex periodic signal with discrete components of identical amplitudes given by $(2N - 1)f$, where M values of N are chosen to give components approximately equally separated in $\omega^{1/2}$ space; frequency components equal to measured frequencies	Same as with PRWN
Phase-varying M-component array	Same as M-component odd-harmonic array	Same as phase-varying PRWN

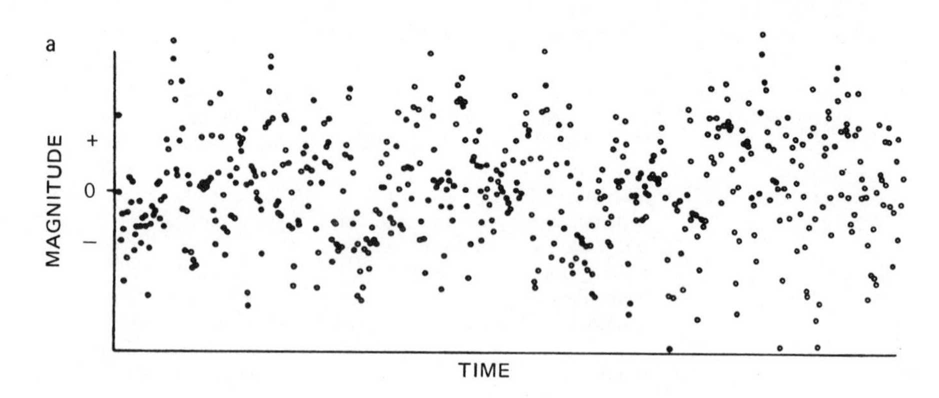

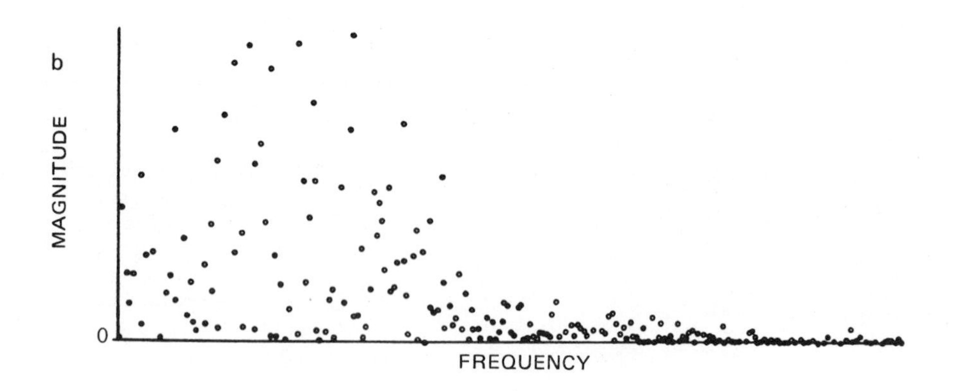

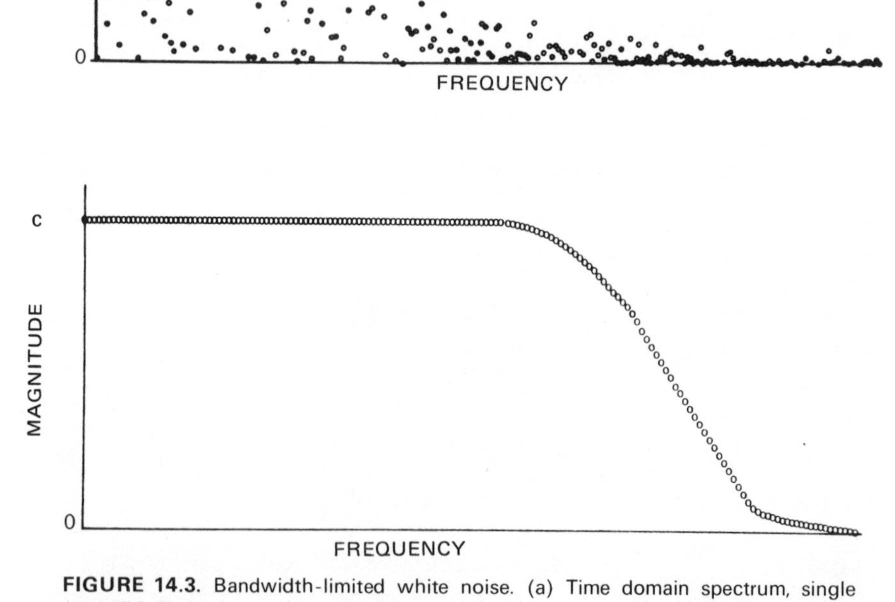

FIGURE 14.3. Bandwidth-limited white noise. (a) Time domain spectrum, single pass; (b) frequency domain spectrum, single pass; (c) frequency domain spectrum at limit of infinite data record length (infinite time). (Reproduced from reference 8 by permission of the Elsevier Publishing Company and the author; copyright 1973.)

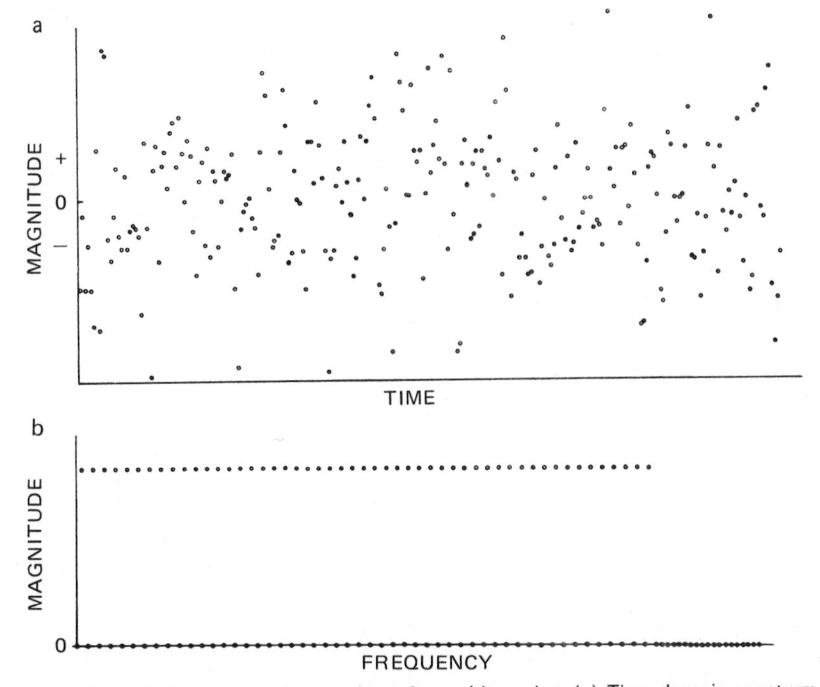

FIGURE 14.4. Odd-harmonic pseudorandom white noise. (a) Time domain spectrum single pass; (b) Frequency domain spectrum, single pass. (Reproduced from reference 8 by permission of the Elsevier Publishing Company and the author; copyright 1973.)

drop formation can be achieved through computer-controlled "drop-knockers."[15,16]

(c) Simultaneous sampling of the perturbing potential and the cell current response through dual sample-and-hold amplifiers.

(d) Rapid repetition of sampling sequences and data storage into two separate arrays representing the time-averaged perturbing potential and cell current time domain waveforms.

(e) Digital data processing—in particular the FFT—at the completion of data acquisition.

Waveforms may be generated either by the computer or by an external source such as a signal generator or a noise generator. Some of the waveforms that have been used are classified in Table 14.1 and summarized in more detail in Table 14.2. It can be seen that several of these waveforms are different types of noise; the discrete time domain and frequency domain records for bandwidth-limited white noise (both for a single pass and for a very large number of averaged passes) and for odd-harmonic pseudorandom white noise are shown in Figures 14.3 and 14.4, respectively.

14.4. KINETICS OF ELECTRODE PROCESSES

The use of the FFT in electrochemical analysis is particularly important when electrochemical response is not diffusion controlled. Many analytically promising electrode processes do not achieve the diffusion-controlled ideal under accessible assay conditions. This is particularly true when one is using a technique with a relatively short observation time (e.g., pulse polarography, fundamental and second harmonic AC polarography, and coulostatic analysis), and when one is using a solid electrode where heterogeneous charge transfer rates are notably slower than at mercury.[17] Measurement of the on-line frequency domain admittance response can allow variations in quantitative data to be minimized by monitoring the magnitude of these kinetic fluctuations. Thus the status of the response characteristic of the electrode can similarly be monitored and appropriate action taken if necessary.

Smith[1,18] has given a detailed example of how this type of procedure can be applied to a quasi-reversible ERM admittance response (controlled both by diffusion and by heterogeneous charge-transfer kinetics) of a single-step electrode reaction, which may be represented by

$$O + ne^- \rightleftarrows R$$

For such reactions, as the standard heterogeneous charge-transfer rate constant k_s becomes larger, the reaction rate becomes progressively more diffusion controlled. When k_s is very large, the reaction is said to be "reversible" or "Nernstian." When the measurement of k_s is of interest and the heterogeneous charge-transfer kinetics have not been observable under a given measurement condition, the usual strategy has involved minimizing the effect of diffusion by raising the frequency (or lowering the time scale) of the observation. In this way it is hoped that the heterogeneous charge-transfer kinetics can be observed before reaching the upper frequency limit imposed by instrumental restrictions. These kinetics are usually studied in terms of the fundamental harmonic phase angle cotangent, cot Φ, obtained in faradaic admittance (usually AC polarographic) measurements.

The rate law for the small-amplitude frequency domain response of the faradaic admittance may be written

$$A(\omega t) = A_{\text{rev}} F(t)\, G(\omega) \sin(\omega t + \Phi) \tag{14.3}$$

A_{rev} is the diffusion-controlled response given by

$$A_{\text{rev}} = \frac{n^2 F^2 A C_O^* (\omega D)^{1/2}}{4RT \cosh^2(j/2)} \tag{14.4}$$

where A is the surface area of the electrode, C_O^* the initial concentration of the oxidized species, and D a diffusion coefficient given by

$$D = D_O^{\beta} D_R^{\alpha} \tag{14.5}$$

In this expression α is the charge transfer coefficient, representing the fraction of the potential that favors the reduction reaction, and

$$\beta = 1 - \alpha \tag{14.6}$$

The subscripts O and R refer to the oxidized and reduced species, respectively. Finally, the term j in equation (14.4) is given by

$$j = \frac{nF}{RT}(E_{dc} - E_{1/2}^{r}) \tag{14.7}$$

where

$$E_{1/2}^{r} = E^{O} - \frac{RT}{nF} \ln \left(\frac{f_R}{f_O}\right)\left(\frac{D_O}{D_R}\right)^{1/2} \tag{14.8}$$

$F(t)$ and $G(\omega)$ represent the effect of heterogeneous charge transfer on the DC and AC time scales, respectively, and are both unity in a diffusion-controlled experiment. $G(\omega)$ is given by

$$G(\omega) = \left[\frac{2}{1 + \cot^2 \Phi}\right]^{1/2} \tag{14.9}$$

where

$$\cot \Phi = 1 + 2^{1/2}/\lambda \tag{14.10}$$

$$\lambda = \frac{k_s}{D^{1/2}}(e^{-\alpha j} + e^{\beta j}) \tag{14.11}$$

The quantity $(2\omega/\lambda)^{1/2}$ in equation (14.10) can be viewed as the contribution of the heterogeneous charge-transfer kinetics to the $\cot \Phi$ magnitude. When $E_{dc} = E_{1/2}^{r}$ (the admittance peak potential for very rapid processes), j is very small, and $e^{-\alpha j}$ and $e^{\beta j}$ are both approximately equal to unity, so that

$$\cot \Phi - 1 = (\tfrac{1}{2}D)^{1/2} k_s^{-1} \omega^{1/2} \tag{14.12}$$

Thus a plot of $\cot \Phi$ vs. $\omega^{1/2}$ should be linear, with a gradient from which k_s can be calculated if D is known, and an intercept of unity at zero frequency.

If k_s is very large, it may require very high frequencies in order that $(2\omega/\lambda)^{1/2}$ exceed the experimental noise. Use of frequencies greater than 10 kHz may mean that the response is limited by the double-layer charging current, which then becomes predominant. Faradaic response measurements at extremely high frequencies have in the past often employed observables resulting from faradaic nonlinearity, such as the faradaic rectification effect,[6,7] where double-layer charging current contributions are greatly minimized. This is accomplished at the expense of greater theoretical complexity and more stringent demands on instrument and cell design.[6,7,19,20]

However, it may be noted that with the reduced noise in admittance spectra measured using FFT strategies, the additional use of signal-averaging allows the noise to be further reduced relative to the $(2\omega/\lambda)^{1/2}$ term, allowing fairly large values of k_s to be measured.

The first application of FFT data processing to measurement of $A_f(\omega)$ involved measurement of the response to purely sinusoidal perturbations.[21,22] Even though this type of measurement is a less far-reaching change in measurement philosophy than applications involving multiple-frequency perturbing waveforms, significant benefits are still realized since the FFT provides a basis for the simultaneous measurement of both linear and nonlinear faradaic response components.

Glover and Smith[15,22] studied the reaction

$$Cr(CN)_6^{-3} + e^- \overset{Hg}{\rightleftarrows} Cr(CN)_6^{-4}$$

in 1 M KCN. With this process the relaxation response reflects mixed rate control by diffusion and the heterogeneous charge-transfer step. Some of the fundamental harmonic AC results are shown in Figure 14.5. The agreement between theory and experiment is excellent, and the rate parameters that were calculated are in excellent agreement with the literature values.

The more fascinating FFT applications involving measurement of the relaxation response to complex multiple frequencies have also been studied by Creason and Smith.[8,23,24] These workers have reported FFT analysis

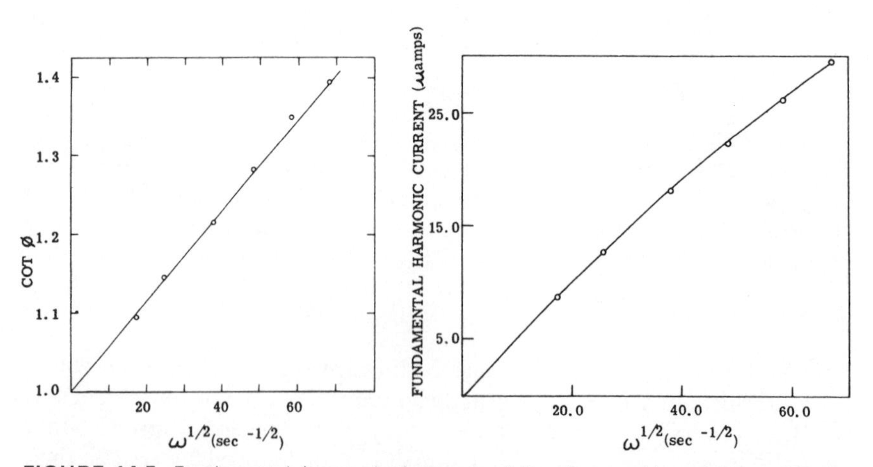

FIGURE 14.5. Fundamental harmonic frequency AC polarographic response data for 2.00×10^{-3} M $Cr(CN)_6^{3-}$ in 1.00 M KCN. Applied: 10 mV peak-to-peak sine wave at frequencies indicated; computer-controlled incremental DC scan. Measured: fundamental harmonic peak cot Φ and peak total current amplitude vs. $\omega^{1/2}$. The solid line represents the theoretical responses for $\alpha = 0.55$, $k_s = 0.35$ cm s^{-1}, $t = 5.0$ s, $A = 0.020$ cm^2, $n = 1$, $D_O = D_R = 8.2 \times 10^{-6}$ cm^2 s^{-1}, $T = 25°$C, $\Delta E = 5.00$ mV, $C_o^* = 2.00 \times 10^{-3}$ M. (Reproduced from reference 22 by permission of the American Chemical Society and the author; copyright 1973.)

TABLE 14.3. Properties of $\cot\Phi - \omega^{1/2}$ Data at Admittance Peak from the Average of 64 Measurement Replicates

Applied AC potential waveform type	Applied AC potential amplitude properties (mV/f)[b]	k_s (cm s^{-1})[a]	Intercept
BLWN	2	0.386	1.012
BLWN	1	0.406	1.030
BLWN (Gaussian window)	1	0.364	0.992
Combed BLWN	1	0.363	0.990
Almost periodic (frequencies at 43.1, 108.8, 230.5, 382.6, 444.6, and 864.3 Hz)	5	0.367	0.998
Almost periodic (same set of frequencies)	2.5	0.373	1.005
Unfiltered rectangular pulse	+ pulse 2 mV × 2 ms	0.362	1.053
Filtered rectangular pulse	+ pulse 2 mV × 2 ms	0.416	0.987
Filtered rectangular pulse	− pulse 2 mV × 2 ms	0.325	0.987
PRWN with frequencies = 10.07N Hz; $N = 1, 2, 3, \ldots, 100$	0.5	0.403	1.003
Phase-varying PRWN	0.5	0.392	0.992
Odd-harmonic PRWN, with frequencies = 10.07(2N − 1) Hz; where $N = 1, 2, 3, \ldots, 50$	1.5	0.367	0.998
Phase-varying, odd-harmonic PRWN	1.5	0.369	0.997
15-Component, odd-harmonic array, with frequencies = 10.07(2N − 1) Hz; where N = 1, 2, 3, 4, 5, 7, 10, 13, 17, 21, 26, 31, 37, 44, and 50	2	0.407	1.012
Phase-varying, 15-component, odd-harmonic array	2	0.400	1.009

[a] Calculated from $\cot\Phi - \omega^{1/2}$ slope using $\alpha = 0.59$, $n = 1$, $T = 25°C$, $D_O = D_R = 8.20 \times 10^{-6}\ s^{-1}$
[b] mV/f: amplitude of individual frequency components (averages in case of BLWN and combed BLWN).

of the fundamental harmonic response of the ten different waveform types summarized in Table 14.2 and their results for the $Cr(CN)_6^{-3}$ reduction are given in Table 14.3. For this system k_s has been previously measured as 0.35 cm s^{-1}, and it can be seen that all these results are self-consistent and that each one of the perturbing waveforms allows a good estimate of the rate constant to be made. The composite k_s value calculated from the 15 entries in Table 14.3 is 0.380 ± 0.033 cm s^{-1}, where the uncertainty represents one standard deviation. The corresponding result for the $\omega = 0$ intercept is 1.005 ± 0.024.

Not shown in the table are the relative slope and intercept standard deviation for each method, which vary considerably; the largest standard deviation is found in the three measurements for which the perturbing potential was a rectangular pulse and the smallest is found for the various classes of pseudorandom white noise. These results have been rationalized[8] in terms of the relative importance of extraneous noise, leakage (*vide infra*), and faradaic nonlinearity with the various waveforms; the following general conclusions were drawn:

(a) Those waveforms whose individual frequency component *amplitudes* are constant as a function of measurement pass and frequency have advantages, relative to those which do not possess this property, with regard to extraneous noise susceptibility.

(b) Waveforms containing components only at the frequencies examined by the measurement system are advantageous relative to others with regard to leakage-induced error.

(c) Waveforms whose individual frequency component *phase angles* are randomized with regard to frequency and measurement pass are advantageous with regard to faradaic nonlinearity error, since phase angle randomization as a function of frequency causes the faradaic nonlinearity error to vary randomly with frequency, and hence to be suppressed by ensemble averaging.

(d) Waveforms with fewer frequency components are superior regarding faradaic nonlinearity error.

14.5. RELEVANT PROPERTIES OF THE FFT FOR ELECTROCHEMICAL RELAXATION MEASUREMENTS

In FT–IR and FT–NMR spectrometry, limited optical retardation or limited data acquisition time, respectively, may cause the instrument line shape function to be broader than the spectral features being measured (see, for example, Chapters 2 and 5). The fact that measurements of cell response can also be taken over a finite period of time can naturally cause the computed frequency domain spectrum to be broadened relative to the true spectrum in electrochemistry. In the nomenclature of electrical engineering, this effect is known as *leakage*.

If the sinusoidal components of the signal do not decay appreciably over the time of data acquisition, each line in the frequency domain spectrum shows large sinc x side-lobes. However, this *leakage error* can be made to disappear in the special case where the waveform is periodic (i.e., described by a Fourier series) and the time domain record length is made *exactly equal* to an integral multiple of the waveform repetition period. This technique makes the measured frequencies coincide with the frequencies of the periodic waveform Fourier components, and is valid when the frequencies of the waveform are known multiples of a certain fundamental. With signals that have continuous spectra or are almost periodic (see Table 14.1), it is impossible to synchronize the majority of frequency components to the data acquisition period in this manner. In this case leakage can be totally eliminated only in the impractical limit of an infinite data record.

Smith and his co-workers[8] have written programs using the classical Fourier transform algorithm to minimize leakage effects in the case of an almost periodic waveform. The algorithm only affects the Fourier transform at those frequencies actually applied to the cell. The applied frequency values are deduced empirically by the measurement system in a preliminary operation in which data arrays are acquired from individual sinusoidal oscillators and analyzed. Prior to performing the Fourier transform at each frequency, the applied potential and cell current waveform arrays are converted to two new arrays, which are truncated to a length corresponding to an integral number of cycles of the transform frequency to be invoked. The classical Fourier transform is then calculated for each of the component frequencies in the waveform.

The FFT algorithm has been used for purposes other than revealing the frequency domain spectrum of a signal for cell admittance measurements. As discussed in earlier chapters in this book, the FFT can also be used for convolution and correlation (and, of course, for the inverse operations deconvolution and decorrelation). Simple multiplication or division of two signals in the frequency domain implies convolution or deconvolution, respectively, in the time domain. Frequency domain multiplication or division using the complex conjugate of one spectrum is equivalent to time domain correlation or decorrelation, respectively. Thus one can recognize that the master equation for computation of the cell admittance from equation (14.2) is equivalent to deconvoluting the time domain applied potential waveform (with the effects of potentiostat nonideality) from the observed time domain current response to obtain the time domain impulse response of the cell. If the latter waveform were desired for data analysis, it is obtainable from $A(\omega)$ simply through the inverse Fourier transform.

Convolution and deconvolution effects on electroanalytical signals have been discussed by Smith[1] with special respect to voltammetry. The stationary electrode linear sweep or cyclic voltammogram provides a promising area for application of the FFT for deconvolution of instrumental effects. A diffusion-controlled voltammogram can be considered to be distorted by a broadening function, originating in the diffusion process, which is proportional to $t^{-1/2}$ (with planar diffusion). *Deconvolution* of $t^{-1/2}$ from a reversible linear sweep voltammogram (which is equivalent to "semidifferentiation") produces for each component the much sharper and symmetrical $1/[\cosh^2(j/2)]$ shape function, where j is given by equations (14.7) and (14.8). The effect of this operation is shown in Figure 14.6a, c.

Convolution of the original voltammogram with $t^{-1/2}$ has also been suggested, in order to obtain a sigmoidal signal similar to a conventional DC polarogram, as shown in Figure 14.6a, b. However, it is apparent that the deconvolution operation yields a more readily interpretable record.

One may consider the $1/[\cosh^2(j/2)]$ line shape obtained through deconvolution of the original voltammogram from $t^{-1/2}$ to be due to the Nernstian broadening of an impulse response, which is characteristic of, and whose height is proportional to the concentration of, each component in solution. In principle it is possible to deconvolve this broadening function from the profiles of Figure 14.6c to obtain these impulse functions, which

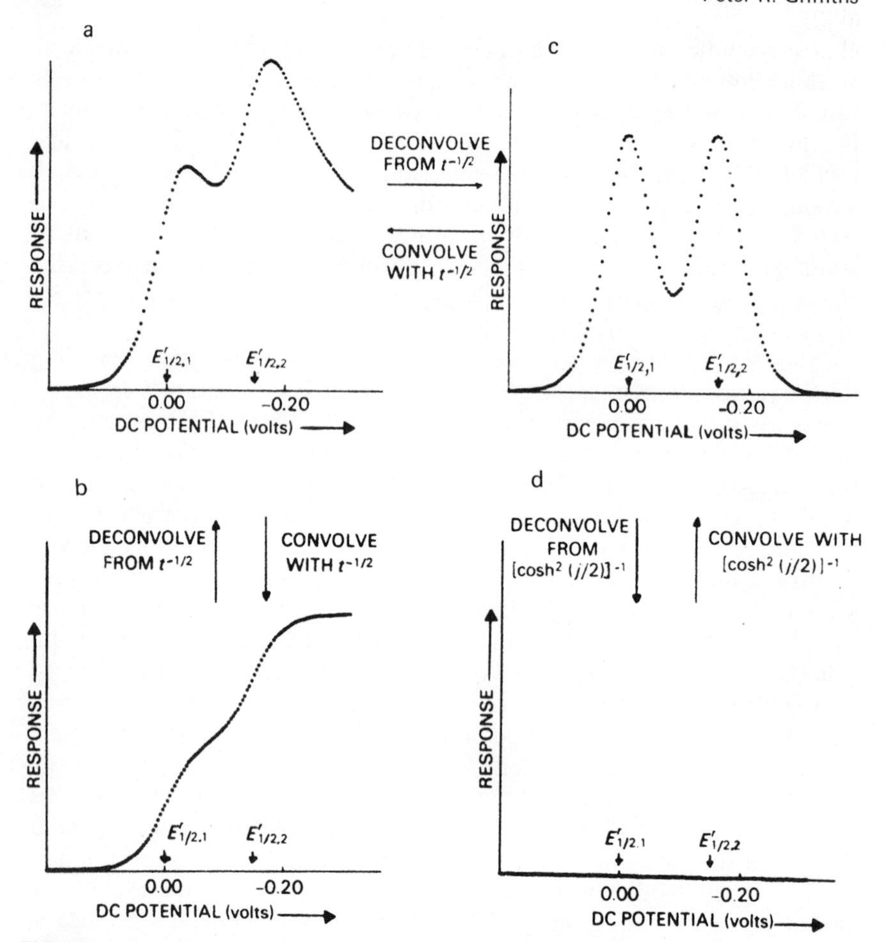

FIGURE 14.6. Illustration of the effects of some convolution and deconvolution operations on the stationary electrode linear sweep voltammetric response of a two-component reversible system with planar diffusion. Curve (a) represents the original voltammogram, assuming two reversibly reduced components with equal bulk concentrations and diffusion coefficients, with $E^r_{\frac{1}{2},1} = 0.000$ V and $E^r_{\frac{1}{2},2} = -0.150$ V. (Reproduced from reference 1 by permission of the American Chemical Society and the author; copyright 1973.)

will be located at the reversible half-wave potential $E^r_{1/2}$, *i*, for each component *i*, as illustrated by the transformation from Figure 14.6c to 14.6d. However, this last step is rather impractical because it amounts to working with very high-order derivatives and will therefore magnify the effects of system noise considerably.

Finally, of course, digital smoothing of any electrochemical signal obtained by conventional analog methods (such as DC polarography or cyclic voltammetry) can be achieved through application of Fourier transform methods, as discussed earlier in this book by Lephardt. The signal is digitized and the resultant array is converted to Fourier space. This array is then multiplied by a rectangular truncating function (or some type of apodization function) to eliminate the high spatial frequencies, and finally the inverse Fourier transform is performed to obtain the smoothed record. A typical example for the smoothing of AC polarographic data[25] is shown in Figure 14.7.

The use of rectangular, or boxcar, smoothing functions is adequate for smoothing data sets that begin and end with zero or near-zero values. However, when the data sets begin and/or end with values whose magnitude is significantly different from zero, the Fourier spectrum may have an appreciable magnitude at high spatial frequencies. Thus after this data set is multiplied by the rectangular filter function, a discontinuous transient is put into the set, which manifests as sinc x side-lobes when the inverse Fourier transform of the array is computed. This effect is illustrated for an exponential function in Figure 14.8a–d.

To get around this problem, Hayes *et al.*[25] have developed a method in which the original data are modified *prior* to Fourier transformation by rotating and translating the original data record so that the initial and final data points have zero values. This process is illustrated in Figure 14.8a, e, in which the rotation–translation operation can be seen to involve subtracting from each successive datum point a quantity Δ_n, which varies linearly between the magnitudes of the first and last points, A_1 and A_k, respectively. Thus the magnitude of the nth point in the rotated–translated array is calculated from the expression

$$R_n = A_n - \Delta_n \tag{14.13}$$

where

$$\Delta_n = A_1 + \frac{(A_k - A_1)(n - 1)}{(k - 1)} \tag{14.14}$$

for $n = 1, 2, 3, \ldots, k$.

The effect of this translation–rotation operation can be seen by comparing the Fourier spectra in Figure 14.8b and 14.8f. By multiplying the array in Figure 14.8f by a rectangular filter function, and performing first the inverse Fourier transform and then the inverse of the rotation–translation operation, the original exponential array is obtained with smoothing of any high-frequency noise component but without significant distortion.

14.6. PUBLISHED AND FUTURE APPLICATIONS OF THE FFT IN ELECTROCHEMISTRY

Since the first papers describing the *feasibility* of applying the FFT for electroanalytical data processing only appeared at the start of this

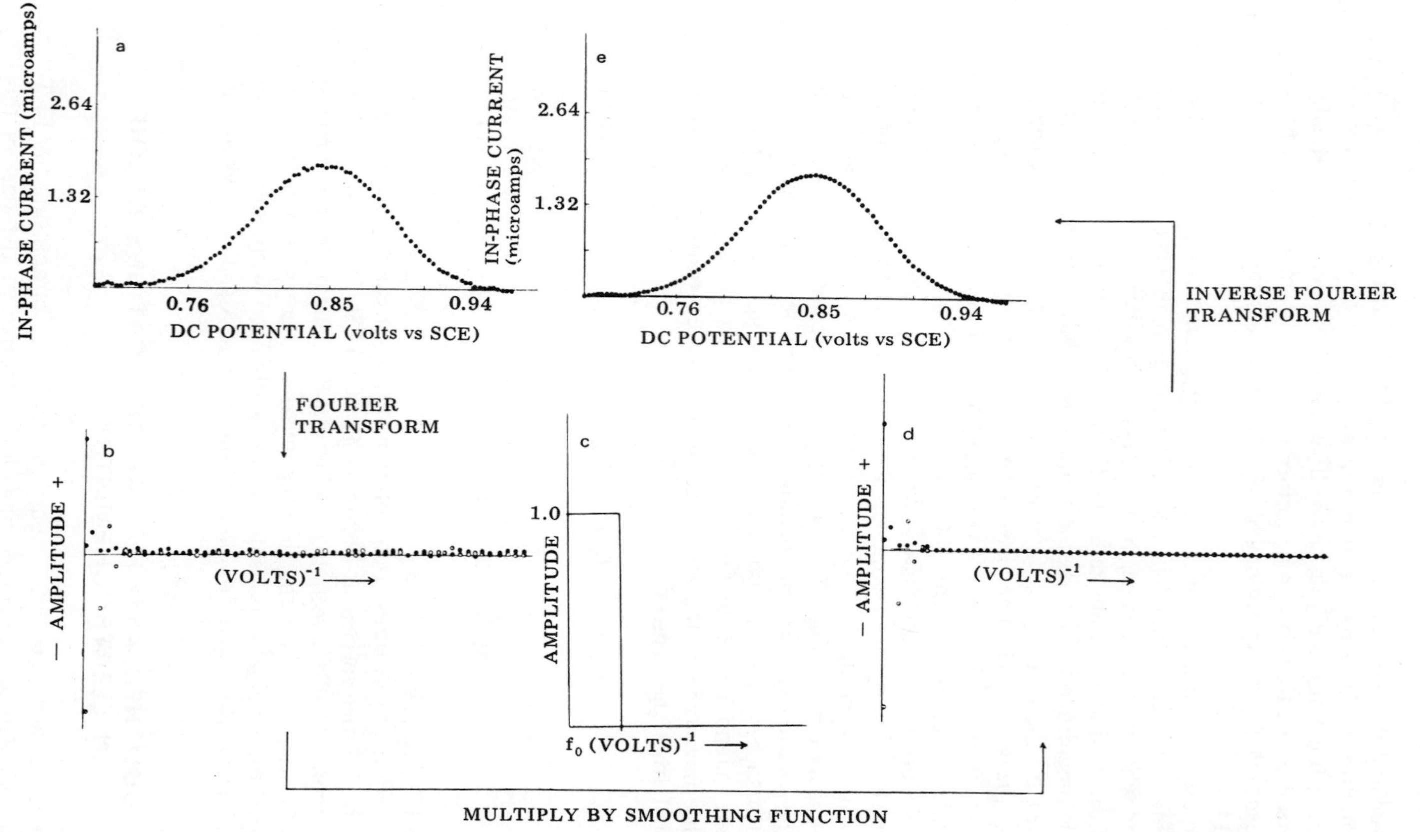

FIGURE 14.7. Fourier transform smoothing of fundamental harmonic AC polarographic data. (a) Original polarogram for 1.0×10^{-3} *M* Cd^{2+} in 1.0 *M* KNO_3, 0.1 *M* sodium acetate, 0.04 *M* nitrilotriacetic acid, pH = 4.95; (b) Fourier transform of (a); ○, real components; ●, imaginary components, or both when they are equal; (c) rectangular smoothing function; (d) result of multiplying (b) by (c); (e) inverse Fourier transform of (d). (Reproduced from reference 25 by permission of the American Chemical Society and the author; copyright 1973.)

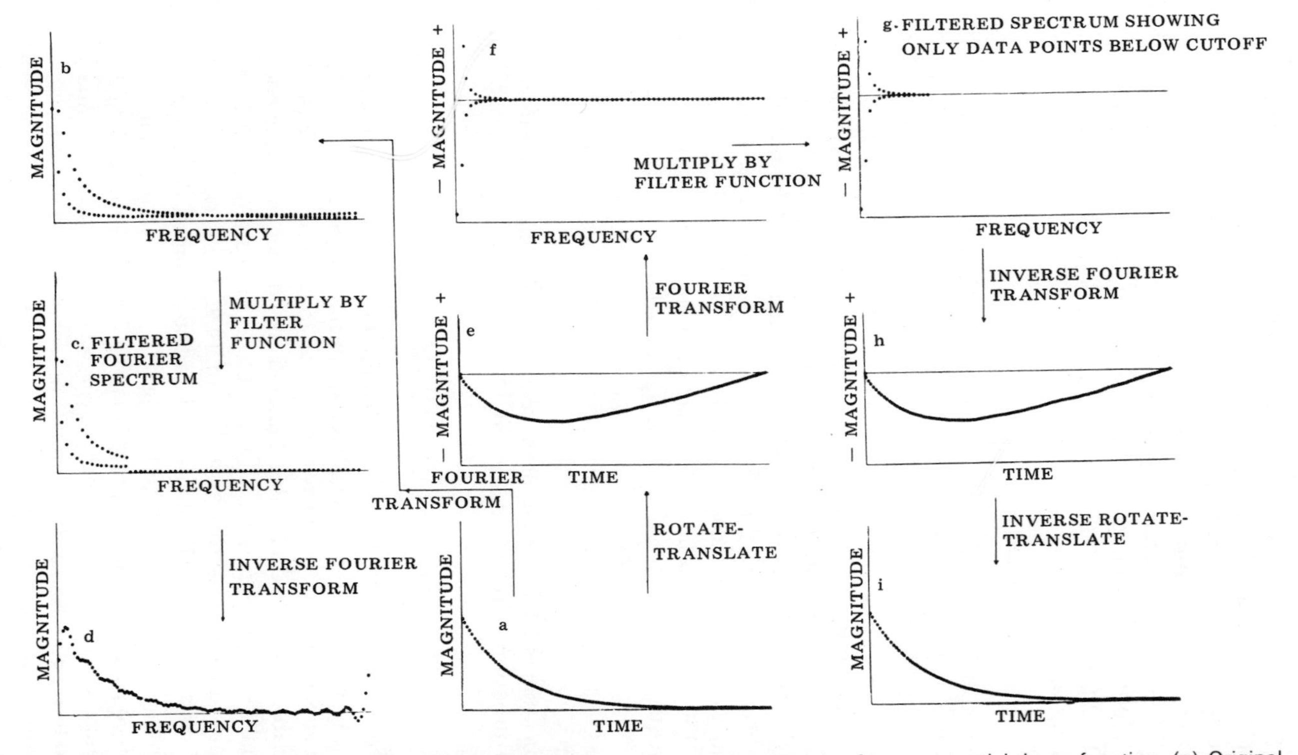

FIGURE 14.8. Effects of conventional and modified Fourier transform smoothing algorithms on a noise-free exponential decay function. (a) Original exponential decay data; (b) Fourier spectrum (real and imaginary points) of (a); (c) result of multiplying (b) by a rectangular smoothing function similar to that in Figure 14.7c; note the discontinuity in the curve; (d) inverse Fourier transform of (c); note the introduction of spurious "wiggles" into the original exponential decay data due to the discontinuity in (c); (e) result of rotation–translation of (a) to give zero values at the start and end of the time domain spectrum; (f) Fourier spectrum (real and imaginary points) of (e); (g) result of multiplying (f) by a rectangular smoothing function (showing only the nonzero data points); (h) inverse Fourier transform of (g); (i) inverse rotation–translation of (h) to give the exponential decay data with high-frequency components removed. Comparison of curves (d) and (i) shows the benefit accrued through the rotation–translation operation to give zero values at the start and end of the original data array. (Reproduced from reference 25 by permission of the American Chemical Society and the author; copyright 1973.)

decade, it is not surprising that the number of published applications to date has been relatively small, and primarily limited to determination of the heterogeneous charge-transfer rate constants for systems involving homogeneous chemical reactions coupled to the heterogeneous charge-transfer step.[2,6,7,26,27] For example, the sinusoidal faradaic admittance response of solutions containing the cadmium–nitrilotriacetate complex in 1 M KNO_3, 0.1 M acetate buffer, and a mercury electrode follows rate laws for the mechanism:

$$\mathrm{Y} \rightleftarrows \mathrm{O} + ne^- \rightleftarrows \mathrm{R}$$

Measurements on this system were used as the basis for obtaining the dissociation rate constant of the complex (2.9×10^5 M^{-1} s^{-1}) and its heterogeneous charge-transfer rate constant ($k_s = 1.5 \times 10^{-7}$ cm s^{-1}).[26] These results are supported by previous NMR line-broadening studies.[28]

Analogous measurements have also been carried out on other systems. For example, reduction of benzaldehyde at mercury in 75% ethanol–water with 0.1 M tetrabutylammonium hydroxide is followed by rapid dimerization of the ketyl radical to the pinacol. FFT faradaic admittance measurements[27] on this system gave the dimerization rate constant as 8.0×10^6 M^{-1} s^{-1} and the k_s as 0.40 cm s^{-1}. Another experiment evaluated rate constants for the reaction sequence in which reduction of the uranyl ion is followed by a disproportionation process. The rate constant in 6 M $HClO_4$ was assessed at 1.4×10^4 M^{-1} s^{-1} and the electrode reaction k_s value was found to be 2.5×10^{-2} cm s^{-1}.[27]

One application of FFT analysis of the relaxation response to multiple-frequency perturbations has been described.[18] The system

$$\mathrm{Fe(C_2O_4)_3^{-3}} + e^- \rightleftarrows \mathrm{Fe(C_2O_4)_3^{-4}}$$

is characterized by a k_s value that is so large that until recently it has not been possible to measure it without special high-frequency observations.[20,29] However, the superior precision provided by the 15-frequency waveform response described at the end of Table 14.3 and the further improvement found on signal averaging has made k_s accessible to measurements at low and moderate frequencies, as shown in Figure 14.9. The measured rate constant of 1.5 cm s^{-1} agrees well with earlier faradaic rectification measurements using frequencies in the megahertz range.[20] Smith[9] has estimated that the combined effects of enhanced precision and frequency range should extend the range over which electrochemical relaxation measurements may be applied to kinetic studies by about two orders of magnitude in terms of accessible kinetic parameters.

Like so many other areas in which Fourier transform techniques have given the analytical chemist new measurement capabilities, the number of problems that can be solved in electroanalytical chemistry far exceeds the resources of the few groups working in the field. The next few years are sure to

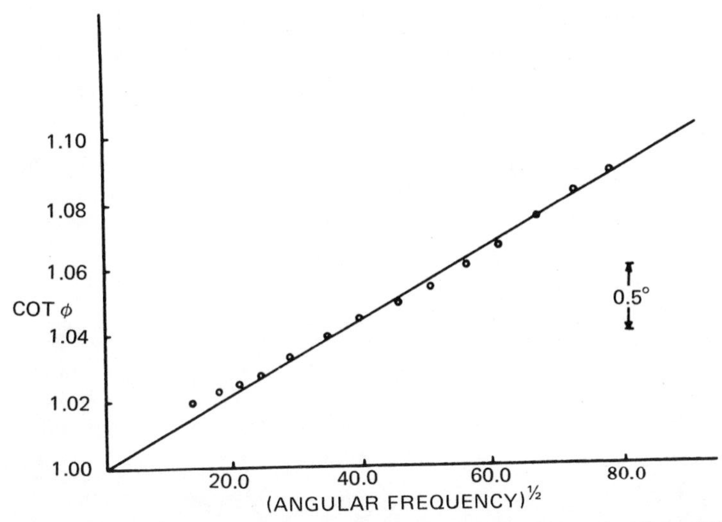

FIGURE 14.9. Fundamental harmonic cot Φ frequency response data from 100-pass Fourier transform measurement, for 5.0×10^{-3} *M* $Fe(C_2O_4)_3^{3-}$ in 1.00 *M* $K_2C_2O_4$ and 0.05 *M* $H_2C_2O_4$ at Hg. A computer-generated, phase-varying, 15-component odd harmonic waveform (see last entry of Table 14.3) was used, with a DC potential of −0.180 V vs. Ag/AgCl. The solid line represents a least-squares best fit to the experimental points. (Reproduced from reference 18 by permission of the American Chemical Society and the author; copyright 1973.)

bring many new applications in which the techniques described in this chapter will yield previously unobtainable solutions. Some of these are forecast to be[10]:

(a) Analytical applications based on kinetic parameter magnitudes, or assay procedures in which undesirable kinetic effects are detected and corrected for in calculating the analytically relevant observable, made possible through the nearly instantaneous FFT kinetic analysis of an electrode process.

(b) Acquisition of time-resolved admittances during electroplating, corrosion, or other processes involving temporal changes in the faradaic admittance occurring on the millisecond time scale.

(c) Acquisition of faradaic admittance data on systems where both forms of a redox couple are transiently stable, but at least one can be generated photochemically, electrolytically, or chemically.

(d) Applications to spectroelectrochemistry in which light absorption or emission is used as one of the observed responses to the input perturbation.

(e) Electroanalytical faradaic admittance instruments that do not rely on analog potentiostats and galvanostats, but consist only of a cell, minicomputer or microcomputer system and very simple analog signal-conditioning networks.

REFERENCES

1. D. E. Smith, *Anal. Chem.* **48**, 517A (1976).
2. D. E. Smith, *Crit. Rev. Anal. Chem.* **2**, 247 (1971).
3. E. R. Brown, T. G. McCord, D. E. Smith, and D. D. DeFord, *Anal. Chem.* **38**, 1119 (1966).
4. E. R. Brown, D. E. Smith, and G. L. Booman, *Anal. Chem.* **40**, 1411 (1968).
5. E. R. Brown, H. L. Hung, T. G. McCord, D. E. Smith, and G. L. Booman, *Anal. Chem.* **40**, 1424 (1968).
6. D. E. Smith, *in: Electroanalytical Chemistry*, Vol. 1, pp. 1–155 (A. J. Bard, ed.), Marcel Dekker, New York, 1966.
7. M. Sluyters-Rehbach and J. H. Sluyters, *in: Electroanalytic Chemistry*, Vol. 4, pp. 1–128 (A. J. Bard, ed.), Marcel Dekker, New York, 1970.
8. S. C. Creason, J. W. Hayes, and D. E. Smith, *J. Electroanal. Chem.* **47**, A1 (1973).
9. D. E. Smith, *in: Information Chemistry: Computer Assisted Chemical Research Design*, pp. 125–142 (H. B. Mark, Jr., and S. Fujiwara, eds.), University of Tokyo Press, Tokyo, Japan, 1975.
10. D. E. Smith, *in: Topics in Pure and Applied Electrochemistry*, pp. 43–67, SAEST, Karaikudi, India, 1975.
11. A. A. Pilla, *J. Electrochem. Soc.* **117**, 467 (1970).
12. K. Doblhofer and A. A. Pilla, *J. Electroanal. Chem.* **39**, 91 (1971).
13. N. Weiner, *Nonlinear Problems in Random Theory*, MIT Press, Cambridge, Massachusetts, 1958.
14. D. E. Smith, *Anal. Chem.* **48**, 221A (1976).
15. D. E. Glover, Ph.D. Dissertation, Northwestern Univ., Evanston, Illinois, (1973).
16. S. C. Creason, Ph.D. Dissertation, Northwestern Univ., Evanston, Illinois, (1973).
17. J. E. B. Randles and K. W. Somerton, *Trans. Faraday Soc.* **48**, 937, 951 (1952).
18. S. C. Creason and D. E. Smith, *Anal. Chem.* **45**, 2401 (1973).
19. R. deLeeuwe, M. Sluyters-Rehbach, and J. H. Sluyters, *Electrochim. Acta* **12**, 1593 (1967).
20. R. deLeeuwe, M. Sluyters-Rehbach, and J. H. Sluyters, *Electrochim. Acta* **14**, 1183 (1969).
21. H. Kojima and S. Fujiwara, *Bull. Chem. Soc. Japan* **44**, 2158 (1971).
22. D. E. Glover and D. E. Smith, *Anal. Chem.* **45**, 1869 (1973).
23. S. C. Creason and D. E. Smith, *J. Electroanal. Chem.* **36**, A1 (1972).
24. S. C. Creason and D. E. Smith, *J. Electroanal. Chem.* **40**, A1 (1972).
25. J. W. Hayes, D. E. Glover, D. E. Smith, and M. W. Overton, *Anal. Chem.* **45**, 277 (1973).
26. K. R. Bullock and D. E. Smith, *Anal. Chem.* **46**, 1069 (1974).
27. J. W. Hayes, D. E. Smith, I. Ruzik, J. R. Delmastro, and G. L. Booman, *J. Electronal. Chem.* **51**, 245, 269 (1974).
28. D. L. Rabenstein and R. J. Kula, *J. Am. Chem. Soc.* **91**, 2492 (1969).
29. T. Rohko, M. Kogoma, and S. Aoyagi, *J. Electroanal. Chem.* **38**, 45 (1972).

Index